T0191922

Advances in Terahertz Technology and Its Applications

Sudipta Das · N. Anveshkumar · Joydeep Dutta ·
Arindam Biswas
Editors

Advances in Terahertz Technology and Its Applications

Editors
Sudipta Das
Department of Electronics
and Communication Engineering
IMPS College of Engineering
and Technlogy
Malda, West Bengal, India

Joydeep Dutta
Department of Computer Science
Kazi Nazrul University
Asansol, West Bengal, India

N. Anveshkumar
Department of Electronics
and Communication Engineering
VIT Bhopal University
Bhopal, Madhya Pradesh, India

Arindam Biswas
Department of Computer Science
School of Mines and Metallurgy
Kazi Nazrul University
Asansol, West Bengal, India

ISBN 978-981-16-5733-7 ISBN 978-981-16-5731-3 (eBook)
https://doi.org/10.1007/978-981-16-5731-3

This Springer imprint is published by the registered company Springer Nature Singapore Pte Ltd.
The registered company address is: 152 Beach Road, #21-01/04 Gateway East, Singapore 189721, Singapore

Preface

In these modern times, the growing applications in communication, sensing, security, safety, spectroscopy, manufacturing, biomedical, agriculture, imaging, etc., demand for higher resolution, greater speeds, and wider bandwidth support. Thanks to THz technology that makes it possible due to its enormous advantages like non-ionizing signal nature, compactness, higher resolution, spatial directivity, high-speed communication, and greater bandwidth. Since the THz radiation covers frequencies from 0.1 THz to around 10 THz and is highly attenuated by atmospheric gases, it is mainly used in short-distance applications.

This book is aimed to bring the emerging application aspects of THz technology and various modules used for its successful realization. It gathers scientific technological novelties and advancements already developed or under development in the academic and research communities. This work focuses on recent advances and different research issues in terahertz technology and would also seek out theoretical, methodological, well-established, and validated empirical work dealing with these different topics. The chapters cover a very vast audience from basic science to engineering, technology experts, and learners. This could eventually work as a textbook for engineering and communication technology students or science master's programs and for researchers as well. These chapters also serve the common public interest by presenting new methods for application areas of technology, its working, and its effect on the environment and human health, etc.

In particular, this textbook covers the broad categories of THz technology aspects, advantages, disadvantages, applications, communication trends, challenges and security, various modules like antennas and absorbers under different material considerations, solid-state devices, THz generators, realization of optical NOT and NAND gates, de-multiplexers and parity generator, THz e-healthcare system, THz metasurface, oversampled OFDM modulation technique in THz communication systems. The antenna trends include planar antennas, horn geometries, SIW structures, THz arrays, fabrication techniques, etc. The solid-state devices cover wide bandgap semiconductors, IMAPTT diodes, etc.

Once the readers finish the study of this book, then definitely they will get to know about the importance of THz technology, advancement in the field, applications,

THz modules, and future scope. It also leads to enhancement in their knowledge in THz technology and gives a platform to future technology and novel application realization.

Malda, West Bengal, India — Sudipta Das
Bhopal, Madhya Pradesh, India — N. Anveshkumar
Asansol, West Bengal, India — Joydeep Dutta
Asansol, West Bengal, India — Arindam Biswas

Contents

About the Editors

Dr. Sudipta Das obtained his Ph.D. degree from the University of Kalyani, India. He is currently working as Associate Professor in the Department of Electronics and Communication Engineering at IMPS College of Engineering & Technology, West Bengal, India. He is having 10 years of teaching and 7 years of research experiences. His research interests are microstrip antennas for microwave, mm-wave and THz communication systems, flexible antenna design, filter design, FSS, RFID, microwave components, and THz systems.

Dr. N. Anveshkumar is currently working as Assistant Professor in the Department of ECE, VIT Bhopal University, Bhopal, India. He pursued his doctoral degree from VNIT Nagpur, in 2019 under the guidance of Dr. A. S. Gandhi from the field of antennas, RF circuits, and wireless communications. He is having 2 years of teaching and research experience. Till now, he published his research papers in seven SCI Journals, two Scopus Indexed Journals, five international journals and 11 in international and national conferences. His research interests include antennas, RF circuits, wireless communications, and embedded systems. Currently, he is working in the area of high directional UWB antennas, reconfigurable antennas, cognitive radio antennas, RF energy harvesting, THz antennas, and mobile antennas.

Dr. Joydeep Dutta completed his Ph.D. from the Department of Computer Science and Engineering, NIT Durgapur, in 2020. He has more than 9 years of teaching experience and 2 years of industrial experience. His current research interests include network optimization problems, artificial intelligence, evolutionary algorithms, combinatorial optimization, etc.

Dr. Arindam Biswas received his M.Tech. in radio physics and electronics from the University of Calcutta, India, in 2010, and Ph.D. from NIT Durgapur, in 2013. He was Postdoctoral Researcher at Pusan National University, South Korea, under the prestigious BK21PLUS Fellowship. He was Visiting Professor at Research Institute of Electronics, Shizuoka University, Japan. Currently, he is working as Associate Professor in the School of Mines and Metallurgy, Kazi Nazrul University, Asansol, West Bengal, India. His research interests are in the areas of carrier transport in low-dimensional systems and electronic devices, nonlinear optical communications, and THz semiconductor sources.

Recent Trends in Terahertz Antenna Development Implementing Planar Geometries

V. Adhikar, A. Karmakar, B. Biswas, and C. Saha

Abstract This chapter summarizes design and development of antenna using micro-fabrication technology for Terahertz application. Two different types of antennas have been discussed here with fabrication process details along with design strategies. Effect of imbibed process related imperfections on final device performance are highlighted here with the help of mathematical models. Surface roughness becomes a bottleneck for realizing any device in THz-regime, which has been explained with various analytical models. FEM-based 3D simulator is used extensively for analyzing all designs. Parametric simulations were carried out to get the optimized design values. The first design is complete of planar type. It is based on a flexible microwave substrate named LCP (Liquid Crystal Polymer), which is very thin (~100 μm), hardly moisture absorbent (<0.04%), and moreover bio-compatible in nature. A 100 GHz design specification is chosen for the targeted medical imaging antenna array structure offers a directive gain of around 19.3 dBi. Electrical equivalent circuit modeling of the whole array structure is also chalked out in this chapter, which eases to understand the interaction between various radiating elements with adjacent parasitic counterparts. Second portion of the chapter introduces design and simulation of a silicon micro-machined W-band sectoral H-plane horn antenna with proposed fabrication process details. Selection of design parameters has been summarized with the help of analytical models and parametric simulations in a 3D-EM environment. Proposed design offers around 13 dBi gains.

Keywords THz antenna · Circuit modeling · LCP · Sectoral horn · Superstrate

---

V. Adhikar
COMSOL Multi Physics Software Limited, Bengaluru, India

A. Karmakar (✉)
Semi-Conductor Laboratory (SCL), Department of Space, Government of India, Chandigarh, India

B. Biswas
Indian Institute of Science Education and Research (IISER), Mohali, Chandigarh, India

C. Saha
Indian Institute of Science and Technology (IIST), Trivandrum, India

S. Das et al. (eds.), *Advances in Terahertz Technology and Its Applications*,
https://doi.org/10.1007/978-981-16-5731-3_1

1 Introduction

In the recent era, remarkable research interests are being paid to the Terahertz (THz) band (0.1 to 10 THz) in various applications for wireless communication sectors, medical diagnostics, radio astronomy, imaging, and many other instrumentation circuits because of its several unique features including free-license wide spectral bandwidth, high spectral resolution, non-ionization property, tranquil from EMI/EMC issue, etc. [1–6]. However, like other technology it has also some demerits in context with high path-loss factors and attenuation in the presence of water vapor or hydroxyl ions. To eradicate these issues, high gain antennas are inevitable for effective establishment of communication links between the transmitter and receiver. Simultaneously, several atmospheric windows co-exist across the THz frequency spectrum, where the path-loss factors are reduced significantly [7]. THz-communication can be established at these low loss frequencies to enhance the wireless link reliability factor. On the contrarily, since the dimension of wavelength becomes very short (0.03 to 3 mm) at THz frequencies, manufacturing antenna at this frequency band becomes a challenging task while employing cost-effective standard fabrication processes. Furthermore, the integration of the antenna with other building blocks of communication sub-system plays a pivotal role to attain the overall figure of merit.

This chapter demonstrates the design and development of two different types of THz-antennas implementing planar geometries. The antennas are based on standard micro-fabrication technologies. One of the antennas is relied upon silicon micromachining process, whereas the other candidate is focused on organic type microwave substrate called 'Liquid Crystal Polymer (LCP)'. Simple design with enriched performance is the attractive feature of the present structures. In the present chapter, the detailed design, simulation, effect of various design parameters, fabrication process intricacies, and finally the developed architectures have been demonstrated for two specified antenna types. Finally, the electrical equivalent circuit modeling aspects are also briefly touched upon for a better understanding of the device physics in association with various parasitic effects.

2 Designing of THz Antenna

Antenna being the electronic eye of whole communication system, its design is always an exigent task to an RF/antenna engineer. For the THz frequency, some more additional constraints act as supplementary with the generic requirements. In this section, the design aspects of two different types of THz antenna are discussed in detail.

2.1 Microstrip Patch Antenna Array

As compared to other antenna candidates for THz frequency ranges, fortunately, microstrip patch antenna inherits potential capabilities to satisfy mandatory requirements of link establishment. In this sub-section, series-fed rectangular shaped microstrip patch antenna (RMAA) array design is discussed. Ease of design and fabrication suitability with a cost-effective way is the attractive feature of this antenna module. As previously discussed in the introduction part, due to higher atmospheric attenuation at the THz frequency range, high gain antennas are essential for THz technology. Herein, the gain for RMAA is enhanced up to 19 dBi either by introducing parasitic elements or by appending the superstrate layer. LCP has been used as a base substrate material for the antenna design. It is a non-toxic, bio-compatible substrate with a permittivity value of 3.14 and dielectric loss tangent of around 0.002 [8]. It is a kind of organic polymer. Surface-wave and leaky wave problems are inherently eliminated/suppressed at higher frequency ranges for this substrate, which is usually of 4 mil (~100 μm) thickness and of flexible in nature. It can be wrapped over outside of any 3D cylindrical/circular shaped object. Conformal design can be made comfortably out of this, which is desirable for several bio-medical usages and air-borne applications [9, 10], targeting for THz frequency regime.

Proposed antenna structure comprises series-fed five patch elements, each having length 'L' and width 'W' and loaded with parasitic patch elements of length and width 'L' and 'W_p' on both sides of non-radiating edges of the five elements. Figure 1 shows the schematic of the proposed antenna. Dimension of the patch width 'W' and length 'L' is evaluated using Eq. (1) and (2) [11] and given by,

$$W = \frac{c}{2f_r}\sqrt{\frac{2}{1+\epsilon_r}} \tag{1}$$

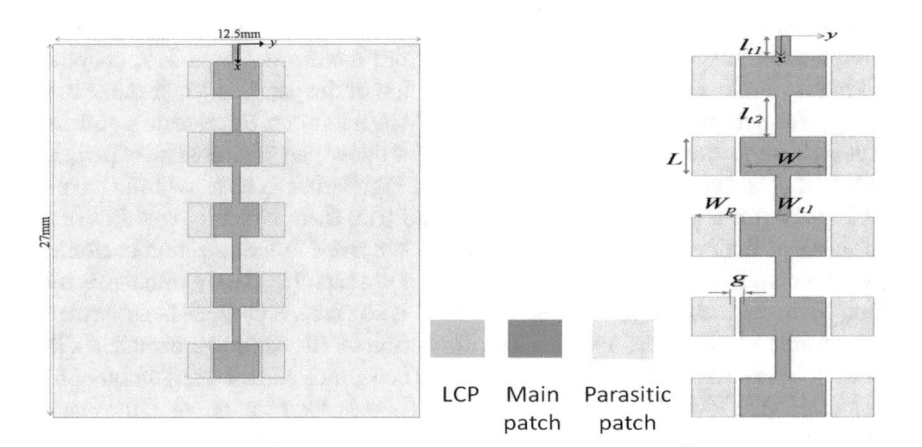

Fig. 1 Top view and dimensions of microstrip patch array with parasitic patches

$$L = \frac{c}{2f_r\sqrt{\epsilon_{eff}}} - 2\Delta L \tag{2}$$

where, ϵ_{eff} is an effective dielectric constant, and 'ΔL' is extended length on both sides of the radiating patch due to the fringing field [11]. Segments of the transmission line in between the patches are responsible for phase matching. The dimension of these transmission lines, i.e., length 'l_{t1}' and 'l_{t2}' and width 'W_{t1}', are computed as follows [12]

$$l_{t1} = (2P+1)\frac{\lambda_0}{4} \tag{3}$$

$$l_{t2} = (2Q+1)\frac{\lambda_0}{4} + 2\Delta L \tag{4}$$

$$W_{t1} = \frac{7.475h}{e^x} - 1.25t \tag{5}$$

$$x = \frac{Z_0\sqrt{1.41+\epsilon_r}}{87} \tag{6}$$

$$Z_0 = \sqrt{50 \times \frac{11.96\lambda_0}{W}} \tag{7}$$

where, P and Q are non-negative integer numbers (here, $P = Q = 1$).

Series-fed array structure is chosen to achieve a high gain antenna profile at the cost of a little bit increased layout area. Further, the directive gain can be enhanced remarkably by placing parasitic patch structures at the non-radiating edges of the main radiating elements. Effectively, the aperture size of the whole planar array antenna is enhanced, which further causes an increase in gain. The gap between the primary antenna and parasitic patch decides the amount of E-field coupling. To ensure strong coupling and optimal excitation of the parasitic patch element, its gap 'g' from the main radiating element should be minimum. Systematic parametric studies have been carried out to get an optimum value of gap 'g' and width of parasitic patch 'W_p' and are indicated in the S_{11} plot of Fig. 2. As can be noted from the plot of Fig. 2b, for a gap dimension of 50 to 200 μm, there is hardly any difference in resonant frequency, but a slight change is observed in the refection coefficient curve. Actually, the flux linkage phenomenon is not captured clearly with the results. But, when the width of parasitic patches 'W_p' is changed, it changes the impedance and current vector of the antenna. The dimension of all the design parameters has been optimized and summarized in Table 1. The surface current distribution of this enhanced gain RMAA is depicted in Fig. 3. As revealed from the plot, the energy associated with each patch element monotonically decreases from the directly fed array element to the last element. This is because most of the energy is radiated by

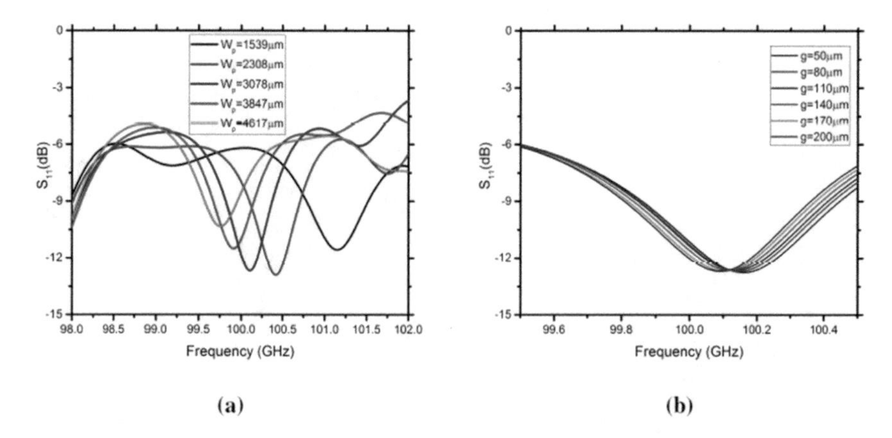

Fig. 2 Variation in reflection coefficient of parasitic patch loaded RMAA with **a** width of the parasitic patch **b** gap between the main and parasitic patch element

Table 1 Optimized dimensions of the RMAA with parasitic patches

Parameter	Values (in μm)
W	3078
L	2489
l_{t1}	1293
l_{t2}	2685
w_{t1}	500
W_p	1539
g	200
h	100

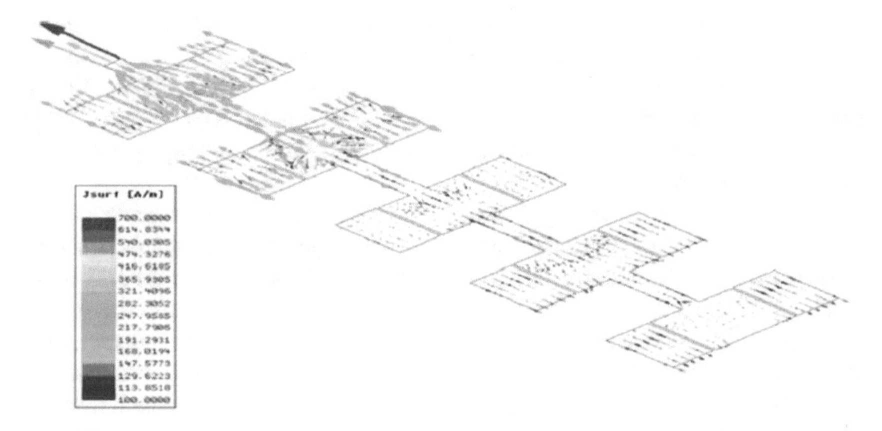

Fig. 3 Surface current distribution for RMAA structure

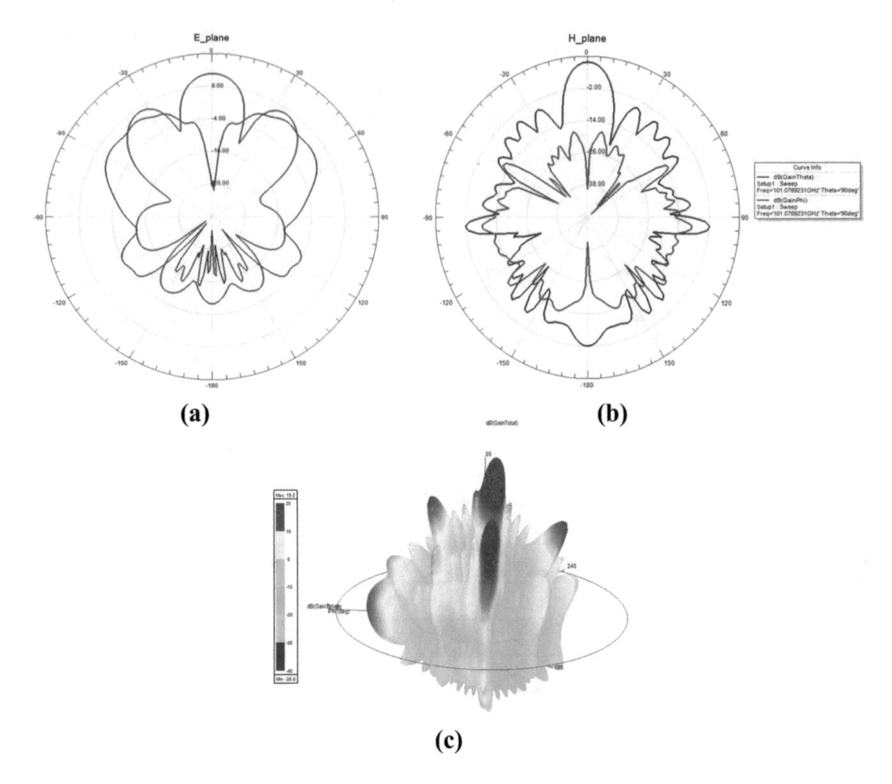

Fig. 4 Simulated far-field radiation patterns for RMAA at 100 GHz in **a** E-plane; **b** H-plane; **c** 3D-plot of the radiation pattern

the first patch itself, and the remaining is passed to the next patch, which radiates the partial amount of energy and partially transfers it to the next.

Simulated far-field radiation pattern of the antenna is shown in Fig. 4. H-plane indicates the broadside profile at the broadside direction and X-pol level is 38 dB below than Co-pol. E-plane pattern is like a flower with multi-shape petals. Peak simulated gain is obtained as 19dBi.

Another option of enhancing the gain of the RMAA structure is utilizing a superstrate structure instead of parasitic patch loading. Figure 5 demonstrates structural description of superstrate structure loaded antenna architecture with detailed cross-sectional and isometric views. Here, h_s is the height of the superstrate material, h_a is the height of air column and h_{sub} is the thickness of the substrate. In this composite structure, in practice, air column is realized by placing a honeycomb structure, whose electrical property greatly resembles air. It is a carbon fiber based material made by Ultracor Company which offers thermal stability of near zero CTE. Part number of the structure is UCF-126-3-8-2.0 [13]. Figure 6 shows the real honeycomb structure used for THz antenna realization.

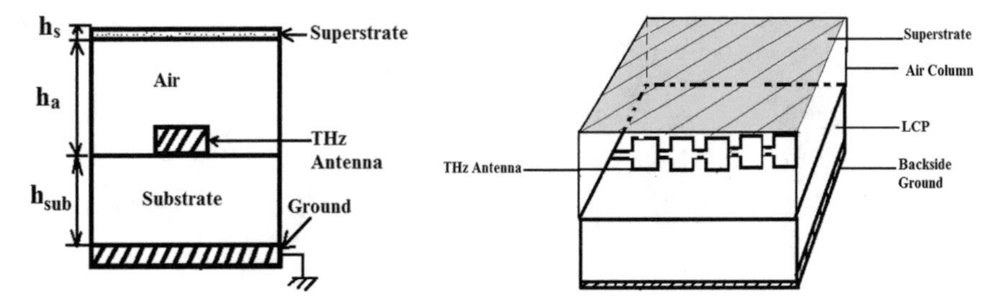

Fig. 5 Superstrate loaded THz antenna for gain enhancement in **a** cross-sectional view; **b** isometric view

Fig. 6 Ultracor honeycomb structure (Part No. UCF-126-3-8-2.0)

This method works on the structural resonance method, and it is also known as the resonance gain method. Earlier, this method has been used to design load-bearing antenna structures [14]. The detailed analysis of this structure with the help of the transmission line model is done in [15]. Co-relation between the substrate thickness and superstrate height is obtained with respect to operating wavelength for a high gain antenna module. The position and thickness of the superstrate are the important parameters for exciting the structural resonance, leading to a higher gain of the whole antenna structure. Three different types of microwave materials are used as superstrate materials, such as-FR-4, quartz glass, and high resistive silicon. Two cases are studied to improve the gain of basic RMAA by the superstrate layer. In the first case, the position of the Silicon superstrate of thickness $h_s = 675$ µm is varied from 0.25 λ_0 to 1.5 λ_0. Corresponding S-parameters are shown in Fig. 7a. We can observe that for $f_0 = 100$ GHz; the resonance occurred for every 0.5 λ_0 step size starting from $h_a = 0.75$ mm. While in the second case, we have varied the superstrate material. The position and the thickness of each superstrate material are determined by the high gain curve for 100 GHz [15]. These dimensions are optimized and listed in Table 3. The S parameter performance for this case is shown in Fig. 7b, and we can observe that structure is resonating at desired 100 GHz frequency for all superstrate

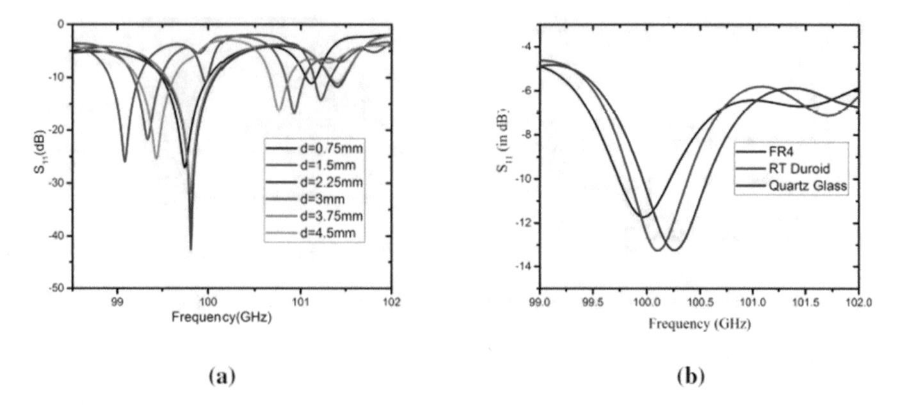

Fig. 7 S_{11} parameter for **a** silicon superstrate with different 'd' **b** different superstrate

layers. The 3D radiation patterns for different superstrates are shown in Fig. 8. It is observed that due to the superstrate, the gain is improved by 3 dB as compared to the stand-alone basic RMAA structure. It has also been observed from the FEM analysis that, the antenna with glass superstrate has very high cross-polarization in H-plane, as shown in Fig. 9 (Table 2).

2.2 *Sectoral H-plane Horn Antenna*

Horn antenna is a widely used simplest and robust candidate of engineers' choice from decades ago. Because of its several inherent characteristics, like wide bandwidth, stable radiation pattern, etc. its demand in the antenna world still remains unaltered as compared to its counterparts. Like lower frequency gamut of EM spectrum, horn antenna becomes indispensable for realizing the feeding network or as a combiner in monopulse tracking radar even in THz frequency bands. It drives the engineers to develop an efficient way of manufacturing such antennas out of advanced fabrication methodologies.

Literature reveals that, one such promising technology in 'Silicon micromachinng' [16–18]. Implementing DRIE (Deep Reactive Ion Etching) method along with wafer bonding makes it possible to realize. Here, in this chapter, we discuss such a silicon micro-machined *H*-plane sectoral horn antenna. Two different frequency bands (75 to 110 GHz and 220–330 GHz) have been targeted here for a 13 dBi gain horn antenna. Flaring the dimension of rectangular waveguide in direction of *H*-field keeping other dimension constant forms *H*-plane sectoral horn. It is expected that, radiation patterns in *H*-plane will be much narrower than *E*-plane because of flaring and larger dimension is in that direction. Design of this horn, as explained in the following subsection, is systematically accomplished using 'normalized directivity v/s aperture size curve'. The directivity and aperture size are related as follows [19]:

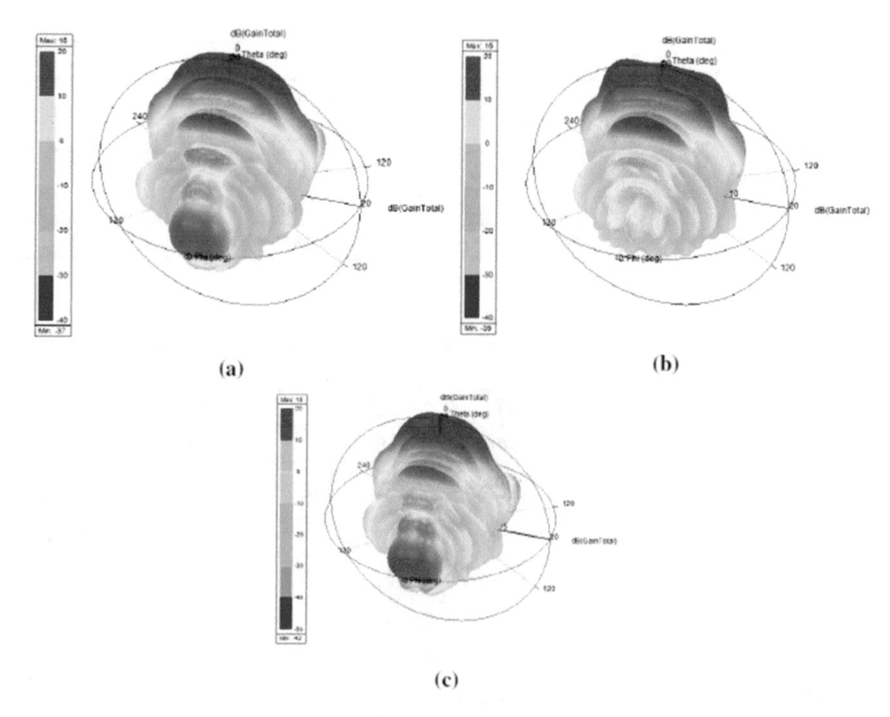

Fig. 8 3D far-field pattern for **a** RT duroid superstrate **b** Quartz glass superstrate **c** FR4 epoxy superstrate

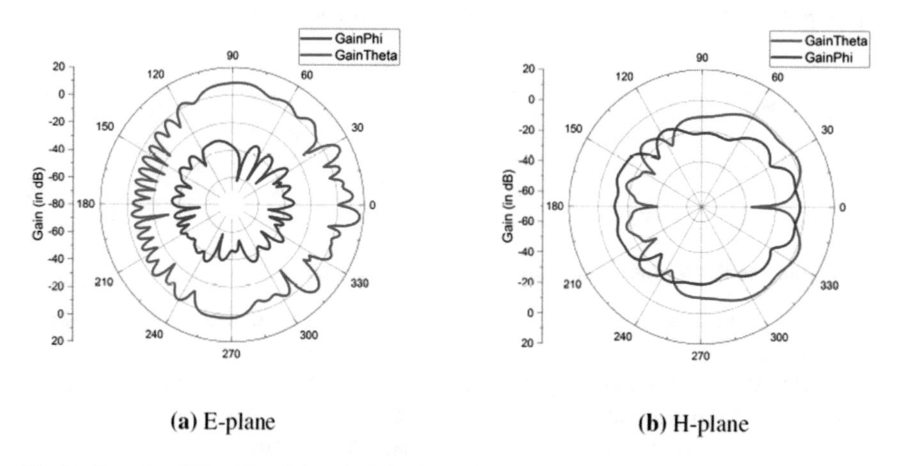

(**a**) E-plane (**b**) H-plane

Fig. 9 Co-pol and X-pol levels in principle planes for glass superstrate RMAA

Table 2 Optimized dimensions of the RMAA with superstrate

Parameter	Values (inμm)
W	3078
L	2489
l_{t1}	1293
l_{t2}	2685
W_{t1}	690

Table 3 Dimension of the composite antenna array with superstrate materials

Superstrate material	Thickness, h_s (in mm)	Height of air column, h_a (in mm)
RT duriod	0.51	1.5
FR4 epoxy	0.35	1.5
Quartz glass	0.55	1.5

$$D_H = \frac{b}{\lambda} \frac{G_H}{\sqrt{\frac{50}{\lambda \rho_h}}} \tag{8}$$

where, 'D_H' is the directivity for H-plane horn antenna, 'b' is the height of the rectangular waveguide, 'λ' is the operating wavelength, 'a' is the width of the feeder waveguide, 'a_1' is the flaring width of the horn antenna, and 'ρ_h' is hypotenuse of right angle triangle form in H-plane given by Eq. (9) and (10).

$$\rho_h = \sqrt{\rho_2^2 + \left(\frac{a_1}{2}\right)^2} \tag{9}$$

$$G_H = \frac{32}{\pi} \frac{a_1}{\lambda} \sqrt{\frac{50}{\lambda \rho_h}} \tag{10}$$

H-plane sectoral horn is designed using flowing steps:

- Defining the desired gain (if it is in dBi convert it to absolute value) of horn antenna.
- Finding the corresponding optimum ρ_1 from the standard normalized directivity versus aperture size graph for H-plane sectoral horn antenna.
- The aperture size a_1 is correspondingly obtained from optimum $\rho_1$value.

Using above mentioned steps the 13 dBi, WR-10, and WR-3.4 Waveguide based H-plane horn antenna is designed and simulated. Figures 10 and 11 show the schematic diagram of the proposed horn. The optimized dimension of antenna structure is given in Table 4. The S-parameter performance is shown in Fig. 12. As revealed from the plot, the S_{11} is below −10 dB in the desired band. Figures 13 and 14 show the far-field plots for these antennas in the principle plane. As it can be observed

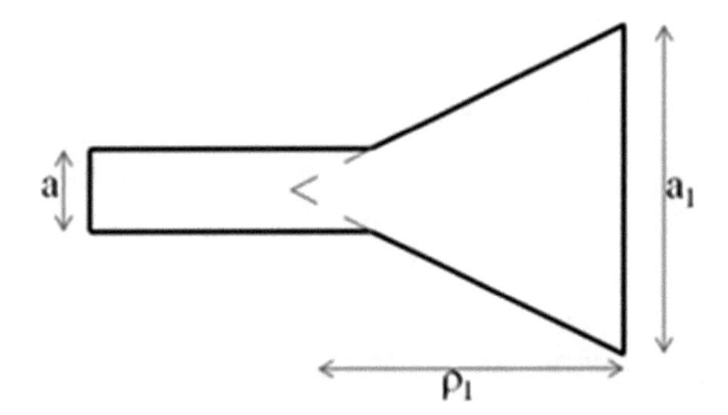

Fig. 10 H-plane view of horn antenna

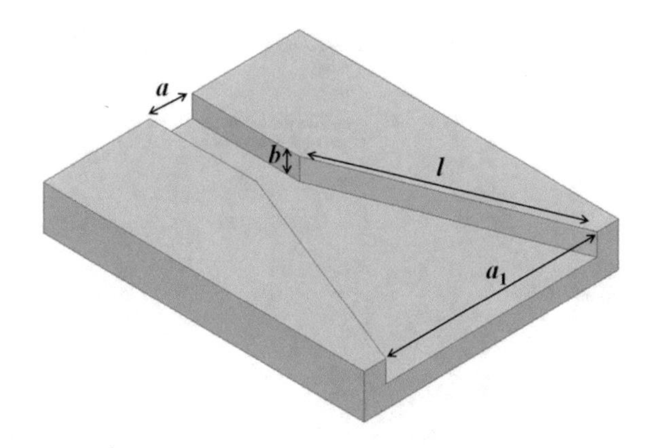

Fig. 11 Isometric view of *H*-plane horn structure

Table 4 Optimized dimensions of the *H*-plane horn antenna

Parameter	WR-10 based structure (in mm)	WR-3.4 based structure (in mm)
a	2.54	0.864
b	1.27	0.432
l	34.63	14.85
a_1	20.12	7.3

that, the *H*-plane plot is much narrower than *E*-plane plot and has simulated gain of 13 dBi. Thus simulated results have close relevance to theory. In addition to this, *X*-pol level is 37 dB below the Co-pol level in principle.

3 Circuit Modeling of THz Antenna

Equivalent circuit model is used for representing the antenna impedance. This helps in extracting features of the antenna. An RLC resonator model has been widely used in modeling antenna impedance. Similar model is attempted here. Out of two antenna categories, circuit modeling for complicated array structure is tried out here.

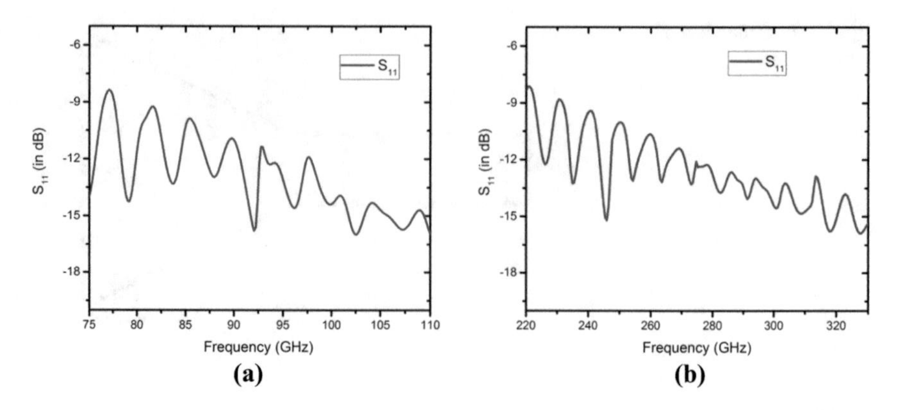

Fig. 12 S_{11} Parameter for **a** WR-10 waveguide based **b** WR-3.4 waveguide based *H*-plane horn structure

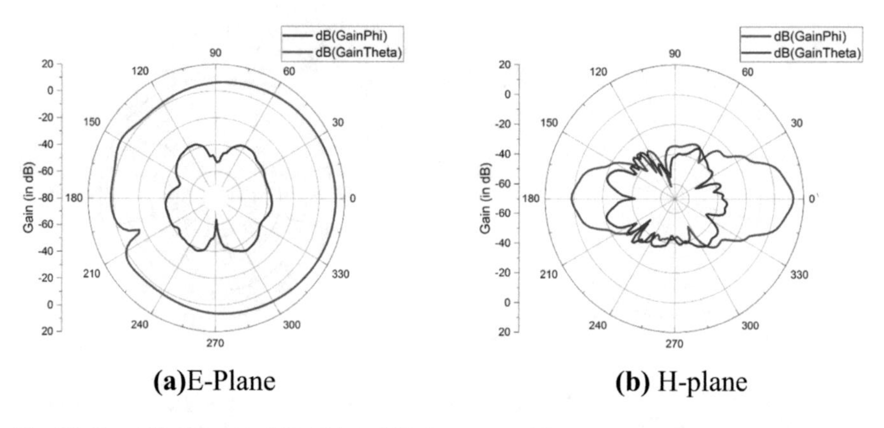

Fig. 13 Far-field pattern for WR-10 based *H*-plane sectoral horn

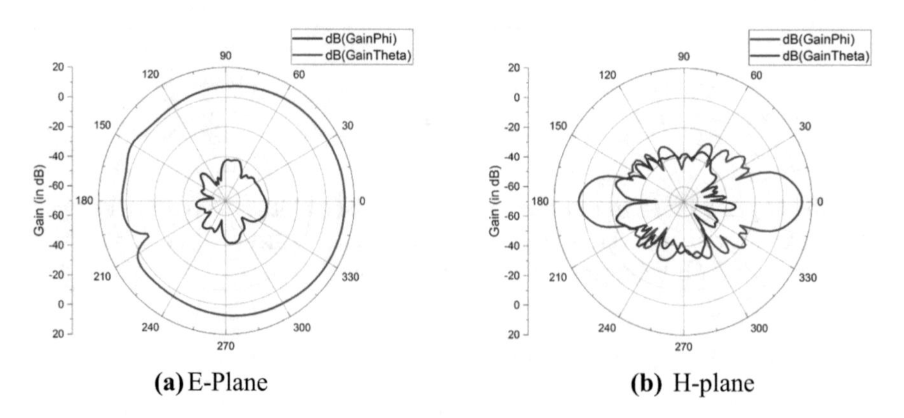

Fig. 14 Far-field pattern for WR-3.4 based *H*-plane sectoral horn

The proposed array structure consists of five main radiating elements, transmission line segments in between them, ten parasitic patches, and feed line as shown in Fig. 15. Main radiating elements are represented by lossy tank circuit (L_{10}; R_{10} and C_{10}) corresponding to TM_{10} mode. Feeding network is represented by series combination of L_f and R_f due to finite conductivity of metal used. Parasitic patches are nothing but capacitive or dielectric loading depending upon substrate electrical properties. Hence, it can be modeled as a leaky capacitor. Transmission line segments between the patches are represented by a combination of lossy inductor and capacitor, corresponding to the parasitic capacitance effects. This lumped circuit is simulated in ADS and tuned to 100 GHz. The obtained lumped components values are listed in Table 5. The S_{11} parameter from lumped circuit and RMAA structure is shown in Fig. 16.

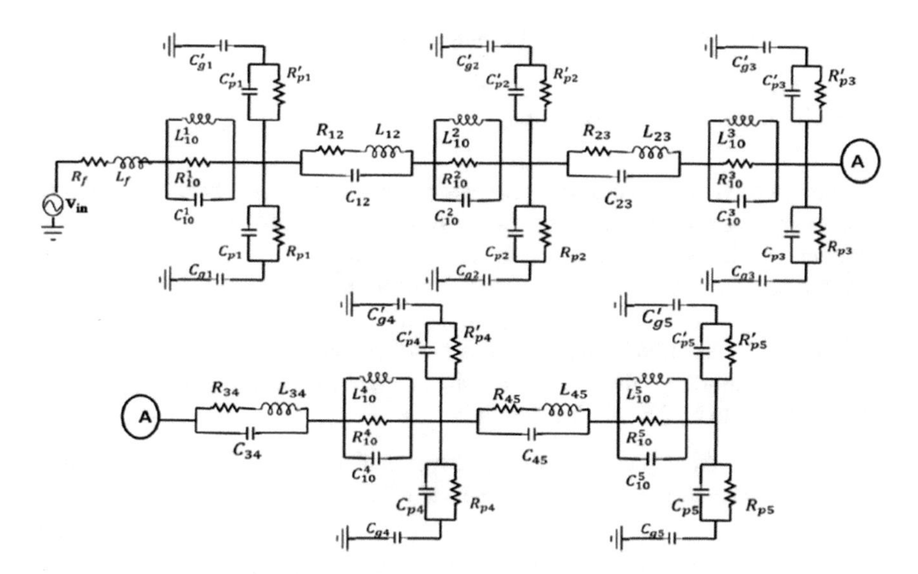

Fig. 15 Transmission line model for the RMAA with parasitic patches

Table 5 Lumped components values for a simulated equivalent model

Variables	Values	Variables	Values
R_f	0.01Ω	R_f	161.12Ω
L_f	54pH	C_{gi}	16.45pF
L^i_{10}	4.28pH	C_{pi}	16.45pF
R^i_{10}	44Ω	R_{ij}	55.56Ω
C^i_{10}	0.591pF	R_{pi}	161.12Ω
C^i_{gi}	16.45pF	L_{ij}	1100pH
C^i_{pi}	16.45pF	C^i_{ij}	7.4pF

Note i varies from 1 to 5 and $j = i + 1$ varies from 2 to 5

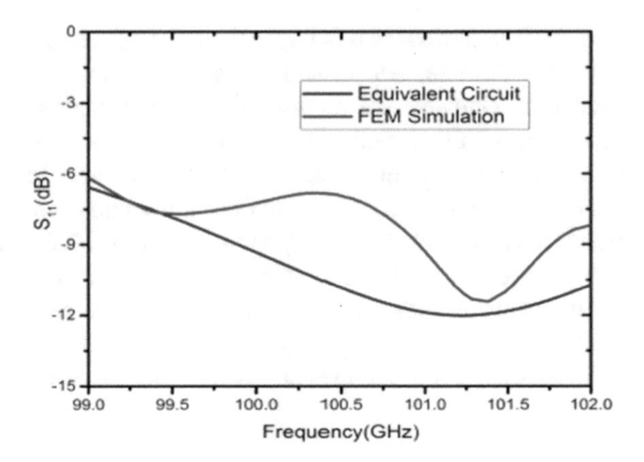

Fig. 16 S_{11} comparison of T-RMAA with parasitic patches and equivalent model

4 Fabrication of Prototypes

Two prototype structures of microstrip patch antenna array have been fabricated on 4 mil thick Liquid Crystal Polymer ($\epsilon_r = 3.14$, tan $\delta = 0.002$) substrate. Figure 17 shows the fabricated prototypes (front view), explaining the flexibility of substrate. Standard wet etching chemistry has been used to pattern the metal layer (25 μm of copper). The backside of the substrate is fully metalized to realize the common ground plane. Currently, the testing process of these antennas is in progress.

The fabrication steps for horn antenna are schematically explained in Figs. 18 and 19. The steps can be described as follows:

(i) It is two wafer processes. Initially, two double-side polished silicon wafers are taken and given requisite chemical cleaning (RCA + SPM) to eliminate

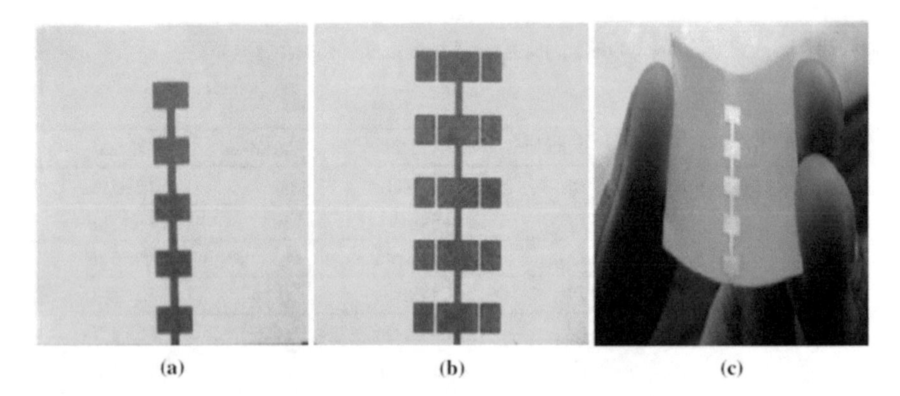

Fig. 17 Fabricated prototypes of the **a** antenna array **b** array with parasitic patches **c** illustration of the conformal shaped antenna

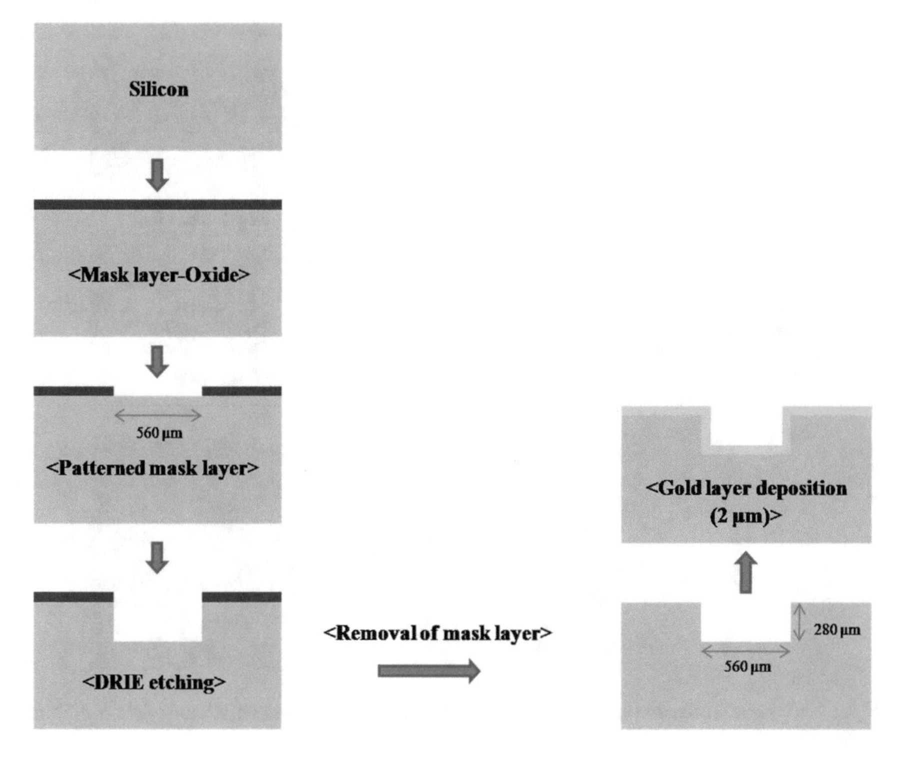

Fig. 18 Preparation of first silicon wafer for making the horn antenna

all possible organic and inorganic contaminants. After that, on the first wafer following process steps have been performed:

(a) A thick oxide layer is deposited on its front side.
(b) The oxide layer is patterned to form a channel through DRIE etching.
(c) DRIE process is applied up to the depth of lateral dimension of the rectangular waveguide.
(d) Oxide layer is removed.
(e) A conformal gold layer is evaporated to metalize the trench.

(ii) Whereas, on the second wafer gold is deposited on the front-side by E-beam evaporation.
(iii) Second wafer is flipped to make the metal layer at its bottom side, and then both of the wafers are bonded together adopting the Eutectic bonding at 363°C. Thus the waveguide channel is formed.

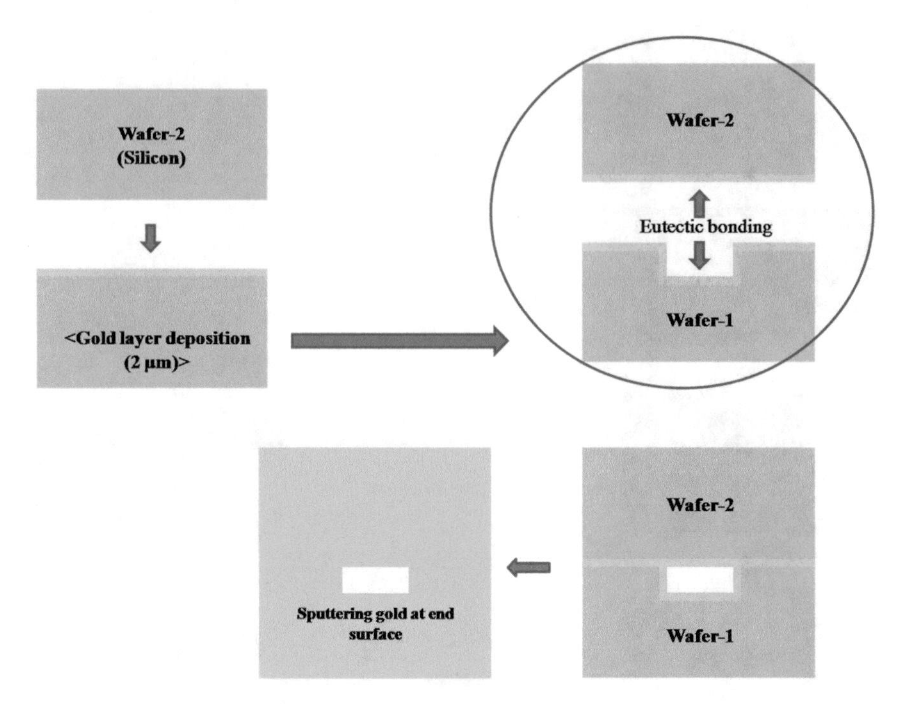

Fig. 19 Generic fabrication steps of horn antenna realization implementing silicon micromachining technique

5 Conclusion

This chapter highlights the essence of high gain antenna module for communication systems in Terahertz regime. Two different types of antennas (horn antenna and microstrip patch array) have been discussed with planar geometries. Ease of design and fabrication simplicity is the main attractive features of these antennas. Detailed design along with circuit modeling and fabrication process details have been covered in this chapter. Parametric simulations were carried out to explore the optimum design parameters. One of the antennas is demonstrated over a bio-compatible, flexible type substrate at 100 GHz, which can further be utilized for medical imaging applications. Effect of superstrate structure is efficiently implemented to enhance the gain of the planar antennas. Another structure involves silicon micromachining technology, where DRIE is extensively used in two silicon wafers separately. Finally, to get an integrated module, two such wafers are bonded. Real time testing of the antenna is underway.

References

1. Z.D. Taylor et al., THz medical imaging: in vivo hydration sensing. IEEE Trans. Terahertz Sci. Technol. **1**(1), 201–219 (2011)
2. D.M. Mittleman et al., Recent advances in terahertz imaging. Appl. Phys. B **68**(6), 1085–1094 (1999)
3. Fukunaga et al., Terahertz spectroscopy applied to the analysis of artists' materials. Appl. Phys. A **100**, 591–597 (2010)
4. C.M. Armstrong, The truth about terahertz. IEEE Spectr. **49**(9), 36–41 (2012)
5. T.G. Phillips, J. Keene, Sub-millimeter astronomy (heterodyne spectroscopy). Proc. IEEE **80**(11), 1662–1678 (1992)
6. M. Rahaman et al., THz communication technology in India present and future. Defense Sci. J. **69**(5), 510–516 (2019)
7. G. Chattopadhyay et al., Terahertz antennas and feeds, in *Aperture Antennas for Millimeter and Sub-Millimeter Wave Applications*. ed. by A. Boriskin, R. Sauleau (Springer International Publishing, Cham, 2018), pp. 335–386
8. N. Kingsley, "Liquid crystal polymer: enabling next-generation conformal and multilayer electronics". Microw. J. 188–200 (May 2008)
9. D.C. Thompson, O. Tantot, H. Jallageas, G.E. Ponchak, M.M. Tentzeris, J. Papapolymerou, "Characterization of liquid crystal polymer(LCP) material and transmission lines on LCP substrates from 30 to 110 GHz". IEEE Trans. Microw. Theory Tech. **52**(4), 1343–1352 (April 2004)
10. Georgia Tech Report. Available: https://phys.org/news/2006-08-georgia-tech-liquid-crystal-polymer.html
11. C.A. Balanis, "*Antenna Theory Analysis and Design*", 3rd edn. (Wiley, 2005)
12. M.S. Rabbani, H. Gi-Shiraz, "Improvement of microstrip antenna's gain, bandwidth and fabrication tolerance at terahertz frequency bands". *Processing Wideband Multi-Band Antennas Arrays Civil, Security Military Applications Conference* (2015)
13. www.ultracorinc.com
14. C.S. You, W. Hwang, S.Y. Eom, "Design and fabrication of composite smart structures for communication, using the structural resonance of radiated field." *Smart Materials and Structures*, 14.2 (Mar. 2005), pp. 441–448
15. C. You, M.M. Tentzeris, W. Hwang, Multilayer effects on microstrip antennas for their integration with mechanical structures. IEEE Trans. Antennas Propag. **55**(4), 1051–1058 (2007)
16. G.M. Rebeiz, "*RF MEMS Theory, Design, and Technology*". (Wiley, 2003)
17. J. Iannacci, RF-MEMS for high-performance and widely reconfigurable passive components—a review with focus on future telecommunications, internet of things (IoT) and 5G applications. J. King Saud Univ.-Sci. **29**(4), 436–443 (2017)
18. A. Karmakar, K. Singh, *"Si-RF Technology", ISBN: 978-981-13-8051-8*, 1st edn. (Springer, Singapore, 2019)
19. L. Jun, A universal method for directivity synthesis. IEEE Trans. Antennas Propag. **35**(11), 1199–1205 (1987). https://doi.org/10.1109/TAP.1987.1144014

Element Failure Correction Techniques for Phased Array Antennas in Future Terahertz Communication Systems

Hina Munsif, B. D. Braaten, and Irfanullah

Abstract The terahertz frequency band is the range of frequencies from 300 GHz to 10 THz (above the microwave range and below the optical spectrum) in the electromagnetic spectrum. The future terahertz-range indoor communication system will require a highly directive terahertz beamforming antenna array to switch between different devices (mobile, home appliances, etc.). As such, there will be a high probability of antenna elements failure, which will disrupt the high-volume wireless communications. Therefore, study and analysis of antenna array failure correction techniques are important. In the literature, only a few methods have been reported for array element failure correction. This chapter provides an overview and tutorial on the methods that have been reported in the literature for array element failure correction in the context of terahertz beam control. A pattern recovery technique of low sidelobe antenna arrays operating at 3.5 THz, using the iterative Fourier technique (IFT) is discussed. The occurrence of defective elements across the array is included by setting the associated excitation coefficients to zero. The presented results are related to 1×8 linear patch antenna array failure correction model with half-wavelength inter-element spacing operating at 3.5 THz in MATLAB and CST simulators. The array failure correction using this method is evaluated for various sidelobe levels (SLLs), where the failed elements are randomly dispersed across the array. The analysis of array failure correction algorithm on the sidelobe level (SLL), directivity, half-power beamwidth (HPBW), and taper efficiency of the recovered radiation patterns are reported in detail, and guidelines are provided on the performance parameters.

Keywords Phased arrays · Radiation pattern · Optimization · Microstrip arrays

H. Munsif · Irfanullah (✉)
Electrical and Computer Engineering Department, COMSATS University Islamabad, Abbottabad, Pakistan
e-mail: eengr@cuiatd.edu.pk

B. D. Braaten
Electrical and Computer Engineering Department, North Dakota State University, Fargo, USA

S. Das et al. (eds.), *Advances in Terahertz Technology and Its Applications*,
https://doi.org/10.1007/978-981-16-5731-3_2

1 Introduction

The antenna is the main component in wireless communication, which was first revealed by Marconi in 1901 [1–3]. Terahertz (THz) frequency band (0.3 THz to 10 THz) is a part of sub-millimeter EM (electromagnetic) spectrum, shown in Fig. 1a. Recently in wireless communication, the terahertz frequency band has been reserved for real life applications [4], as shown in Fig. 1b. In the literature, recently there is growing research on THz applications, for example in communications, sensors, radar, imaging, and other research applications [5–7]. Some of the major reasons for the increasing interest in THz frequency band are given below [8, 9]:

1. THz radiations are able to penetrate through luggage and covering materials like corrugated cardboard, packaging, book bags, clothes, and shoes, etc. to discover any dangerous or unsafe materials concealed within.
2. The THz systems do not damage the suspect or human body that is scanned under THz radiations.
3. Terahertz imaging systems can detect corrosion or deterioration on metallic exterior of aerospace structure and any fault in composite materials of aerospace structure as well [10, 11].
4. The ECRs imaging and terahertz spectroscopy have been explored intensively for possible biomedical, security, and defense applications.

As antenna is the backbone for every wireless application and therefore antenna arrays are used for the purpose of beamforming and increasing the gain of antenna. Usually, the antenna arrays are made of large number of antenna elements. Hence, there is a high probability of failure of radiating elements. Phased antenna arrays used for radar and satellite applications can have element failure typically due to amplifiers. As antenna arrays are used in battlefield, aerospace, and other communication applications, so the damaged components in antenna arrays are hard to be repaired in real time. The failure of radiating elements can result in an increase in sidelobe level (SLL), filling nulls position, beam widening, and loss in gain, depending on the number of failed elements. To recover these performance parameters of array

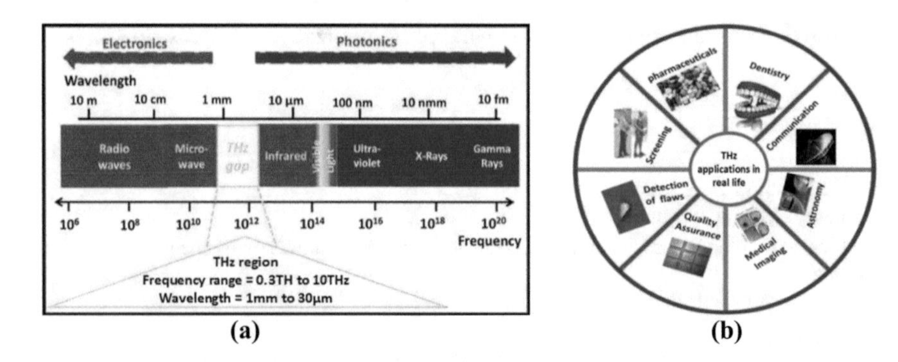

Fig. 1 THz frequency **a** Frequency range and wavelength **b** Applications in real life

by replacing non-radiating elements is quite impossible in situations, where antenna array is on a platform in space. The hardware replacement is a costly way to recover the array availability. While the presence of faulty elements makes antenna array non-symmetrical, because of this issue, the traditional analytical techniques are difficult to implement. Recently, the researchers have explored some self-repairing techniques based on optimization and these optimization techniques can recover the array pattern in the presence of failed elements. Thus, researchers sort out an approach through which the properties of array radiation patterns can be maintained: this approach is to re-synthesize or recalculate the excitation coefficients of the working elements instead of replacing them. In the literature, there are two approaches for antenna array failure correction: (1) software solutions: some of them are optimization algorithms, for example, PSO (particle swarm optimization algorithm), GA (genetic algorithm), BFO (bacteria foraging algorithm), and IFT (iterative Fourier technique), and (2) hardware solutions: rarely used due to its complicated designs (discussed in section I.II).

1.1 Antenna Array Failure Correction Software Techniques

In the software techniques presented in this section, majority algorithms work to recalculate the amplitudes/phases of the array antenna elements for radiation pattern recovery. In [12], the proposed algorithm calculates the new phases of non-defective phase shifters to restore the broad null (20° to 30°) with −66 dB depth in the desired direction of the 28 isotropic elements linear array pattern. It is shown that the null can be recovered at the cost of increasing SLL for random elements failures. A rectangular 5800 elements planar array was investigated in [13] for SLL (−45 dB) and gain by defecting 290 random elements using inverse Fourier transform. In [14], 50 and 80 elements linear isotropic array was investigated for low SLL (−45 dB) synthesis using 8 elements random failure. In [15], the genetic algorithm (GA) can recover the best pattern with SLL of −20 dB for 18 and 44 elements linear isotropic arrays from the damaged pattern by re-synthesizing the excitation for the working elements. The work in [16] has presented a novel real-time pattern correction approach based on sparse-recovery technique on 17 × 17 planar arrays for SLLs of −20 dB and −40 dB. The IFFT approach in [17] has investigated single and dual wide nulls along with specified SLL value for 1 × 34 linear isotropic arrays. The DE (Differential Evolution) algorithm in [18] is used to investigate the 1 × 30 linear half-wave dipole array by optimizing spacing among antenna elements for pattern recovery in case of antenna elements failure. In [19], the performance of two optimization algorithms particle swarm optimization (PSO) and bacteria foraging optimization (BFO) is analyzed. By using the above algorithms, the excitations were recalculated for the array with elements failure and the radiation patterns are restored. The main emphasis of [19] was to recover the SLL and steering nulls in the specified directions, using amplitude correction approach. The analysis was performed on a 1 × 32 linear array with element spacing of $\lambda/2$, and the array was designed at 2.33 GHz. The work in [20] has proposed a novel technique CUSUM (cumulative sum scheme) for active phased antenna arrays that are used for correction and detection of failures using a systemic process for the arrays. This technique is considered to monitor phase shifters and alternators inside TRMs (transmitter receiver modules). Two techniques NU-NCLMS (non-uniform norm constrained least mean square) and GA (genetic algorithm) are analyzed in [21] for re-optimization of array pattern. The implementation of PSO algorithm reported in [22] can be used for pattern correction/optimization and additionally considers some other effects, like element pattern, element failure, mutual coupling, and installation platform. A new signal processing algorithm known as RIO (Recursive Intelligent Optimizer) was used in [23] to operate in stochastic and deterministic mode and offers beam steering, fast beamforming, tolerates faulty array element and interference rejection by using digitally controlled phase shifters. The software techniques presented above are tabulated in Table 1.

1.2 Antenna Array Failure Correction Hardware Solutions

The hardware techniques for antenna array failure correction are limited due to its complexity in real-time implementation. Majority of hardware solutions reported in the literature work using the technique in [24]. In the hardware solution, backup/redundant or idle transmit/receive (TR) modules are used in case of its failure. The hardware connections of faulty TR cells are disconnected, and backup TR cells are connected to the input/output switching module, power distribution system, and detection module. The solution in [24] has analyzed different types of faulty TR cells in 1×16 linear antenna array, for example, single faculty TR cell, adjacent faulty TR cells with different sequences.

The detailed literature review of software and hardware approaches for antenna array failure correction is given in Table 1. At this point, the inverse Fourier technique (IFT) presented in this book chapter is closely related to the work in [13] and [14] with multi-functional capabilities of computing SLL, gain, and taper efficiency simultaneously.

2 Proposed IFT Antenna Array Failure Correction Technique

The array pattern of N-elements linear array with inter-element spacing of d can be written as the product of element pattern (EP) and array factor (AF):

$$AP = EP \times AF \tag{1}$$

Table 1 Literature review of antenna array failure correction techniques

Ref. No.	Technique used	Year	Synthesized factor	Array type	Array element	Max defected elements	Taper Efficiency (%)	SLL recovery	Gain recovery	HPBW recovery	Null's restoration
[12]	Computer coding	1994	Phase-only	Linear	28	2	No	No	No	No	**Yes**
[13]	Iterative fourier method	2007	Amplitude-only	planar	5800	1740	No	**Yes**	No	No	No
[14]	Iterative fourier method	2009	Amplitude-only	Linear	50	8	**58.2**	**Yes**	No	No	No
[15]	Genetic algorithm	2011	Amplitude-only + Complex weight	Linear	44	3	No	**Yes**	No	No	No
[16]	Approach based on sparse-recovery technique	2014	Amplitude-only	Planar	17 × 17	3	No	**Yes**	No	No	No
[17]	IFFT base algorithm	2015	Amplitude-only	Linear	34	4	No	**Yes**	No	No	**Yes**
[18]	Differential evolution algorithm	2016	Inter-element spacing	Linear	30	3	No	**Yes**	No	**Yes**	No
[19]	PSO and BFO	2017	Amplitude-only	Linear	32	3	No	**Yes**	No	No	**Yes**
[20]	Novel technique based on CUSUM	2018	Complex weights	Linear Planar	32 8 × 8	10 20	No	**Yes** **No**	No	No **Yes**	No
[21]	NU-NCLMS and GA	2018	Amplitude-only	Linear	29	8	No	**Yes**	No	No	**Yes**

(continued)

Table 1 (continued)

Ref. No.	Technique used	Year	Synthesized factor	Array type	Array element	Max defected elements	Taper Efficiency (%)	SLL recovery	Gain recovery	HPBW recovery	Null's restoration
[22]	PSO	2019	Complex weights	Linear	8	1	No	**Yes**	No	No	No
[23]	RIO	2019	Complex weights	Linear	50	5	No	**Yes**	No	No	No
[24]	Hardware	2020	Backup TR modules used	Linear	16	6	No	**Yes**	**Yes**	**Yes**	**Yes**
Proposed technique	IFT technique		Amplitude-only + complex weights	Linear	64	16	**Yes**	**Yes**	**Yes**	No	No

$$AF = \sum_{n=0}^{N-1} w_n.e^{j2\pi \text{ndsin}\theta} \quad (2)$$

where $w_n = A_n e^{j\Delta\emptyset}$ is the complex excitation weight (amplitude and phase) of nth antenna element and θ is the spherical angular distance measured from array normal (z-axis). A flow chart for antenna array failure correction of the proposed IFT technique based on the work in [13, 14] is shown in Fig. 2 and briefly discussed as follows: Initially, the antenna array is excited with uniform amplitudes A_n = ones(1, N) and zero inter-element phase shift $\Delta\emptyset = 0$. Then AF calculation of (2) is done using M-point IFFT (inverse Fourier transform), where M > N and M–N zeros padding are used. The MATLAB command used is given in (3). Next, the iterative synthesis is applied on the sidelobe region of the AF to achieve the desired SLL, and only those sidelobe samples of the AF are corrected which are exceeded the SLL threshold, while other samples are unchanged. After this correction step, a direct M-point FFT is applied on the modified array factor of (3) to compute the new updated elements excitation coefficients as given using MATLAB command in (4).

$$AF = \text{ifftshift}(\text{ifft}(A_n, M)); \quad (3)$$

$$A_n = \text{fft}(\text{ifftshift}(AF), M); \quad (4)$$

The above steps are repeated for random elements failure by simply setting the weights of defected elements equal to zero, and recalculating the updated weighting coefficients. When the required SLL with random elements failure is achieved, then the updated excitation coefficients are used in the AF formula given in Eq. (5), where for amplitude-only the progressive phase is set to zero, and for complex weights, the progressive phase for the required scan angle can be calculated using Eq. (6).

$$AF = \sum_{n=0}^{N-1} A_n e^{j\Delta\emptyset}.e^{j2\pi \text{nd} \sin\theta} \quad (5)$$

where $\Delta\emptyset$ is progressive phase shift given by

$$\Delta\emptyset = -2\pi \text{nd} \sin\theta_s \quad (6)$$

θ_s is the scan angle for mainbeam.

3 Single Antenna Element Design at 3.5 THz

Using the guidelines in [25], a 3.5 THz microstrip patch antenna was designed in CST on FR4 substrate ($\varepsilon_r = 4.4$ and $\tan\delta = 0.02$) as shown in Fig. 3a and 8 elements linear array shown in Fig. 3b. Various formulae for calculating the effective patch

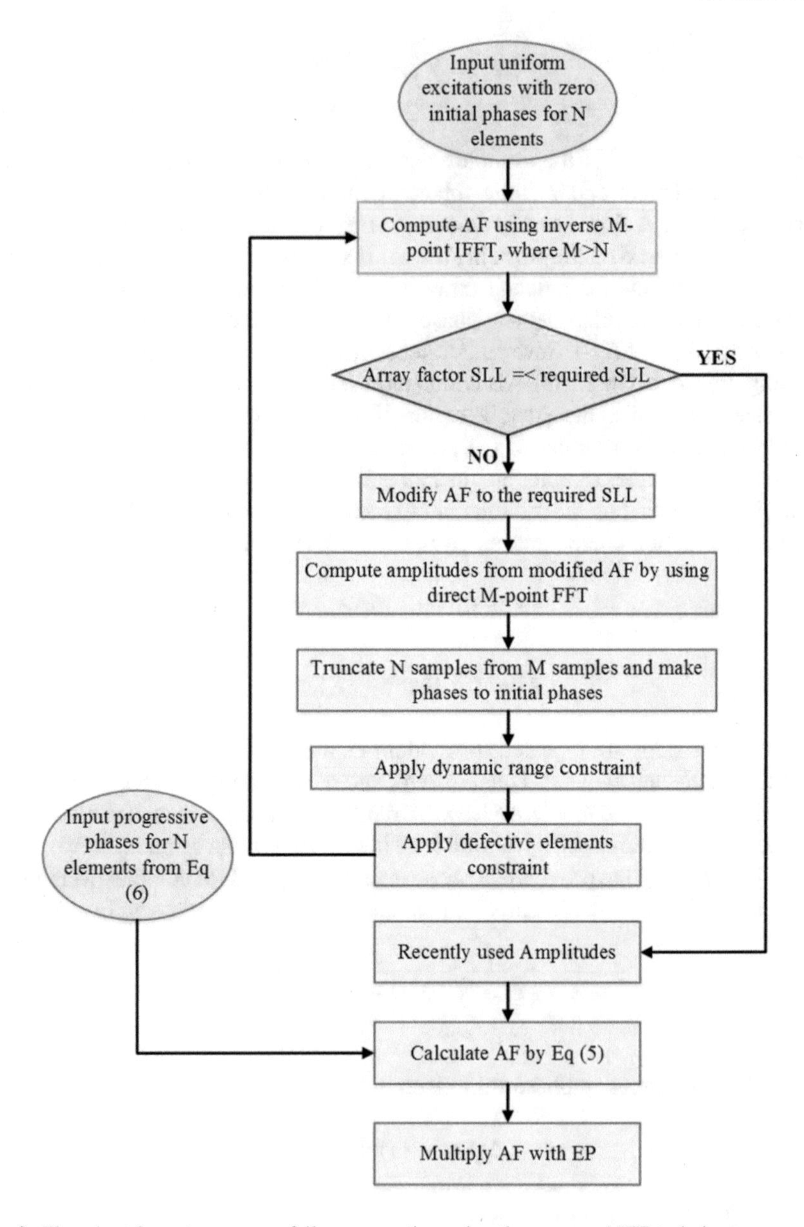

Fig. 2 Flowchart for antenna array failure correction using the proposed IFT technique

dimensions of a microstrip antenna are known in the literature [26], but they are limited to $h/\lambda_o \leq 1/10$. The thickness of the chosen substrate is lowered as a result of this enforced constraint. The dimensions of the single patch element operating at 3.5 THz are listed in Table 2, and the performance parameters of the antenna are given in Table 3. The length of the patch can be calculated from Eq. (7) to (10),

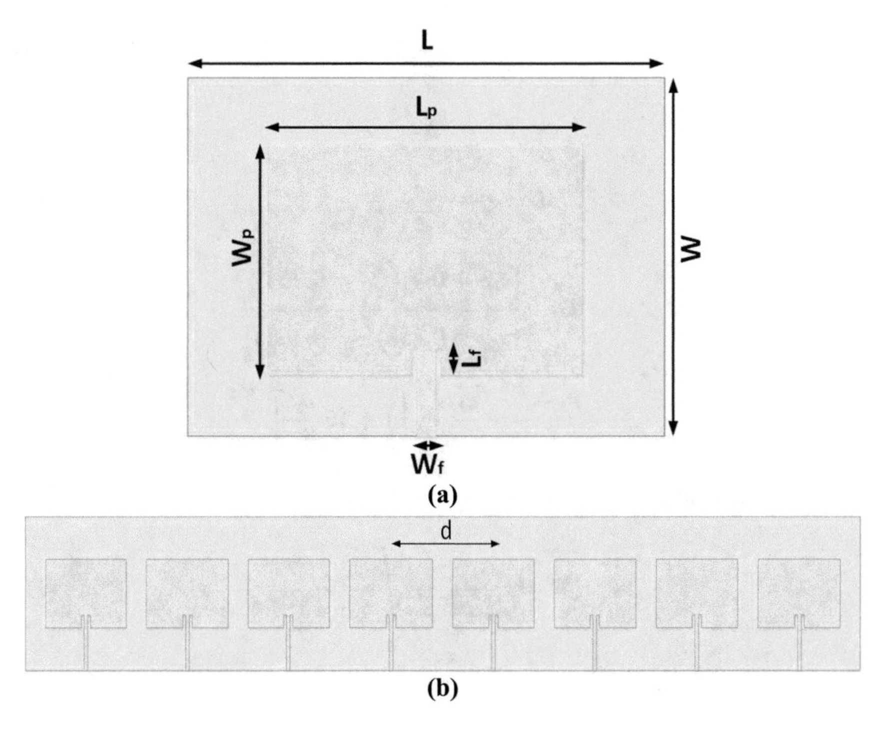

Fig. 3 **a** dimensions of a 3.5 THz microstrip patch antenna **b** 1 × 8 THz microstrip patch linear antenna array

Table 2 Design parameters of the proposed 3.5 THz microstrip patch antenna

Variable	Value (μm)
Patch length (L_p)	18.969
Patch width (W_p)	25
Inset feed length (L_f)	2.349
Microstrip feed width (W_f)	1.368
Ground/substrate length (L)	28.453
Ground/substrate width (W)	37.5
Patch/ground thickness (h)	0.02
Substrate thickness (h_s)	0.45
Normalized inter-element spacing (d)	0.5

Table 3 Performance parameters of the proposed 3.5 THz microstrip patch antenna

Performance parameter	Value
Bandwidth	0.1376 THz
Directivity	6.01 dBi
HPBW	88.2°

$$L_{eff} = L + 2\Delta L \tag{7}$$

where

$$L = \frac{1}{2f_r\sqrt{\varepsilon_{eff}}\sqrt{\varepsilon_o\mu_o}} \tag{8}$$

$$\Delta L = 0.412\frac{\left(\varepsilon_{eff} + 0.3\right)\left(\frac{W_p}{h_s} + 0.264\right)}{\left(\varepsilon_{eff} + 0.258\right)\left(\frac{W_p}{h_s} + 0.8\right)}h_s \tag{9}$$

$$\varepsilon_{eff} = \frac{\varepsilon_r + 1}{2} + \frac{\varepsilon_r - 1}{2}\left[1 + 12\frac{h_s}{W_p}\right]^{-1/2} \tag{10}$$

and width is calculated by Eq. (11).

$$W = \frac{1}{2f_r\sqrt{\varepsilon_o\mu_o}}\sqrt{\frac{2}{\varepsilon_r + 1}} \tag{11}$$

3.1 3.5 THz Linear Antenna Array with Element Failure Correction

THz beamforming with 25% elements failure correction using 1 × 8 microstrip patch linear array is considered. Two cases, adjacent elements (1st and 2nd) and edge elements (1st and 8th) as defected elements are implemented. In both cases, broadside, 20° scanned and 40° scanned patterns for −20 dB, −30 dB and −40 dB SLLs are discussed.

4 Simulation Results

To validate the proposed IFT technique for antenna array failure correction, a 1 × 8 elements microstrip patch antenna array operating at 3.5 THz was simulated in a commercially available CST simulator. The proposed IFT technique for 25% elements failure correction was first implemented in MATLAB to recalculate the array weights (amplitudes and phases) after random elements failure. The array elements in CST were excited with these MATLAB computed complex weights to recover the radiation pattern. Figures 4, 5, 6, 7, 8, 9, 10, 11, 12 and 13, show the pattern obtained by the proposed algorithm in MATALB, and 14 to 13 show patterns achieved in CST. In each array, 25% elements are considered defective elements.

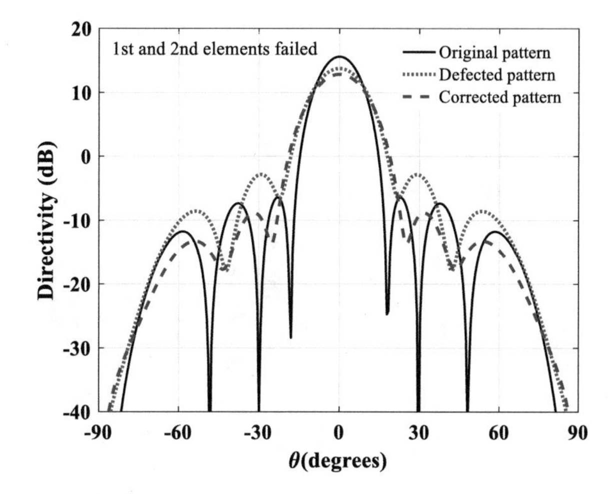

Fig. 4 MATLAB comparison of radiation patterns for broadside ($\theta_S = 0°$) 3.5 THz 1 × 8 linear array failure correction with 1st and 2nd failed elements and SLL = −20 dB

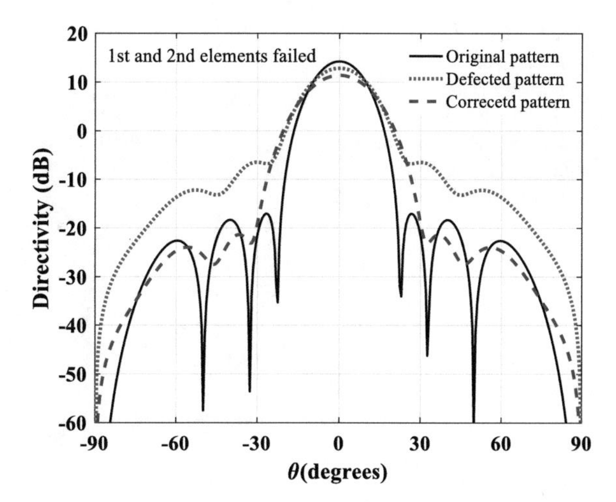

Fig. 5 MATLAB comparison of radiation patterns for broadside $\left(\theta_S = 0^{\circ}\right)$ 3.5 THz 1 × 8 linear array failure correction with 1st and 2nd failed elements and SLL = −30 dB

Each example of linear array failure correction is divided into three cases: broadside, scanned at 20°, and scanned at 40°, and then these three cases are further investigated for 3 sub-cases of −20 dB SLL, −30 dB SLL, and −40 dB SLL.

Dynamic range is the ratio of maximum amplitude value to minimum amplitude value of excitation coefficients of elements ($A_{\max}/A_{\min}$), and its value is limited to 25 dB in the proposed IFT failure correction algorithm to maintain an acceptable taper efficiency. The low sidelobe synthesis has 512-point inverse and direct FFTs. In Tables 1 to 18, OP stands for original pattern which means array pattern without defecting any element, DP stands for defected pattern which means array pattern with defecting 25% elements without applying proposed IFT technique, and CP stands for corrected pattern which means array pattern of array after defecting 25% elements and corrected by the proposed IFT technique.

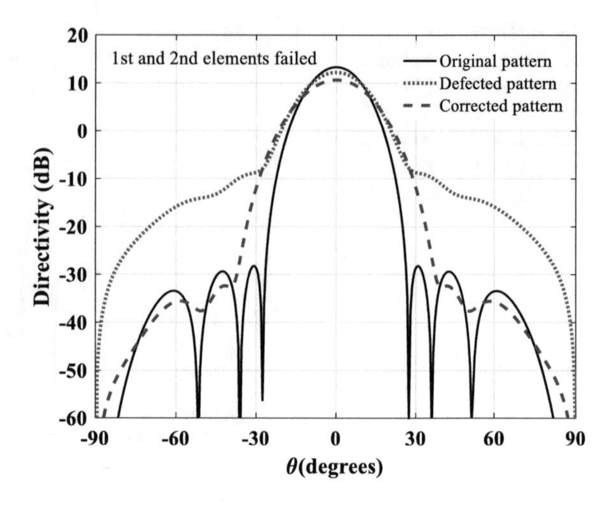

Fig. 6 MATLAB comparison of radiation patterns for broadside ($\theta_S = 0^\circ$) 3.5 THz 1 × 8 linear array failure correction with 1st and 2nd failed elements and SLL = −40 dB

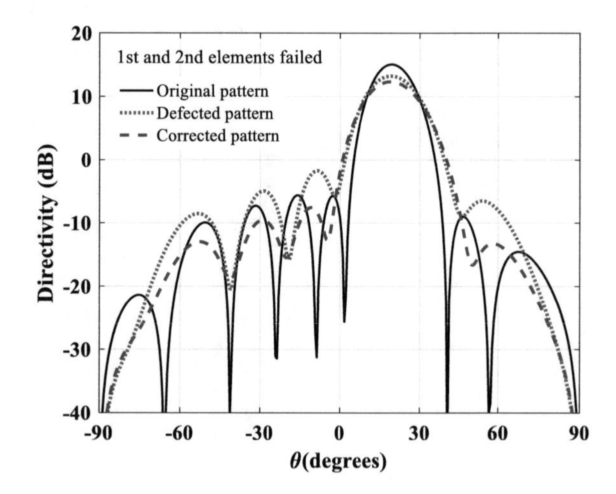

Fig. 7 MATLAB comparison of radiation patterns for scanned ($\theta_S = 20^\circ$) 3.5 THz 1 × 8 linear array failure correction with 1st and 2nd failed elements and SLL = −20 dB

A. MATLAB simulation

Case 1: Adjacent Defected Elements (1st and 2nd)

In the first case of 3.5 THz 1 × 8 microstrip patch linear antenna array failure correction, 1st and 2nd adjacent elements are taken as defected elements. For case 1, the proposed IFT antenna array failure correction technique in section II was implemented in MATLAB to compute the array weights for sidelobe levels (SLL = −20 dB, SLL = −30 dB and SLL = −40 dB) of the broadside array radiation pattern. The original pattern (denoted as OP), defected pattern (denoted as DP), and corrected pattern (denoted as CP)) are given in Figs. 4, 5 and 6, and the achieved SLLs, HPBWs, directivity, and taper efficiency are provided in Table 4. Similarly, original, defected, and corrected patterns for above SLLs using main beam scanned at 20° and at 40° are

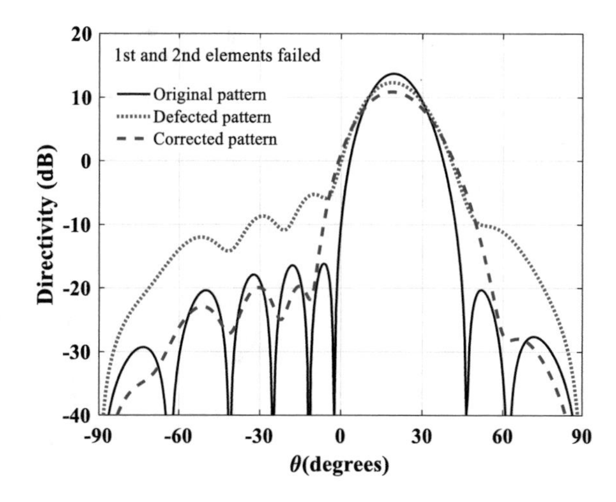

Fig. 8 MATLAB comparison of radiation patterns for scanned ($\theta_S = 20°$) 3.5 THz 1 × 8 linear array failure correction with 1st and 2nd failed elements and SLL = −30 dB

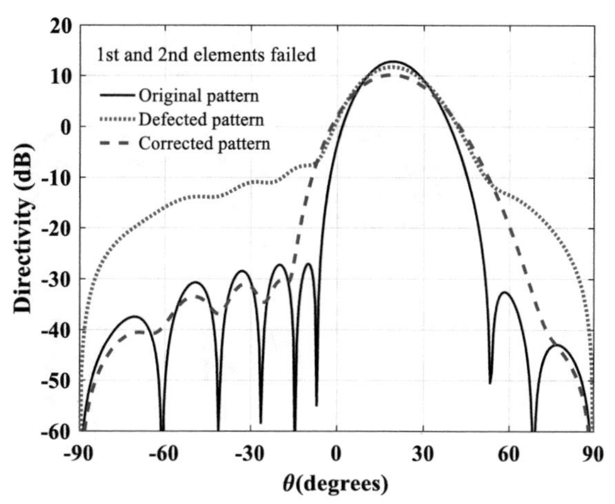

Fig. 9 MATLAB comparison of radiation patterns for scanned ($\theta_S = 20°$) 3.5 THz 1 × 8 linear array failure correction with 1st and 2nd failed elements and SLL = −40 dB

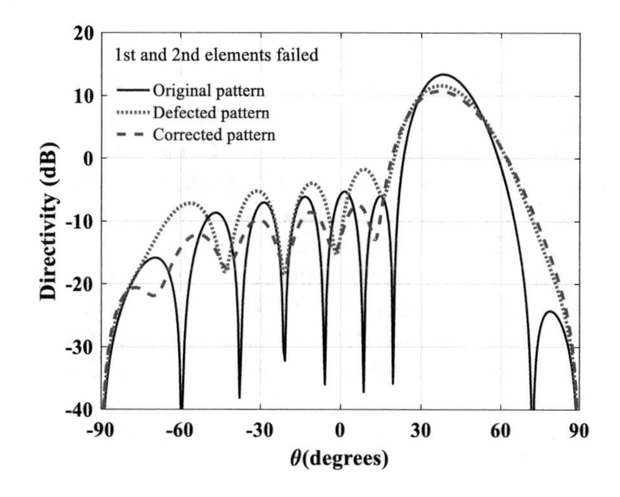

Fig. 10 MATLAB comparison of radiation patterns for scanned ($\theta_S = 40°$) 3.5 THz 1 × 8 linear array failure correction with 1st and 2nd failed elements and SLL = −20 dB

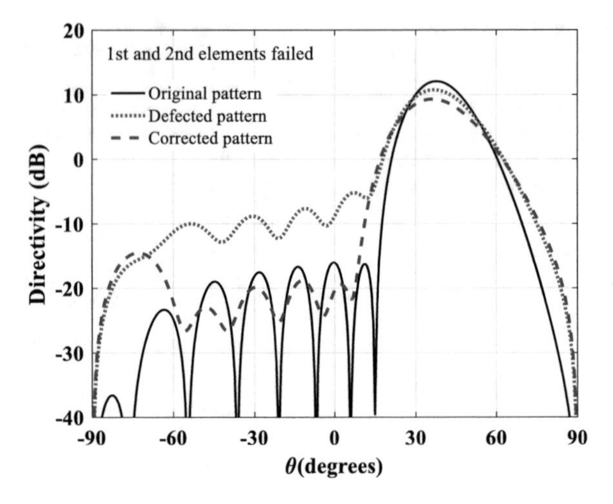

Fig. 11 MATLAB comparison of radiation patterns for scanned ($\theta_S = 40°$) 3.5 THz 1×8 linear array failure correction with 1st and 2nd failed elements and SLL $= -30$ dB

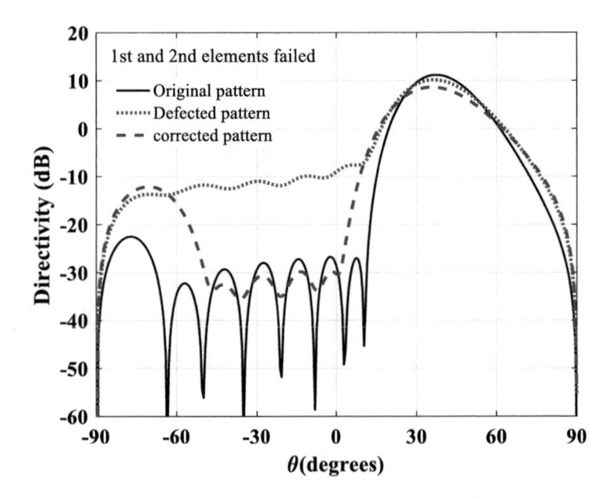

Fig. 12 MATLAB comparison of radiation patterns for scanned ($\theta_S = 40°$) 3.5 THz 1×8 linear array failure correction with 1st and 2nd failed elements and SLL $= -40$ dB

given in Figs. 7, 8 and 9 and Figs. 10, 11 and 12, respectively. Their achieved SLLs, taper efficiency, directivity, and HPBWs are provided in Tables 5 and 6. These figures and tables show that for all desired SLLs with 1st and 2nd defecting elements, the original radiation patterns are recovered fairly by the proposed IFT failure correction algorithm. However, the cost of algorithm is widening the main beam due to inverse relationship between SLL and HPBW. If the desired SLL decreases, the HPBW increases. The decrease in the directivity can also be seen in the corrected patterns; this decrease is due to the 25% of elements taken as defected and therefore cannot contribute to the overall array radiation pattern. This decrease in directivity can also be the cause of widening the HPBW. In phased array applications, when mainbeam is scanned from broadside to 20° and 40°, the directivity of the array decreases due to scan loss. During scanning, the element pattern has maximum value of directivity

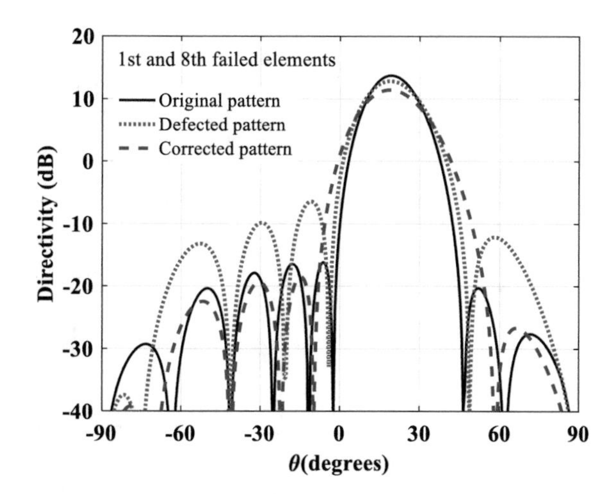

Fig. 13 MATLAB comparison of radiation patterns for scanned ($\theta_S = 20°$) 3.5 THz 1 × 8 linear array failure correction with 1st and 8th failed elements and SLL = −30 dB

at broadside and decreases gradually as moving away from broadside. The taper efficiency of original pattern is greater than the taper efficiency of corrected pattern. As the taper efficiency is related to the amplitudes dynamic range, and the dynamic range constraints of 25 dB are applied in the algorithm to maintain the taper efficiency. If these constraints are relaxed to 30 dB instead of 25 dB, then the taper efficiency of corrected pattern further decreases than the mentioned efficiencies in the Tables. It should be noted that in Figs. 11 and 12, the desired SLL is not achieved; this is most probably because of not fine tuning the algorithm. If the algorithm is further optimized, the problem will be solved, or increasing the dynamic range to 30 dB can also resolve the issue. However, increasing dynamic range can also decrease the taper efficiency. The computed excitation amplitudes for various SLLs using the proposed IFT antenna array failure correction algorithm are given in Table 8 of appendix A.

Case 2: Edge Defected Elements (1st and 8th)

For case 2, the edge elements 1st and 8th in the 3.5 THz 1 × 8 linear antenna array failure correction are considered as defected. For SLLs (SLL = −20 dB, SLL = −30 dB and SLL = −40 dB) of broadside and scanned radiation patterns, the original, defected, and corrected patterns are investigated similarly as in case 1. The required SLLs were successfully obtained in recovered patterns with the help of the proposed IFT failure correction algorithm. The behavior of widening HPBW and decreasing the directivity in 1 × 8 linear arrays with first and eighth defected elements is the same as in case 1(adjacent first and second antenna elements were defected). Therefore, for the edge defected elements (1st and 8th), only results for test case of ($\theta_S = 20°$) with SLL = −30 dB are reported in this chapter and are shown in Fig. 13. For edge defected elements in 1 × 8 linear array, the achieved SLLs, HPBWs, directivity, and taper efficiency are provided in Table 7. The computed excitation amplitudes for case 2 using the proposed IFT antenna array failure correction algorithm are given in Table 9 of appendix A.

Table 4 Performance parameters comparison for broadside 3.5 THz 1 × 8 linear antenna array failure correction with defected elements 1 and 2

Main beam broadside ($\theta_s = 0°$)													
Antenna array type	Defected elements	Required SLL (dB)	Achieved SLL (dB)			Taper efficiency (%)		Directivity (dB)			HPBW (degree)		
			OP	DP	CP	OP	CP	OP	DP	CP	OP	DP	CP
1 × 8 elements linear array	1 and 2	−20	−20.88	−16.6	−20.25	94.2	68.7	15.6	13.74	12.85	15.5	18.4	19.56
		−30	−30.24	−19.27	−30.02	83.8	61.6	14.25	12.84	11.38	16.1	19.6	21.86
		−40	−40.01	−21.3	−40.21	76	57.1	13.33	12.21	10.65	17.84	20.76	23

Table 5 Performance parameters comparison for 20° scanned 3.5 THz 1 × 8 linear antenna array failure correction with defected elements 1 and 2

Main beam scanned ($\theta_S = 20°$)													
Antenna array type	Defected elements	Required SLL (dB)	Achieved SLL (dB)			Taper efficiency (%)		Directivity (dB)			HPBW (degree)		
			OP	DP	CP	OP	CP	OP	DP	CP	OP	DP	CP
1 × 8 elements linear array	1 and 2	−20	−20.3	−15.03	−20	94.2	68.7	15.07	13.23	12.33	23.9	19.2	18.2
		−30	−30	−17.66	−30.62	83.8	61.6	13.73	12.32	10.87	21.6	19	15.2
		−40	−39.82	19.62	−40.95	76	57.1	12.81	11.69	10.5	20.6	15.8	13.66

Table 6 Performance parameters comparison for 40° scanned 3.5 THz 1 × 8 linear antenna array failure correction with defected elements 1 and 2

Main beam scanned ($\theta_S = 40°$)													
Antenna array type	Defected elements	Required SLL (dB)	Achieved SLL (dB)			Taper efficiency (%)		Directivity (dB)			HPBW (degree)		
			OP	DP	CP	OP	CP	OP	DP	CP	OP	DP	CP
1 × 8 elements linear array	1 and 2	−20	−19.4	−13.4	−18.2	94.2	68.7	13.4	11.61	10.73	13.75	22.53	22.92
		−30	−28.33	−16	−28.78	83.8	61.6	12.08	10.73	9.31	19.68	24.07	26.6
		−40	−37.88	−17.87	−38.43	76	57.1	11.18	10.13	8.61	21.3	25.31	26.92

Table 7 Performance parameters comparison for 20° scanned 3.5 THz 1 × 8 linear antenna array failure correction with defected elements 1 and 8

Main beam scanned ($\theta_S = 20°$)													
Antenna array type	Defected elements	Required SLL (dB)	Achieved SLL (dB)			Taper efficiency (%)		Directivity (dB)			HPBW (degree)		
			OP	DP	CP	OP	CP	OP	DP	CP	OP	DP	CP
1 × 8 elements linear array	1 and 8	−30	−30	−19.28	−30	83.8	62.9	13.73	12.83	11.42	17.19	20.55	22.34

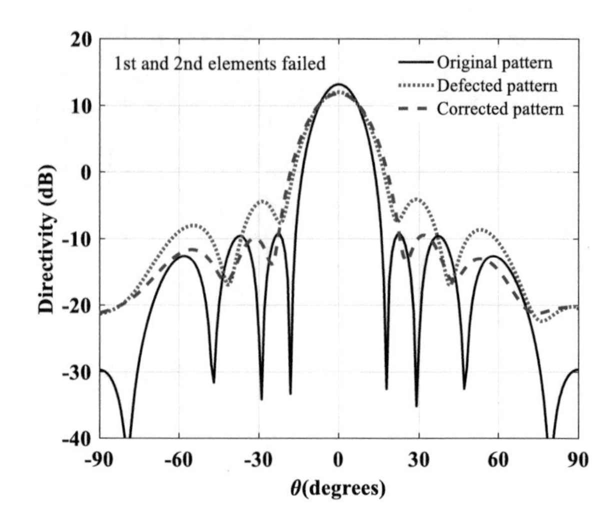

Fig. 14 CST comparison of radiation patterns for broadside ($\theta_S = 0°$) 3.5 THz 1×8 linear array failure correction with 1st and 2nd failed elements and SLL = −20 dB

CST Simulation

Case 1: Adjacent Defected Elements (1st and 2nd)

A 1×8 microstrip patch antenna element array constituted of single THz microstrip patch antenna element operating at 3.5 THz (as discussed in section III) is designed in CST simulator. The required excitation amplitudes are computed from the proposed low sidelobe synthesis IFT failure correction algorithm and are given in Table 8 of appendix A. The designed 3.5 THz 1×8 linear patch antenna element failure correction array in CST was excited with these computed array weights. For corrected patterns, all the defected elements (1st and 2nd) are excited with zero magnitudes. The original, defected, and corrected patterns (denoted as OP, DP, and CP)) for broadside THz 1×8 microstrip patch linear array for −20 dB, −30 dB, and −40 dB SLLs are given in Figs. 14, 15 and 16, respectively. For the above three SLL cases of broadside, the required SLLs are achieved in corrected patterns. Similar to MATLAB patterns, the CST corrected patterns have an increase in HPBW and decrease in directivity from original patterns. The original, defected, and corrected patterns for 20° scanned 3.5 THz 1×8 microstrip patch linear array for −20 dB, −30 dB and −40 dB SLLs are given in Figs. 17, 18, and 19, respectively. For the above three SLL cases of 20° main beam scanning, the required SLLs are achieved, while the corrected patterns have

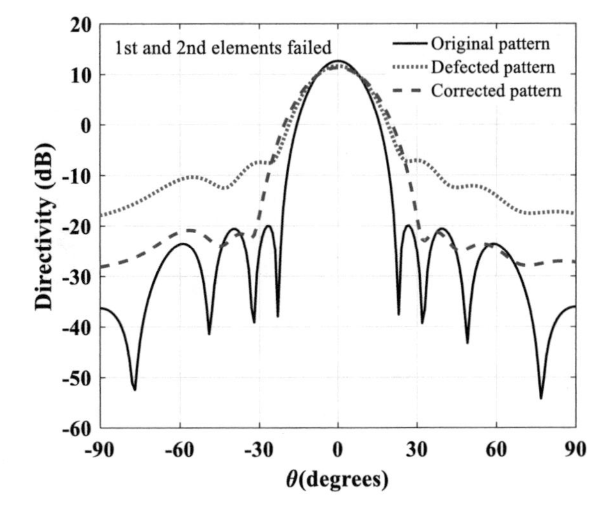

Fig. 15 CST comparison of radiation patterns for broadside ($\theta_S = 0°$) 3.5 THz 1×8 linear array failure correction with 1st and 2nd failed elements and SLL $= -30$ dB

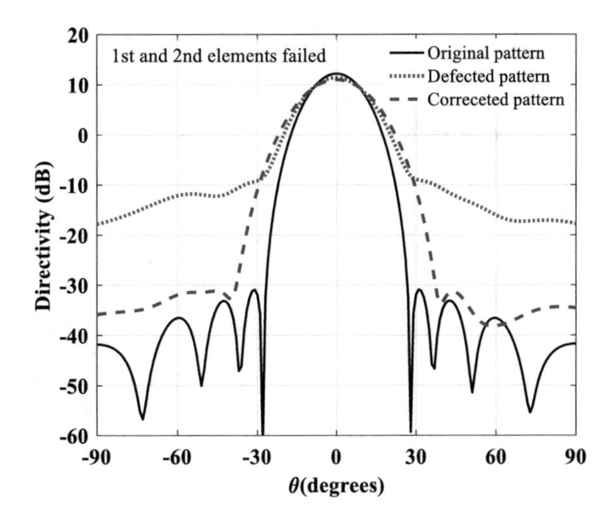

Fig. 16 CST comparison of radiation patterns for broadside ($\theta_S = 0°$) 3.5 THz 1×8 linear array failure correction with 1st and 2nd failed elements and SLL $= -40$ dB

widened HPBW and low directivity than original patterns. The original, defected, and corrected patterns for 40° 3.5 THz 1×8 microstrip patch linear array for -20 dB, -30 dB and -40 dB SLLs are given in Figs. 20, 21, and 22, respectively. For above three SLL cases of 40° scanning, the required SLLs are achieved, while the corrected patterns have widened HPBW and low directivity than original patterns. Due to the influence of individual element patterns, the directivity also decreases as mainbeam is scanned off broadside. At this point, it is worth mentioning that MATLAB and CST results (HPBW, directivity, and SLLs) and patterns are almost same for all cases with slight changes in MATLAB and CST patterns. This is thought to be due to the mutual coupling between elements of arrays. Mutual coupling effects are not included in MATLAB coding, while 3D EM based CST simulator takes into account the mutual

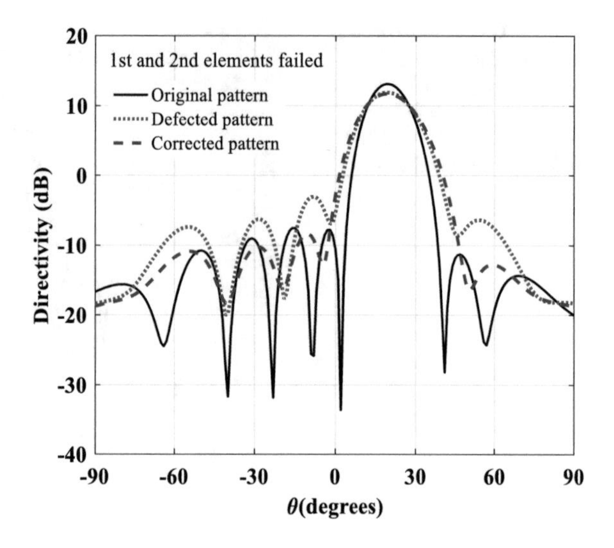

Fig. 17 CST comparison of radiation patterns for scanned ($\theta_S = 0°$) 3.5 THz 1 × 8 linear array failure correction with 1st and 2nd failed elements and SLL = −20 dB

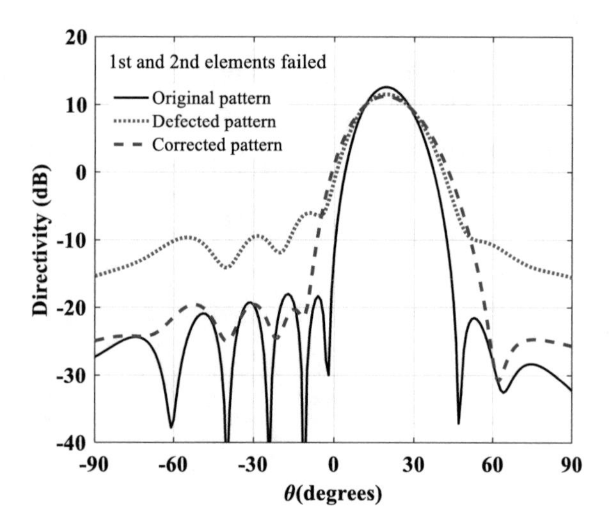

Fig. 18 CST comparison of radiation patterns for scanned ($\theta_S = 20°$) 3.5 THz 1 × 8 linear array failure correction with 1st and 2nd failed elements and SLL = −30 dB

coupling in the 3D simulation of the radiation patterns. This mutual coupling affects the radiation pattern and it can be seen by comparing Figs. 11 with 21 and Figs. 12 with 22.

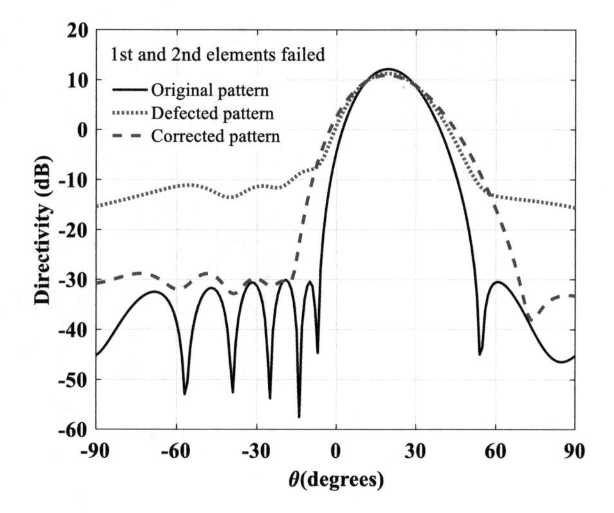

Fig. 19 CST comparison of radiation patterns for scanned ($\theta_S = 20°$) 3.5 THz 1 × 8 linear array failure correction with 1st and 2nd failed elements and SLL = −40 dB

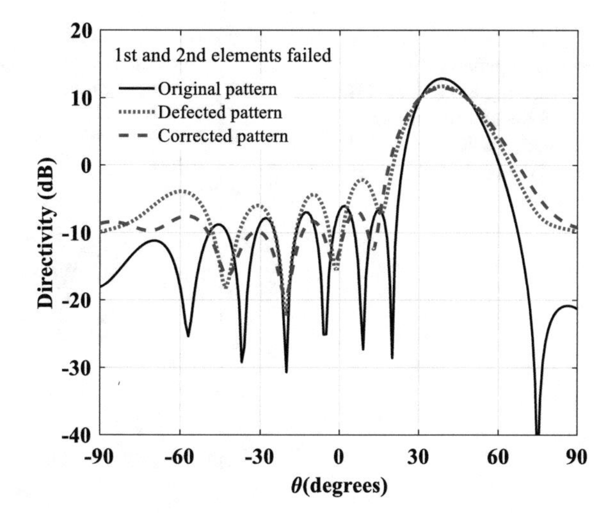

Fig. 20 CST comparison of radiation patterns for scanned ($\theta_S = 40°$) 3.5 THz 1 × 8 linear array failure correction with 1st and 2nd failed elements and SLL = −20 dB

Case 2: Edge Defected Elements (1st and 8th)

For case 2 in CST, only results for test case of ($\theta_S = 20°$) with SLL = −30 dB are reported in this chapter and are shown in Fig. 23. As can be seen, the SLL of −30 dB has been successfully recovered in the corrected pattern by the proposed IFT antenna array failure correction algorithm.

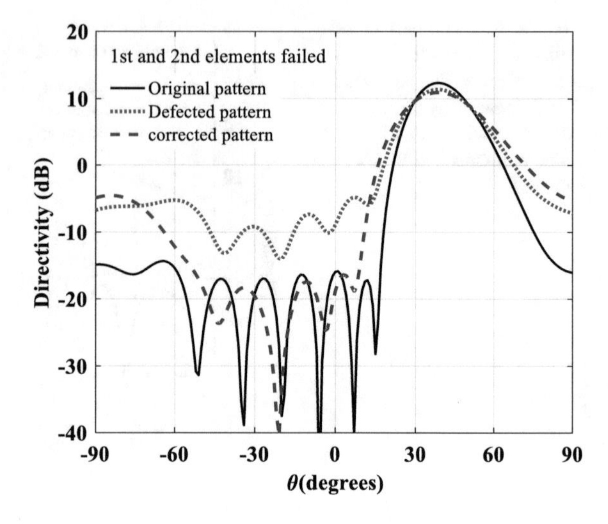

Fig. 21 CST comparison of radiation patterns for scanned ($\theta_S = 40°$) 3.5 THz 1×8 linear array failure correction with 1st and 2nd failed elements and SLL = -30 dB

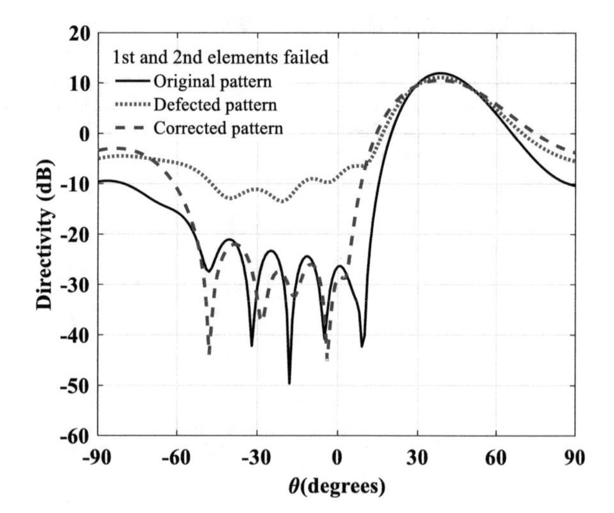

Fig. 22 CST comparison of radiation patterns for scanned ($\theta_S = 40°$) 3.5 THz 1×8 linear array failure correction with 1st and 2nd failed elements and SLL = -40 dB

5 Summary

A linear antenna array failure correction algorithm based on iterative Fourier technique in [13, 14] is discussed in this chapter. Then a 1×8 microstrip patch antenna array with 25% elements failure correction capability operating at 3.5 THz was designed in CST simulator. The required array weights (amplitudes and phases)

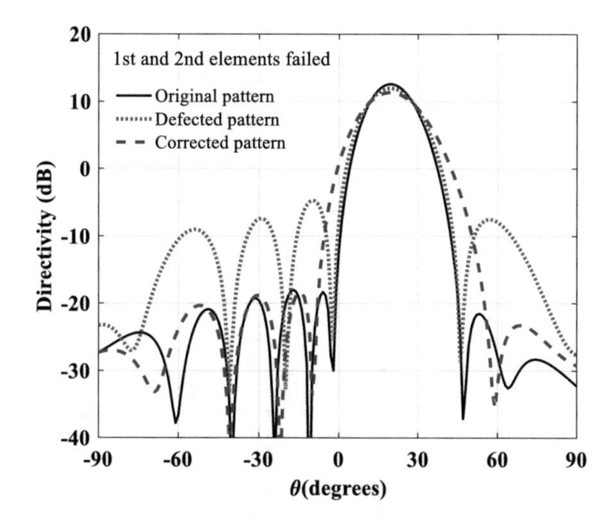

Fig. 23 CST comparison of radiation patterns for scanned ($\theta_S = 20°$) 3.5 THz 1×8 linear array failure correction with 1st and 8th failed elements and SLL = -30 dB

were computed in MATLAB using the proposed IFT failure correction algorithm for random failure of antenna elements. The calculated array weights were put on individual antenna elements (weights of defected elements were set to 0) in CST and effects on directivity, HPBW, SLLs, and taper efficiency were thoroughly investigated. There are two main constraints of the proposed antenna array failure correction algorithm: 1. Dynamic range constraints, 2. Defective elements excitation. The main purpose of this iterative low sidelobe synthesis is that the proposed algorithm is highly robust, simple, and easy for implementation in MATLAB. The calculations are only based on DFT and FFT (inverse fast Fourier transform) calculations. Because of this, it has fast computational speed. The presented examples and results are based on 1×8 linear antenna array failure correction model implemented in MATLAB and CST simulators. MATLAB and CST recovered radiation pattern results for various SLLs are in close agreement with each other. This validates the efficient implementation of iterative IFT based antenna array failure correction algorithm for future use in THz beamforming antenna arrays for various wireless applications.

Appendix A

See Tables 8 and 9.

Table 8 Computed amplitudes using proposed IFT algorithm for various SLLs of 3.5 THz 1 × 8 linear antenna array failure correction (defected elements 1 and 2)

Computed Excitation Amplitudes forCase 1: Adjacent defected elements (1st and 2nd)									
Element No.	For SLL = −20 dB			For SLL = −30 dB			For SLL = −40 dB		
	For original pattern	For defected pattern	For corrected pattern	For original pattern	For defected pattern	For corrected pattern	For original pattern	For defected pattern	For corrected pattern
1	0.5178	0	0	0.2554	0	0	0.1452	0	0
2	0.6401	0	0	0.5152	0	0	0.4167	0	0
3	0.8546	0.8546	0.6388	0.8081	0.8081	0.3638	0.7584	0.7584	0.2477
4	0.9994	0.9994	0.8681	0.9998	0.9998	0.7566	1.0000	1.0000	0.6913
5	1.0000	1.0000	1.0000	1.0000	1.0000	1.0000	1.0000	1.0000	1.0000
6	0.8561	0.8561	0.9111	0.8085	0.8085	0.8845	0.7585	0.7585	0.8789
7	0.6412	0.6412	0.6305	0.5157	0.5157	0.5196	0.4168	0.4168	0.4668
8	0.5154	0.5154	0.3396	0.2541	0.2541	0.1837	0.1448	0.1448	0.1252

Table 9 Computed amplitudes using proposed IFT algorithm for case 2 of 3.5 THz 1 × 8 linear antenna array failure correction (defected edge elements 1 and 8)

Computed excitation amplitudes for case 2: edge defected elements (1st and 8th)			
Element No.	For SLL = −30		
	For original pattern	For defected pattern	For corrected pattern
1	0.2554	0	0
2	0.5152	0.5152	0.2940
3	0.8081	0.8081	0.6822
4	0.9998	0.9998	0.9998
5	1.0000	1.0000	1.0000
6	0.8085	0.8085	0.6829
7	0.5157	0.5157	0.2947
8	0.2541	0	0

Note The required inter-element phase shifts for broadside ($\theta_S = 0°$) and scanned main beam ($\theta_S = 20°, \theta_S = 40°$) for cases 1 and 2 were calculated using Eq. (6) and are not reported here

References

1. J.S. Belrose, "Fessenden and Marconi: their differing technologies and transatlantic experiments during the first decade of this century". in *Proceedings, International Conference 100 Years Radio*, (1995).
2. Z. Ying, Antennas in cellular phones for mobile communications. Proc. IEEE **100**(7), 2286–2296 (2012)
3. P. Hall, P. Gardner, G. Ma, "Active integrated antennas". IEICE Trans. Commun. **E85-B**(9), 1661–1667 (2002)
4. G. Alexander, Y. Cherevko, V. Morgachev, Terahertz antenna modules, in 17th *International Conference on Micro/Nanotechnologies and Electron Devices EDM*. (2016), pp. 145–150
5. P.H. Siegel, Terahertz technology. IEEE Trans. Microw. Theory Technol. **50**, 910–928 (2002)
6. T. Kleine-Ostman, T. Nagatsuma, A review on terahertz communications research. J. Infrared Milli Terahertz Waves **32**, 143–171 (2011)
7. W. Yin, S.K. Khamas, R.A. Hogg, "High input resistance terahertz dipole antenna with an isolating photonic band gap layer". *2016 10th European Conference on Antennas and Propagation* (EuCAP), (Davos, 2016), pp. 1–3
8. T. Löffler, K. Siebert, S. Czasch, T. Bauer, H.G. Roskos, "Visualization and classication in biomedical terahertz pulsed imaging". Phys. Med. Biol. (2002)
9. R.H. Clothier, N. Bourne, "Effects of THz exposure on human primary keratinocyte differentiation and viability". J. Biol. Phys. (2003)
10. D. Petkie, et al., Active and passive imaging in the THz spectral region: phenomenology, dynamic range, modes, and illumination. J. Opt. Soc. Am. B **25**(9), (2008)
11. C. Petkie et al., Active and passive millimeter- and sub-millimeter-wave imaging. Proc. SPIE, (2005)
12. S.C. Liu, "Phase-only sidelobe sector nulling for a tapered array with failed elements". in *Proceedings of IEEE Antennas and Propagation Society International Symposium and URSI National Radio Science Meeting*, vol. 1 (IEEE, 1994), pp. 136–139
13. W.P.M.N. Keizer, "Element failure correction for a large monopulse phased array antenna with active amplitude weighting." IEEE Trans. Antennas Propag. **55**(8), 2211–2218 (2007)

14. W.P.M.N. Keizer, Low-sidelobe pattern synthesis using iterative Fourier techniques coded in MATLAB [EM programmer's notebook]. IEEE Antennas Propag. Mag. **51**(2), 137–150 (2009)
15. J.-H. Han, S.-H. Lim, N.-H. Myung, "Array antenna TRM failure compensation using adaptively weighted beam pattern mask based on genetic algorithm. "IEEE Antennas Wireless Propag. Lett. **11**, 18–21 (2011)
16. M.D. Migliore, D. Pinchera, M. Lucido, F. Schettino, G. Panariello, A sparse recovery approach for pattern correction of active arrays in presence of element failures. IEEE Antennas Wireless Propag. Lett. **14**, 1027–1030 (2014)
17. K. Yadav, A.K. Rajak, H. Singh, "Array failure correction with placement of wide nulls in the radiation pattern of a linear array antenna using iterative fast Fourier transform." in *2015 IEEE International Conference on Computational Intelligence and Communication Technology*. (IEEE, 2015), pp. 471–474
18. S.K. Mandal, S. Patra, S. Salam, K. Mandal, G.K. Mahanti, N.N. Pathak, "Failure correction of linear antenna arrays with optimized element position using differential evolution." in *2016 IEEE Annual India Conference (INDICON)*, pp. 1–5. (IEEE, 2016)
19. O.P. Acharya, A. Patnaik, 'Antenna array failure correction.' IEEE Antennas Propag. Mag. **59**(6), 106–115 (2017)
20. Y.-S. Chen, I-L. Tsai, "Detection and correction of element failures using a cumulative sum scheme for active phased arrays." IEEE Access **6**, 8797–8809 (2018)
21. X. Jiang, P. Xu, T. Jiang, "Comparison between NU-CNLMS and GA for antenna array failure correction." in *2018 IEEE International Symposium on Antennas and Propagation and USNC/URSI National Radio Science Meeting*. (IEEE, 2018), pp. 2205–2206
22. L.A. Greda, A. Winterstein, D.L. Lemes, M.V.T. Heckler, "Beamsteering and beamshaping using a linear antenna array based on particle swarm optimization." IEEE Access **7**, 141562–141573 (2019)
23. Z. Hamici, Fast beamforming with fault tolerance in massive phased arrays using intelligent learning control. IEEE Trans. Antennas Propag. **67**(7), 4517–4527 (2019)
24. S. Zhu, C. Han, Y. Meng, X. Jiawen, T. An, Embryonics based phased array antenna structure with self-repair ability. IEEE Access **8**, 209660–209673 (2020)
25. P. Kalra, E. Sidhu, "Rectangular TeraHertz microstrip patch antenna design for riboflavin detection applications." in *2017 International Conference on Big Data Analytics and Computational Intelligence (ICBDAC)*. (IEEE, 2017), pp. 303–306
26. C.A. Balanis, *Antenna Theory Analysis and Design* (Wiley, NY, 2001)

The Magneto Electron Statistics in Heavily Doped Doping Super-Lattices at Terahertz Frequency

R. Paul, S. Chakrabarti, B. Chatterjee, K. Bagchi, P. K. Bose, M. Mitra, and K. P. Ghatak

Abstract This chapter investigates the magneto electron statistics (ES) in Heavily Doped Doping Super-lattices (HDDS) under terahertz frequency. It is found taking HDDS of Cd_3As_2, $CdGeAs_2$, InAs, InSb, $Hg_{1-x}Cd_xTe$, $In_{1-x}Ga_xAs_yP_{1-y}$ as examples that the Fermi energy (E_F) oscillates with inverse quantizing magnetic field (1/B) and increases with increasing electron concentration n_0 with different numerical magnitudes.

Keywords Electron statistics · Heavily doping super-lattices · Magnetic quantization · Terahertz frequency

R. Paul · S. Chakrabarti
Department of Computer Science and Engineering, University of Engineering and Management, Kolkata 700156, India
e-mail: rajashree.paul@uem.edu.in

S. Chakrabarti
e-mail: satyajit.chakrabarti@iemcal.com

B. Chatterjee
Department of Computer Science and Engineering, University of Engineering and Management, Jaipur, Rajasthan 303807, India
e-mail: biswajoy.chatterjee@iemcal.com

K. Bagchi
237, Canal Street, Sreebhumi, Kolkata 700048, India

P. K. Bose
Department of Mechanical Engineering, Swami Vivekananda Institute of Science and Technology, Dakshin Gobindapur, Sonarpur, Kolkata, West Bengal 700145, India

M. Mitra
Department of Electronic and Telecommunication Engineering, Indian Institute of Engineering Science and Technology, Shibpur, Howrah 711103, India

K. P. Ghatak (✉)
Department of Basic Science & Humanities, Institute of Engineering & Management, 1, Management House, Salt Lake, Sector—V, Kolkata, West Bengal 700091, India

S. Das et al. (eds.), *Advances in Terahertz Technology and Its Applications*,
https://doi.org/10.1007/978-981-16-5731-3_3

1 Introduction

The importance of electron statistics (ES) is well known in the whole arena of quantized structures and Ghatak and his group for the last forty years in their seventeen research monographs [1–17] and more than 300 research papers [18–40] to investigate the different electronic properties of various quantized compounds. From the in-depth study of the said vast literature, it appears that the magneto ES in HDDS has yet to be investigated and in what follows this is done in HDDS of various materials as stated in the abstract.

2 Theoretical Background

The DOS function in HDDS of tetragonal compounds assumes the form [17]

$$N_1 = \beta_1\left[\overline{\delta}\left(E - E_{10,2}\right)\right] \tag{1}$$

where $\beta_1 = \frac{g_v eB}{\pi\hbar d_z}\sum_{n_i=0}^{n_{i\max}}\sum_{n=0}^{n_{\max}}\overline{\delta}$, is the Dirac's delta function, and the other symbols are given in [17].

From Eq. (1), the n_0 can be written as

$$n_0 = \beta_1\left[1 + \exp(-\eta_{10,2})\right]^{-1} \tag{2}$$

where $\eta_{10,2} = (k_B T)^{-1}[E_F - E_{10,2}]$ and $E_{10,2}$ is given in [46].

The DOS function in HDDS of III-V and alloy compounds on the basis of the HD three and two band models of Kane together with wide band gape models assume the forms

$$N_1 = \beta_1\left[\overline{\delta}(E - E_{10,3})\right] \tag{3}$$

$$N_1 = \beta_1\left[\overline{\delta}(E - E_{10,5})\right] \tag{4}$$

$$N_1 = \beta_1\left[\overline{\delta}(E - E_{10,7})\right] \tag{5}$$

where $E_{10,3},\ E_{10,5}\ and\ E_{10,7}$ are defined in [46].

From Eqs. (3), (4) and (5), the n_0 in the respective cases assume the forms

$$n_0 = \beta_1\left[1 + \exp(-\eta_{10,3})\right]^{-1} \tag{6}$$

$$n_0 = \beta_1\left[1 + \exp(-\eta_{10,5})\right]^{-1} \tag{7}$$

$$n_0 = \beta_1 \left[1 + \exp(-\eta_{10,7})\right]^{-1} \tag{8}$$

where $\eta_{10,3} = (k_B T)^{-1}[E_F - E_{10,3}]$, $\eta_{10,5} = (k_B T)^{-1}[E_F - E_{10,5}]$ and $\eta_{10,7} = (k_B T)^{-1}[E_F - E_{10,7}]$.

3 Results and Discussion

Using Eqs. (2), (5)–(8) and taking HDDS of Cd_3As_2, $CdGeAs_2$, InAs, InSb, $Hg_{1-x}Cd_xTe$, $In_{1-x}Ga_xAs_yP_{1-y}$, we have plotted the E_F versus 1/B and E_F versus n_0 in Figs. 1, 2, 3, 4, 5, 6, 7, 8, 9, 10 and 11, respectively. From these figures, we observe the followings:

1. The Fermi energy oscillates with 1/B.
2. Fermi energy increases with increasing n_0 in an oscillatory way.
3. The points 1 and 2 are valid for all types of HDDS with different numerical values due to different band structures.

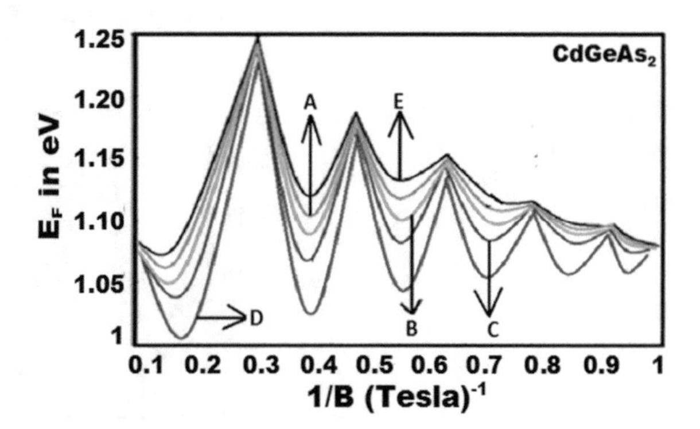

Fig. 1 Plot of E_F for HDDS of the indicated material versus 1/B in five different cases

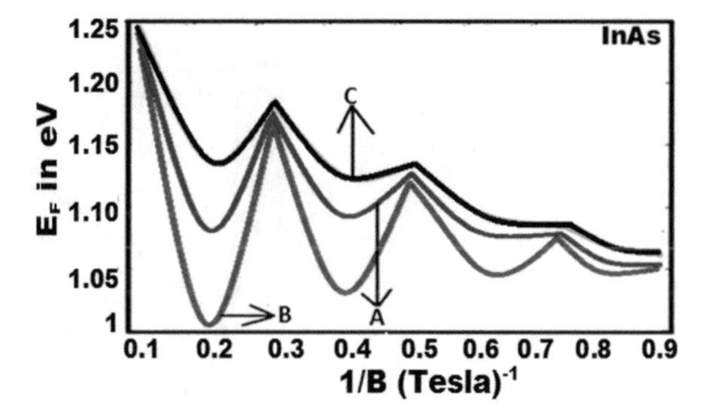

Fig. 2 Plot of E_F for HDDS of the indicated material versus 1/B in three different cases

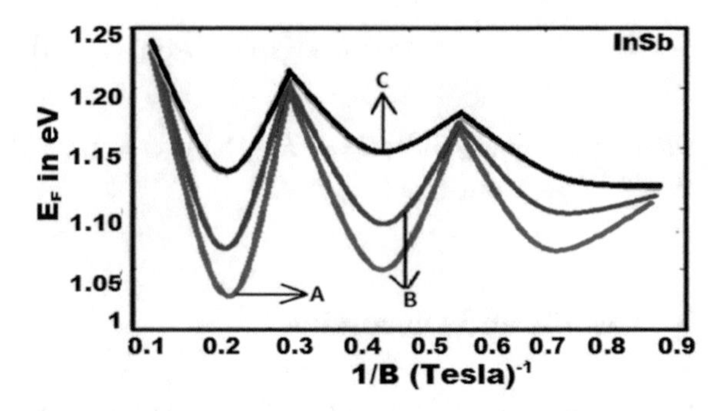

Fig. 3 Plot of E_F for HDDS of the indicated material versus $1/\mathrm{B}$ for all graphs of Fig. 2

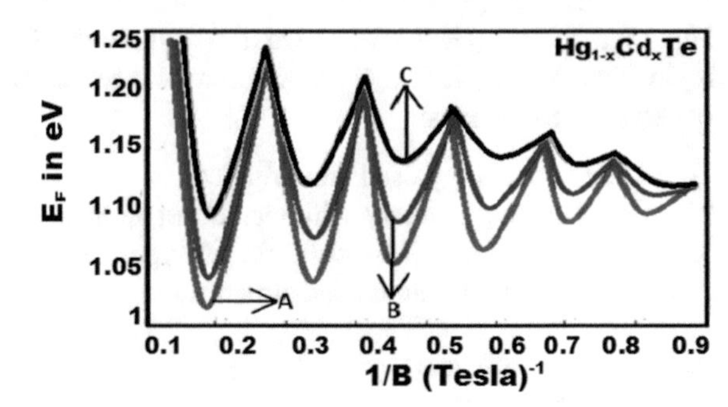

Fig. 4 Plot of E_F for HDDS of the indicated material versus $1/\mathrm{B}$ for all graphs of Fig. 2

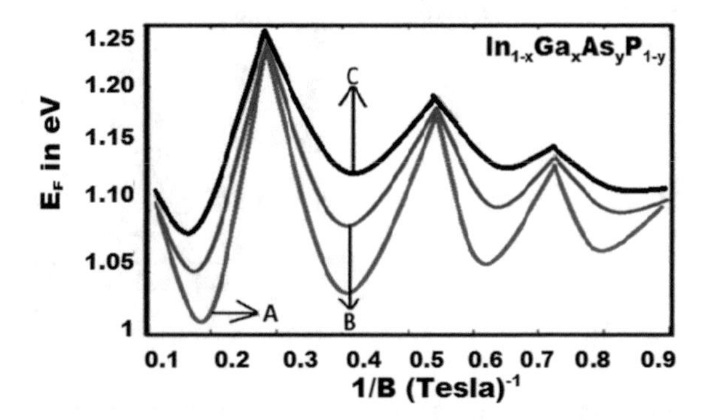

Fig. 5 Plot of E_F for HDDS of the indicated materials versus $1/\mathrm{B}$ for all graphs of Fig. 2

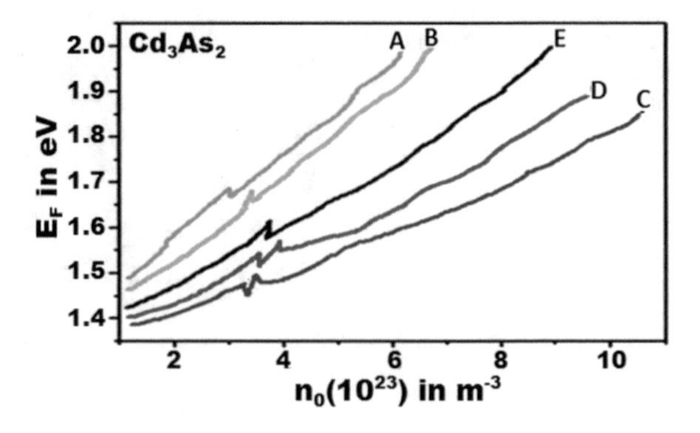

Fig. 6 Plot of E_F for HDDS of the indicated materials versus n_0 for all graphs of Fig. 1

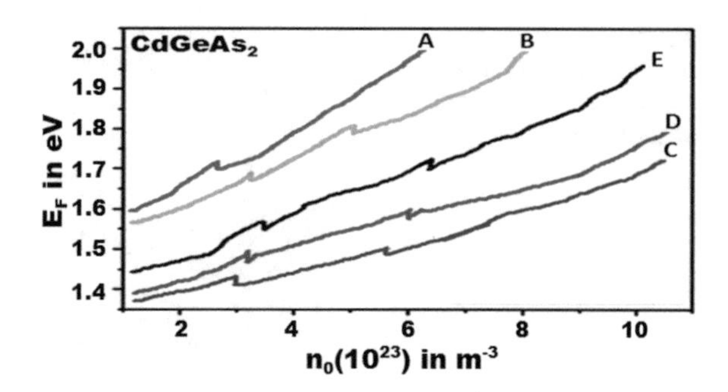

Fig. 7 Plot of E_F for HDDS of the indicated materials versus n_0 for all graphs of Fig. 1

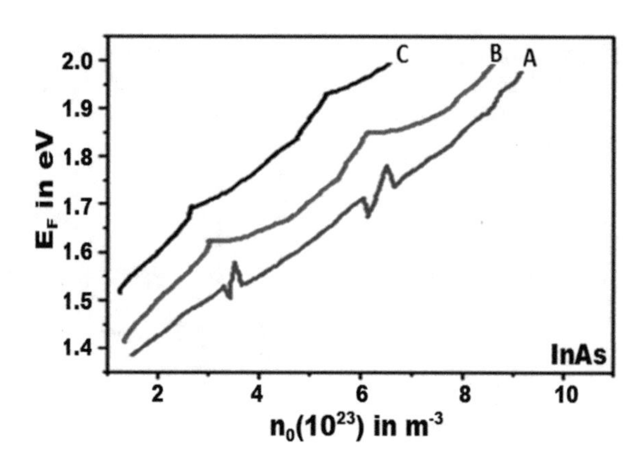

Fig. 8 Plot of E_F for HDDS of the indicated materials versus n_0 for all graphs of Fig. 2

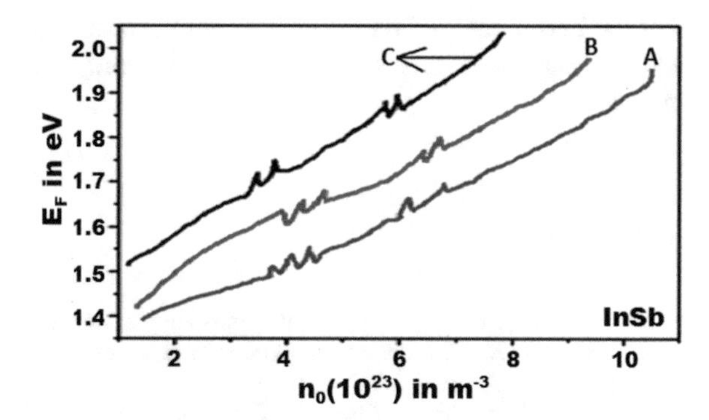

Fig. 9 Plot of E_F for HDDS of the indicated materials versus n_0 for all graphs of Fig. 2

4 Conclusion

This chapter investigates the magneto electron statistics (ES) in Heavily Doped Doping Super-lattices (HDDS) under terahertz frequency. It is found taking HDDS of Cd_3As_2, $CdGeAs_2$, InAs, InSb, $Hg_{1-x}Cd_xTe$, $In_{1-x}Ga_xAs_yP_{1-y}$ as examples

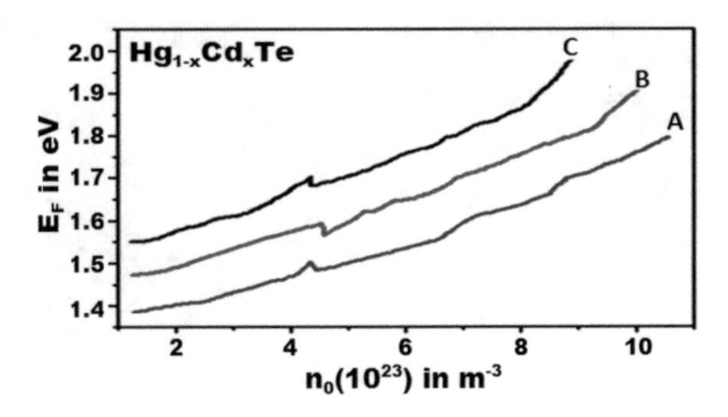

Fig. 10 Plot of E_F for HDDS of the indicated materials versus n_0 for all graphs of Fig. 2

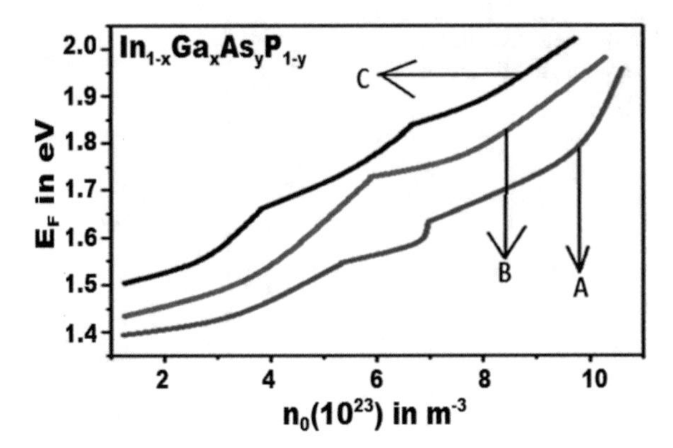

Fig. 11 Plot of E_F for HDDS of the indicated materials versus n_0 for all graphs of Fig. 2

that the Fermi energy (E_F) oscillates with inverse quantizing magnetic field (1/B) and increases with increasing electron concentration n_0 with different numerical magnitudes.

References

1. K.P. Ghatak, M. Mitra, Series on the foundation of natural science and technology, in *Elastic Constants in Heavily Doped Low Dimensional Materials*, vol. 17 (World Scientific Publishing Co. Ltd., Singapore, USA, 2021)
2. K.P. Ghatak, M. Mitra, *Electronic Properties, Series in Nanomaterials*, vol. 1, De Guyter, Germany (2020)
3. K.P. Ghatak, M. Mitra, Series in nanomaterials, in*Quantization and Entropy*, vol. 2, De Guyter, Germany (2020)
4. K.P. Ghatak, Series in nanotechnology science and technology, in *Quantum Wires: An Overview*, Nova, USA (2020)
5. D. De, S. Bhattacharya, K.P. Ghatak, Physics research and technology, in *Quantum Dots and Quantum Cellular Automata: Recent Trends and Application*, Nova, USA (2013)
6. K.P. Ghatak, Series on the foundations of natural science and technology, in *Quantum Effects, Heavy Doping, and The Effective Mass*, vol. 8 (World Scientific Publishing Co. Ltd., Singapore, USA, 2017)

7. K.P. Ghatak, S. Bhattacharya, Springer tracts in modern physics, in *Debye Screening Length: Effects of Nanostructured Materials*, vol. 255 (Springer International Publishing, Switzerland, 2014)
8. K.P. Ghatak, S. Bhattacharya, Springer tracts in modern physics, in *Heavily Doped 2D Quantized Structures and the Einstein Relation*, vol. 260 (Springer International Publishing, Switzerland, 2015)
9. K.P. Ghatak, S. Bhattacharya, in *Bismuth: Characteristics, Production, and Applications*, Materials Science and Technology, Nova, USA (2012)
10. S. Bhattacharya, K.P. Ghatak, Springer series in materials sciences, in *Effective Electron Mass in Low Dimensional Semiconductors*, vol. 167 (Springer, Berlin, 2013)
11. K.P. Ghatak, S. Bhattacharya, D. De, Springer series in materials science, in *Einstein Relation in compound semiconductors and their heterostructures*, vol. 116 (Springer, Berlin, 2009)
12. S. Bhattacharya, K.P. Ghatak, Springer series in solid state sciences, in *Fowler-Nordheim Field Emission: Effects in semiconductor nanostructures*, vol. 170 (Springer, Berlin, 2012)
13. K.P. Ghatak, Springer tracts in modern physics, in *Einstein's Photo-emission: Emission from Heavily Doped Quantized Structures*, vol. 262 (Springer International Publishing, Switzerland, 2015)
14. K.P Ghatak, S. Bhattacharya, Springer series in materials science, in *Thermo Electric Power In Nano structured Materials Strong Magnetic Fields*, vol. 137 (Springer, Berlin, 2010)
15. K.P Ghatak, D. De, S. Bhattacharya, *Photoemission from Optoelectronic Materials and Theirnanostructures* (Springer, Berlin, 2009)
16. K.P. Ghatak, Series on the foundations of natural science and technology, in *Magneto Thermo-electric Power in Heavily Doped Quantized Structures*,vol. 7, World Scientific Publishing Co. Pte. Ltd. Singapore (2016)
17. N.G Anderson, W.D. Laidig, R.M. Kolbas, Y.C. Lo. Y.C., J. Appl. Phys. **60**, 2361–2367 (1986)
18. F. Capasso, Semiconductors Semimetals **22**, 2 (1985)
19. F. Capasso, K. Mohammed, A.Y. Cho, R. Hull, A.L. Hutchinson, Appl. Phys. Letts. **47**, 420–422 (1985)
20. F. Capasso, R.A.J. Kiehl, Appl. Phys. **58**, 1366–1368 (1985)
21. K. Ploog, G.H. Doheler, Adv. Phys. **32**, 285–359 (1983)
22. F. Capasso, K. Mohammed, A.Y. Cho, Appl. Phys. Lett. **48**, 478–482 (1986)
23. R. Grill, C. Metzner, G.H. Döhler, Phys. Rev. B **63**, 235316–235317 (2001)
24. A.R. Kost, M.H. Jupina, T.C. Hasenberg, E.M. Garmire, J. Appl. Phys. **99**, 023501–023503 (2006)
25. A.G. Smirnov, D.V. Ushakov, V.K. Kononenko, Proc. SPIE **4706**, 70–77 (2002)
26. K.P. Ghatak, Springer tracts in modern physics, in *Dispersion Relations in Heavily—Doped Nanostructures*, vol. 265 (Springer, Berlin, 2016)
27. J.Z. Wang, Z.G. Wang, Z.M. Wang, S.L. Feng, Z. Yang, Phys. Rev. B., **62**, 6956–6958 (2000)
28. A.R. Kost, L. West, T.C. Hasenberg, J.O. White, M. Matloubian, G.C. Valley, Appl. Phys. Lett., **63**, 3494–3496 (1993)
29. S. Bastola, S.J. Chua, S.J. Xu, J. Appl. Phys. **83**, 1476–1480 (1998)
30. Z.J. Yang, E.M. Garmire, D. Doctor, Appl. Phys., **82**, 3874–3880 (1997)
31. G.H. Avetisyan, V.B. Kulikov, I.D. Zalevsky, Bulaev, Proc. SPIE, **2694**, 216 (1996)
32. U. Pfeiffer, M. Kneissl, B. Knüpfer, N. Müller, P. Kiesel, G.H. Döhler, J.S. Smith, Appl. Phys. Lett., **68**, 1838–1840 (1996)
33. H.L. Vaghjiani, E.A. Johnson, M.J. Kane, R. Grey, C.C. Phillips, J. Appl. Phys., **76**, 4407–4412 (1994)
34. P. Kiesel, K.H. Gulden, A. Hoefler, M. Kneissl, B. Knuepfer, S.U. Dankowski, P. Riel, X.X. Wu, J.S. Smith, G.H. Doehler, Proc. SPIE, **1985**, 278 (1993)
35. G.H. Doehler, Phys. Scripta. **24**, 430 (1981)
36. S. Mukherjee, S. Mitra, S.N., P.K. Bose, A.R. Ghatak, A. Neoigi, J.P. Banerjee, A. Sinha, M. Pal, S. Bhattacharya, K.P. Ghatak. J. Compu. Theor. Nanosc. **4**, 550–573 (2007)

37. R. Paul, S. Chakrabarti, B. Chatterjee, P.K. Das, T. De, S.D. Biswas, M. Mitra, in *Emerging Trends in Terahertz Engineering and System Technologies* (Springer, Singapore, 2021), pp. 64–110 and pp. 111–143; J. Pal, M. Debbarma, N. Debbarma, P.B. Mallik, K.P. Ghatak, J. Math. Sci. Comput. Math. **1**, 79–118 (2020); M. Mitra, J. Pal, P.K. Das, K.P. Ghatak, J. Math. Sci. Comput. Math. 374–438 (2020); P.K. Bose, K.P. Ghatak, J. Wave Mater. Interact. **12**, 67–75 (1997); K.P. Ghatak, D.K. Basu, B. Nag, J. Phys. Chem. Solids, **58**, 133–138 (1997); KP. Ghatak, P.K. Bose, J. Wave Mater. Interact. **12**, 96–103 (1997); P.K. Chakraborty, S.K. Biswas, K.P. Ghatak, Phys. B Condens. Matter, **352**, 111–117 (2004)
38. P.K. Chakraborty, L.J. Singh, K.P. Ghatak, J. Appl. Phys., **95**, 5311–5315 (2004); K.P. Ghatak, S.K. Biswas, D. De, S. Ghosal, S. Chatterjee, Phys. B Condens. Matter, **353**, 127–149 (2004); K.P. Ghatak, M. Mondal, Phys. Status Solidi (B), **175**, 113–121 (1993); K.P. Ghatak, B. Mitra B, IlNuovoCimento D, **15**, 97–112 (1993); K.P. Ghatak, IlNuovoCimento D, **14**, 1187–1190 (1992); N. Paitya, K.P. Ghatak, Rev. Theor. Sci. **1**, 165–305 (2013)
39. S.M. Adhikari, K.P. Ghatak, Quant. Matter, **2**, 296–306 (2013); S. Bhattachrya, N. Paitya, K.P. Ghatak, J. Comput. Theor. Nanosci. **10**, 1999–2018 (2013); K.P. Ghatak, S. Bhattacharya, S.K. Biswas, A. Dey, A.K. Dasgupta, PhysicaScripta, **75**, 820–835 (2007); S. Mukherjee, D. De, D. Mukherjee, J. Bhattacharya, S. Sinha, A. Ghatak, K.P. Physica B: Condens. Matter **393**, 347–362 (2007); K.P. Ghatak, S. Bhattacharya, J. Appl. Phys. **102**, 73704–73714 (2007); S. Mukherjee, S.N. Mitra, P.K. Bose, A.R. Ghatak, A. Neogi, J.P Banerjee, A. Sinha, M. Pal, S. Bhattacharya, K.P. Ghatak, J. Comput. Theor. Nanosci. **4**, 550–573 (2007)
40. S. Chakrabarti, B. Chatterjee, S. Debbarma, K.P. Ghatak, J. Nanosci. Nanotechnol. **15**, 6460–6471 (2015); S.M. Adhikari, A. Karmakar, K.P. Ghatak, Rev. Theor. Sci. **3**, 273–353 (2015); S.M. Adhikari, K.P. Ghatak, Quant. Matter, **4**, 599–609 (2015); K.P. Ghatak, S. Bhattacharya, K.M. Singh, S. Choudhury, S. Pahari, Phys. B Condens. Matter, **403**, 2116–2136 (2008); K.P. Ghatak, S. Bhattacharya, S. Bhowmik, R. Benedictus, S. Choudhury, J. Appl. Phys. **103**, 094314–094334 (2008); K.P. Ghatak, S. Bhattacharya, D. De, R. Sarkar, S. Pahari, A. Dey, A.K. Dasgupta, S.N. Biswas, J. Comput. Theor. Nanosci. **5**, 1345–1366 (2008); S.L. Singh, S.B. Singh, K.P. Ghatak, J Nanosci. Nanotechnol. **18**, 2856–2874 (2018); K.P. Ghatak, S. Chakrabarti, B. Chatterjee, P.K. Das, P. Dutta, A. Halder, Mater. Focus **7**, 390–404 (2018); K.P. Ghatak, S. Chakrabarti, B. Chatterjee, Mater. Focus **7**, 361–362 (2018); K.P. Ghatak, J. Wave Mater. Interact. **14**, 157–169 (1999); P.K. Chakraborty, K.P. Ghatak, J. Phys. D-Appl. Phys. **32**, 2438–2441 (1999)

Circularly Polarized Dual-Band Terahertz Antenna Embedded on Badge for Military and Security Applications

Sarosh Ahmad, Asma Khabba, Saida Ibnyaich, and Abdelouhab Zeroual

Abstract Terahertz antennas have got much attention in the field of security and military applications. As terahertz antennas are the perfect tools for short-distance communication for the armed forces. Therefore, in this research, a compact size dual-band circularly polarized (CP) microstrip patch terahertz antenna embedded on the badge is designed and simulated for military and security applications. As the badge is being stitched with the uniform or cap of the officers that is why the material used for the badge is Jeans with relative permittivity $\varepsilon_r = 1.78$ and the loss tangent tanδ = 0.085. The proposed antenna consists of three layers, i.e., a flexible substrate, a single element slotted patch, and a full ground plane. The antenna is being operated in the range of 0.126–0.133 THz and 0.184–0.194 THz for security and military applications. As the antenna is dual band so it will receive the data at one frequency and transmit the data to another frequency and the pattern radiation of the antenna is broadside at both frequencies. The overall size of the design is $1.5 \times 1 \times 0.15$ mm^3, and the patch is printed on a polyimide substrate with relative permittivity $\varepsilon_r = 3.5$ and the loss tangent tanδ = 0.0027. The simulated results like realized gain, radiation efficiency, and 10 dB impedance bandwidth at terahertz frequencies are found to be 4.87 dB and 5.59 dB, 89.6% and 77.9%, and 0.651 THz and 0.979 THz, respectively. The antenna gives circularly polarized behavior at both frequencies and has an axial ratio of less than 3 dB. The antenna is designed in computer simulation technology (CST) software and the simulated results proved that this proposed terahertz antenna is well-suited for security and military applications. As the antenna is made up of flexible substrate that is why bending analysis of the antenna along the x-axis and y-axis is also discussed in this article. Furthermore, in the future we will try to increase the gain, make the size more compact and to increase the bandwidth.

Keywords Dual band · Terahertz · Circularly polarized · Military and security applications

---

S. Ahmad
Department of Electrical Engineering, Government College University, Faisalabad, Pakistan

A. Khabba (✉) · S. Ibnyaich · A. Zeroual
Instrumentation, Signals and Physical Systems (I2SP) Team, Faculty of Sciences Semlalia, Cadi Ayyad University, Marrakesh, Morocco
e-mail: asma.khabba@edu.uca.ac.ma

S. Das et al. (eds.), *Advances in Terahertz Technology and Its Applications*,
https://doi.org/10.1007/978-981-16-5731-3_4

1 Introduction

Recently, a terahertz antenna has got the interests of scientists in the area of military and security applications. There are some major requirements for security and military applications that involve wideband greater efficiency, low profile, extreme integration, and conformability to the host platform [1]. The conformability stem and the low profile from the desire to blend the antenna into its environments to avoid easy detection and identification. These applications need novel solutions for antenna designing. In general, the more complicated structure of the antenna, the more it can be affected by the platform and the operating surroundings [2]. This is not always addressed by the researchers even the performance of the antenna tells the system performance. The more the antenna is efficient, the more it will work fine. The other imperative parameter in the designing of the antenna is the choice of the materials for the substrates that should encounter electronic, electromagnetic as well as structural requirements. The selection of the materials plays a vital role in antenna appearance and identification.

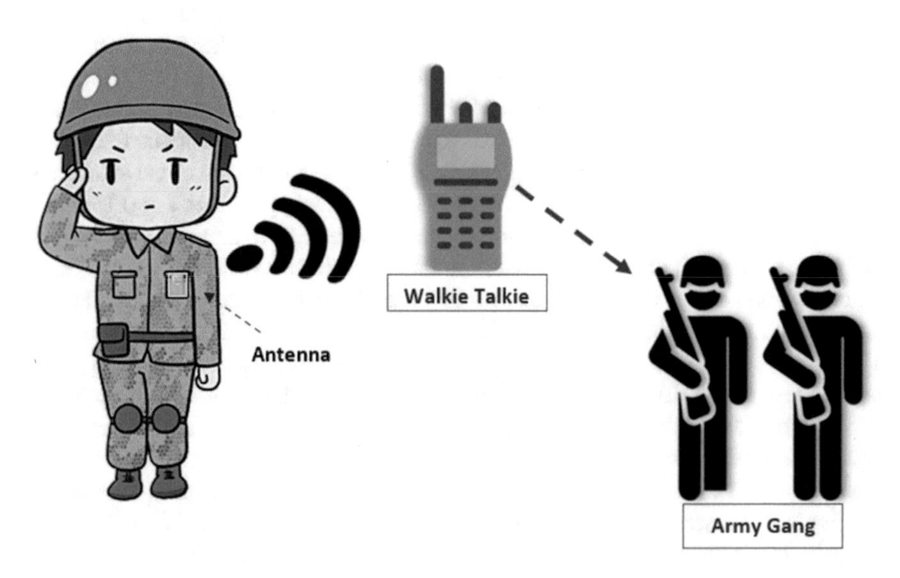

The major component used for communication between wireless systems is antenna. The main advantage of designing a microstrip patch antenna is its compactness and ease to fabricate. High performance, compactness of the size, low cost, easy fabrication, wideband, multiband, and less power consumption are the severe requirements these days for communication engineering [3]. In order to reduce the size of the antenna, slots can be added, and by making modifications in the electrical length of the patch antenna in this way the area of the antenna can be scaled down. Also, in order to achieve dual-band slots can be introduced inside the patch [4].

This paper provides a circularly polarized very compact size dual-band terahertz antenna printed on a polyimide substrate for security and military applications. The peak gain of the proposed design is achieved 4.87 and 5.59 dB at 0.130, and 0.185 THz with an overall size is $1.5 \times 1 \times 0.15$ mm^3. This paper is categorized as CP antenna measurements are discussed in Sect. 2, antenna simulation on badge is presented in Sect. 3, comparison with related works is presented in Sect. 4, and at the end, conclusion part is presented in Sect. 5.

2 Circularly Polarized Antenna Geometry and Simulation Results

The projected design has a very compact size and is suitable for military applications. The projected antenna is configured with a ground plane and a stripline feeding technique, so its response to the greater bandwidth and reduced antenna back radiation. Dimensions of the antenna are small and were constructed and investigated by physical layers, for example, the crust of the external body and its corresponding dielectric conduction, electrical conductivity and weight, and antenna parameters like power loss, voltage standing wave ratio, radioactive pattern, and realized gain are simulated in this article. Therefore, a low-profile terahertz antenna is proposed for military and security utilizations at two different frequencies. To make it flexible, a polyimide material is used as a substrate. It should also be kept in mind that bending of the design must have an insignificant result on the performance. The proposed design is made up of three different layers such as ground plane, substrate, and patch. The conducting patch and the ground plane are printed on polyimide substrate. The precision of the proposed system is improved, reducing the background radioactivity of the antenna, and the proposed system is investigated on badge made up of Jeans material with $\varepsilon_r = 1.78$, and $\tan\delta = 0.085$. The overall size of the antenna is $1.5 \times 1 \times 0.15$ mm^3. The proposed design is shown in Fig. 1 along with the dimensions. The design is small in size having a volume of 0.225 mm^3, and it is flexible by using flexible material, polyimide ($\varepsilon_r = 3.5$, thickness = 0.15 mm, $\tan\delta = 0.0027$), as a substrate. L_s and W_s are the length and width of the substrate and a 50 Ω SMA port is used for the antenna's excitation.

The list of the parameters of the proposed design is presented in Table 1. The CST MW Studio software is utilized to create this type of antenna.

2.1 *Return Loss of the Proposed Design*

The proposed CP antenna is simulated in CST MW Studio software, and the return loss of the design is found to be −22 dB and −44.38 dB at the operating frequency

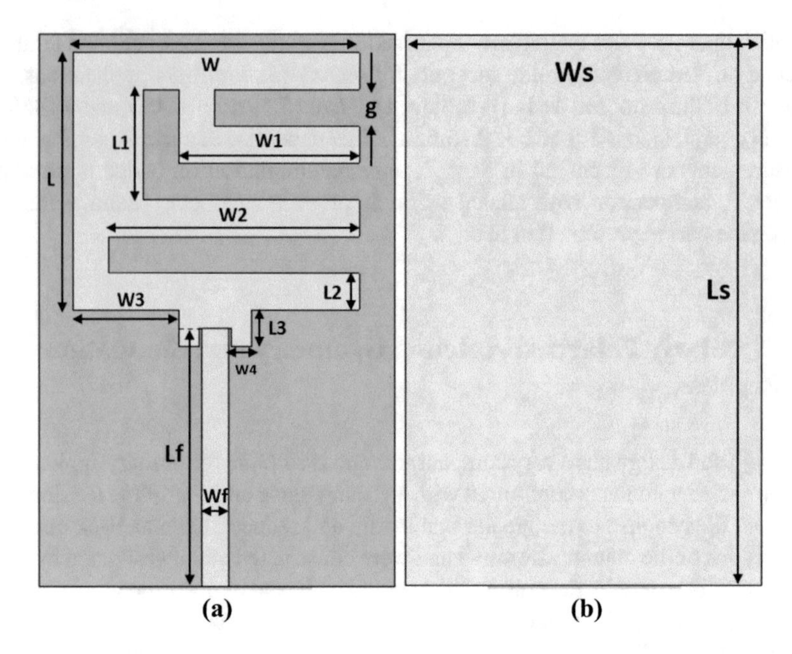

Fig. 1 Proposed CP dual-band antenna

Table 1 Parameters of the proposed design

Parameters	Values (mm)	Parameters	Values (mm)
L_s	1.50	W_s	1.00
L	0.70	W	0.80
L_f	0.70	W_f	0.08
L_1	0.30	W_1	0.50
L_2	0.10	W_2	0.70
L_3	0.10	W_3	0.30
g	0.10	W_4	0.06
L_p	0.80	W_p	0.80

of 0.13 THz and 0.185 THz, respectively. The return loss is shown in Fig. 2. It should be considered that the return loss for an idea antenna is negative infinity [5].

2.2 Design Process and Equations Used

Figure 3 presents the designing process of the proposed dual-band antenna. The basic antenna design (Fig. 3a) consists of a 50 Ω uniform feedline, a ground plane, and a radiating patch. On the basis of the transmission line model for the square-shaped

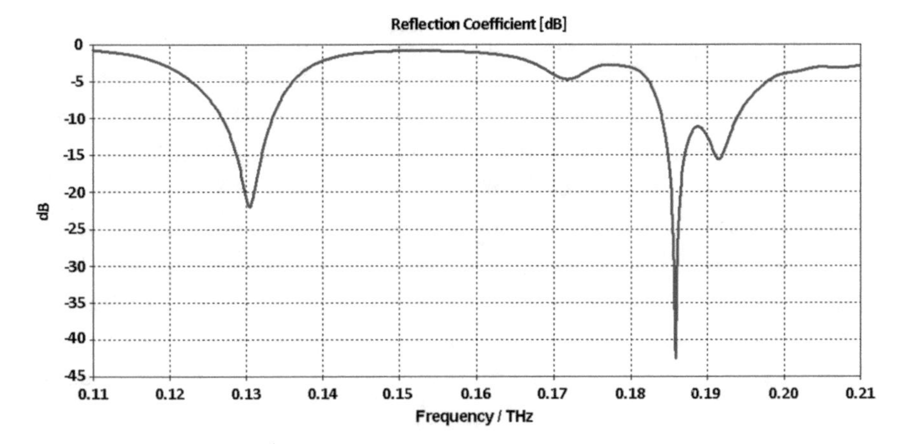

Fig. 2 S11 parameter of the antenna

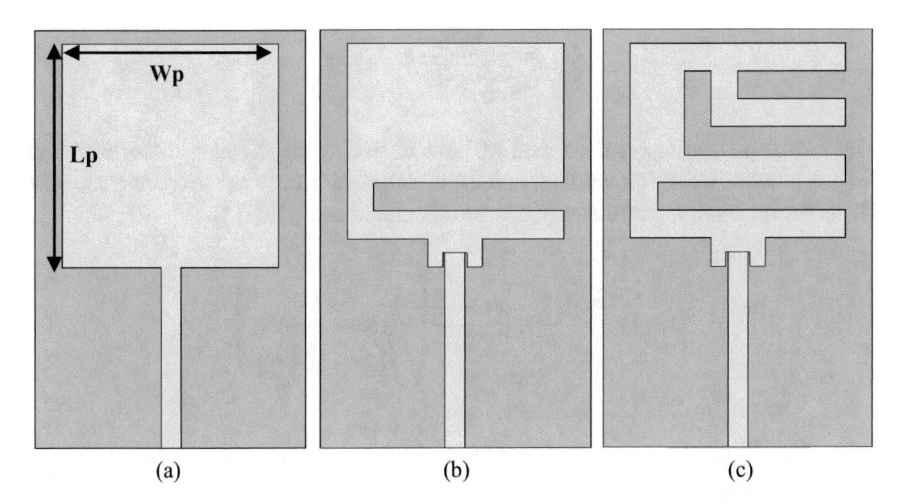

Fig. 3 Radiating patch design procedure; **a** Square patch, **b** Patch with bottom slot, **c** Proposed design

radiating patch at the dominant mode is calculated by using Eqs. (1)–(4). The width of the patch, W_p, is calculated by using Eq. (1).

$$Wp = \frac{\lambda_o}{2\sqrt{0.5(\varepsilon_r + 1)}} \tag{1}$$

where λ_o and ε_r are the wavelength is free space and the relative permittivity of the substrate at the operating frequency. The best choice of *Wp* leads to the perfect impedance matching. The length of the patch can be calculated using Eq. (2).

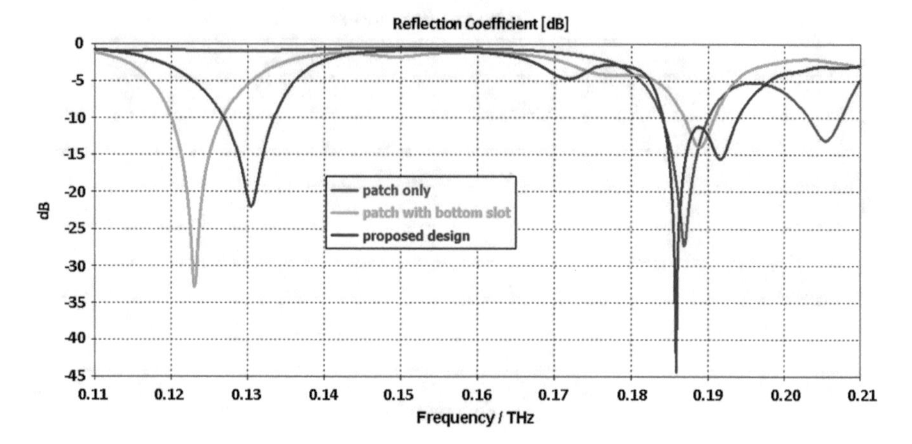

Fig. 4 S11 Parameter of the design procedure discussed in Fig. 3

$$L_p = \frac{c_o}{2f_o\sqrt{\varepsilon_{\text{eff}}}} - 2\Delta L_p \tag{2}$$

where c_o,ΔL_p and ε_{eff} is the speed of light in free space, change in the length of the patch due to its fringing effect, and the effective dielectric constant, respectively. The effective relative permittivity can be calculated using Eq. (3).

$$\varepsilon_{\text{eff}} = \frac{\varepsilon_r + 1}{2} + \frac{\varepsilon_r - 1}{2}\left(\frac{1}{\sqrt{\left(1 + 12\frac{h_{\text{sub}}}{W_p}\right)}}\right) \tag{3}$$

where h_{sub} is the height of the substrate. At the end, the fringing effect can be calculated using Eq. (4).

$$\Delta Lp = 0.421h_{\text{sub}}\frac{(\varepsilon_{\text{eff}} + 0.300)\left(\frac{W_p}{h_{\text{sub}}} + 0.264\right)}{(\varepsilon_{\text{eff}} - 0.258)\left(\frac{W_p}{h_{\text{sub}}} + 0.813\right)} \tag{4}$$

With the placement of $\varepsilon_r = 3.5$, $h_{\text{sub}} = 0.15$ mm in Eqs. (1)–(4), the initial dimensions of the square patch are $L_p = 0.8$ mm and $W_p = 0.8$ mm. Alternatively, the width and length of the patch are adjusted to $L_p = 0.8$ mm and $W_p = 0.8$ mm, based on full-wave optimization. As the size of the antenna substrate is directly proportional to the wavelength at the lower frequency band.

The step-by-step design procedure is clearly presented in this section. The first simple square patch is designed and noticed this simple patch antenna only working for a higher frequency band as can be seen in Fig. 3a. Then one slot at the bottom

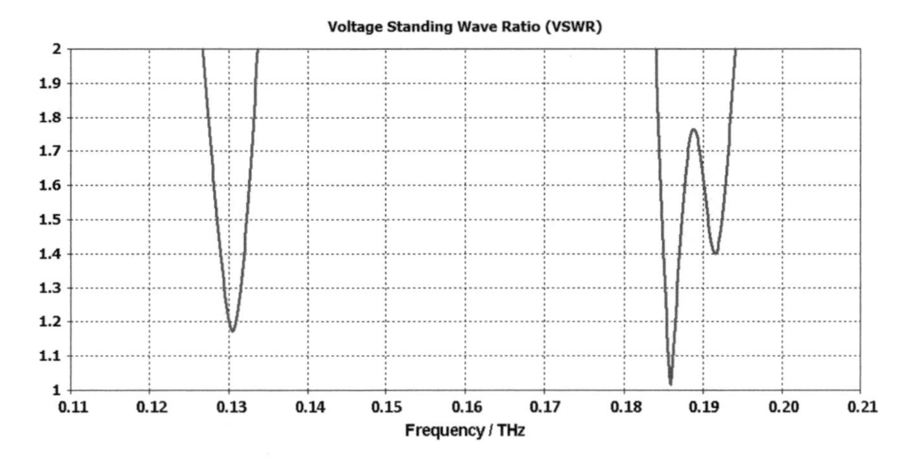

Fig. 5 VSWR of the proposed design

of the patch is created in the patch (see Fig. 3b) to generate another lower-frequency band but it is also observed that by creating this slot at the bottom, a lower-frequency band is generated but a higher band is disturbing. Then two upper slots are created to set both the frequencies at the desired operating bands, and it can be seen that this slotted patch antenna is working at the desired bands, i.e., at 0.13 and 0.185 THz.

2.3 VSWR of the CP Antenna

The voltage standing wave ratio (VSWR) is a computation of how proficiently a radio frequency power is communicated from a power source, through a communication line into a load. The value of the VSWR of the proposed design is found to be 1.17 and 1.01 at 0.13 and 0.185 THz, as presented in Fig. 5.

2.4 Farfield of the Proposed CP Antenna

The gain of the proposed design is found to be 4.87 and 5.59 dB at 0.13 THz and 0.185 THz, respectively, as can be seen in Fig. 6. The simulated 2D radiation pattern of the antenna at 0.13 and 0.185 THz is also presented.

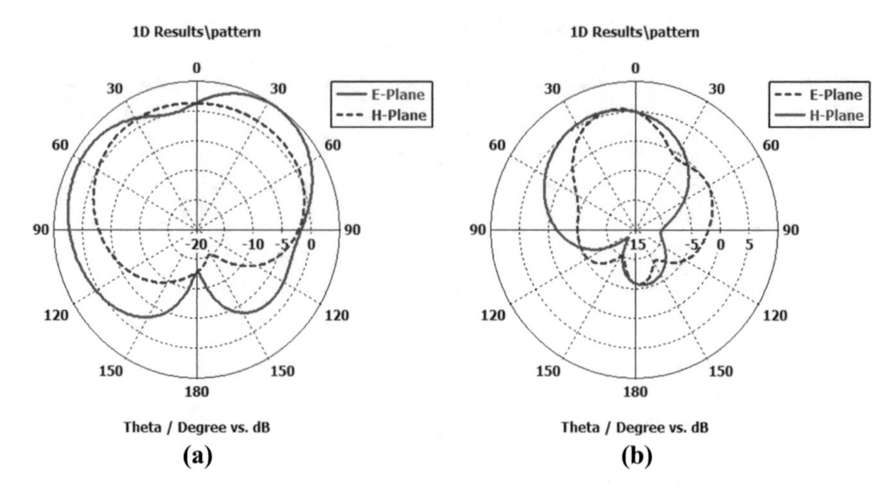

Fig. 6 Farfield of the proposed design; **a** at 0.13 THz, **b** at 0.185 THz

2.5 Circularly Polarized Behavior

Circularly polarization behavior of the electromagnetic (EM) wave is a circular polarization in which the EM field of the antenna has constant magnitude, and it rotates at a continuous rate in a plane orthogonal to the wave direction. The axial ratio of the proposed design is found to be below 3 dB as can be seen in Fig. 7.

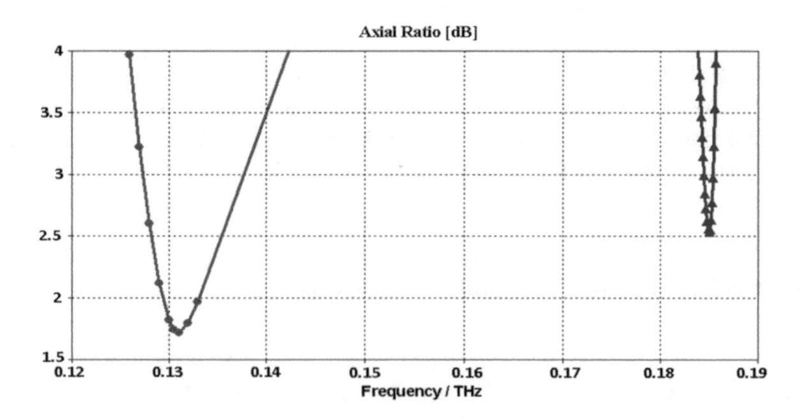

Fig. 7 Proposed CP antenna axial ratio (dB)

2.6 Bending Analysis

As the antenna will be embedded on the badge so it must be flexible along x- and y-axes situations depends on the twisting while in practice. In this portion, we learned how the act of receiving wire is changed by bending antenna on y-, x-directions in a vacuum. Unlike twisting radii ($b_y = b_x =$ 10–100 mm) were selected in y- and x-directions to study the S11. The configuration of the different bending is shown in Fig. 8a, and the S11 of the receiving wire for different bending radii on the x-axis is presented in Fig. 8b. It is well-known, there are unnecessary changes in all twisting radii. In the same way, return loss for different twisting radii in the y-axis is presented in Fig. 8c. From the figures, it is clear that bending of the antenna along the x-axis and the y-axis has negligible effects on the return loss of the dual-band patch antenna. Different bending radii of the antenna along the x- and y-axes are analyzed, i.e., at 10, 40, 70, and 100 mm. Return loss behavior at both frequencies can be seen in Fig. 8b, c. It is clear that bending analysis has a minor effect on the return loss just the higher band is shifting a little bit. Also, the comparison graph of the realized gain (dB) versus frequency (THz) is presented in Fig. 8d when the antenna is bent along the x-axis and y-axis. There is a very minor reduction in the gain, when it is bent along the x-axis. And the comparison of the total radiation efficiency of the antenna along the x- and y-axes is also presented in Fig. 8e. The total radiation efficiency in both of the cases is more than 75%. Hence, all these results indicate that the antenna is stable in all cases even it is bent along the x-axis and y-axis because of its very compact size. This antenna is a good candidate for military and security applications.

2.7 Surface Current Distribution

In metal type of antennas, the surface current distribution is the main electric current that flows in the radiating patch and is induced by an applied electromagnetic field. In Fig. 9, surface current distribution is given at 0.13 and at 0.185 THz frequencies. And the red portion indicates that portion of the radiating patch has contribution to resonate the antenna at the required frequency.

2.8 Parametric Study

In order to optimize the results, the analysis of the different parameters of the projected dual-band CP antenna plan is achieved when placed on the three-dimensional box that has the Jeans properties. By varying the patch width at one side "$W1$," the frequency shift is presented in Fig. 10a. By increasing the value of $W1$ from 0.1 to 0.6 mm, then the frequency varies. Now, again by changing the width

"$W2$" in radiating patch, the projected antenna frequency also varies. By enhancing the value from 0.1 to 0.4 mm, the frequency moves, as presented in Fig. 10b.

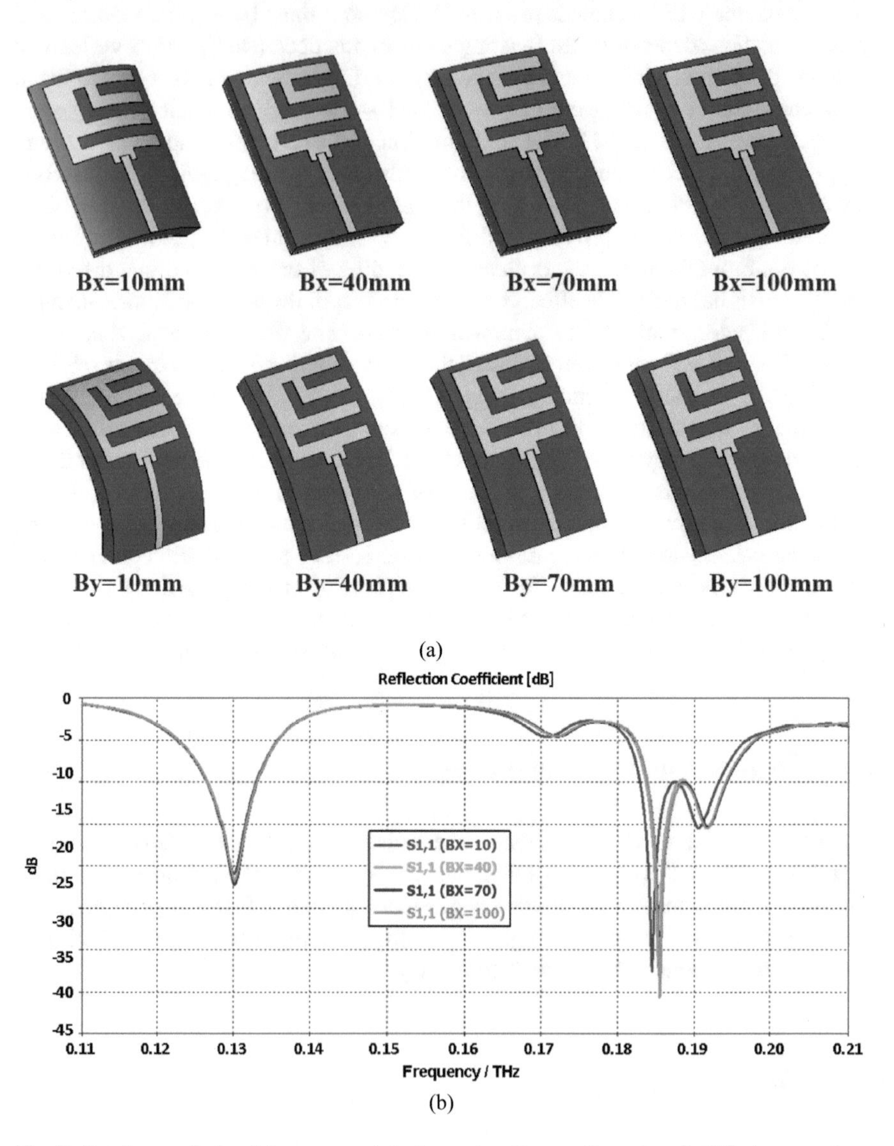

Fig. 8 Bending analysis of the proposed design; **a** bending configuration; **b** S11 changes along *x*-axis; **c** S11 changes along *y*-axis; **d** realized gain (dB) versus frequency (THz); **e** total radiation efficiency along *x*- and *y*-axes

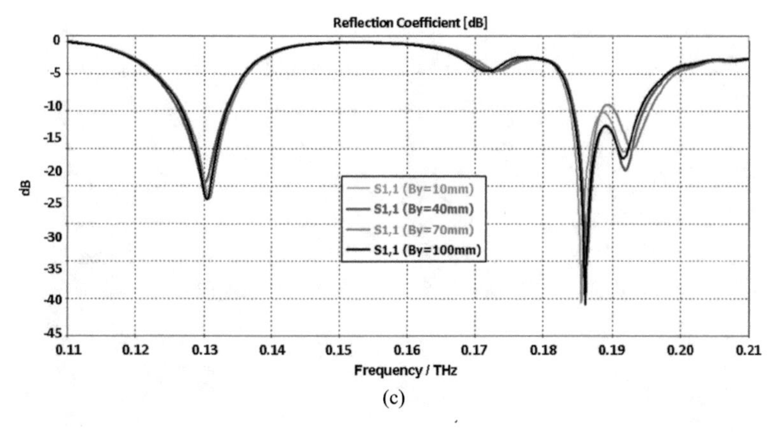

(c)

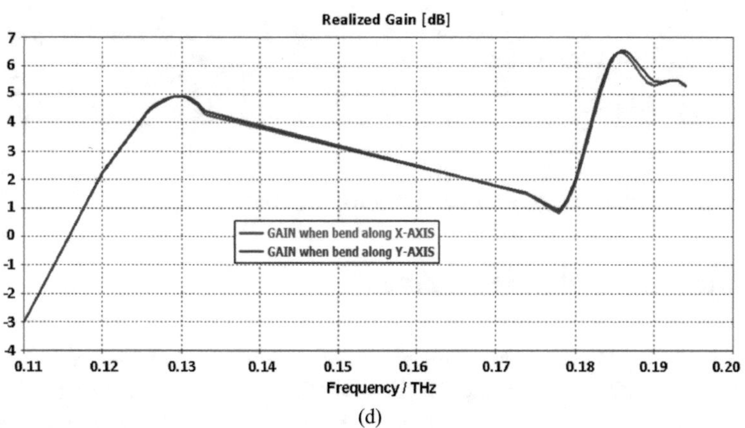

(d)

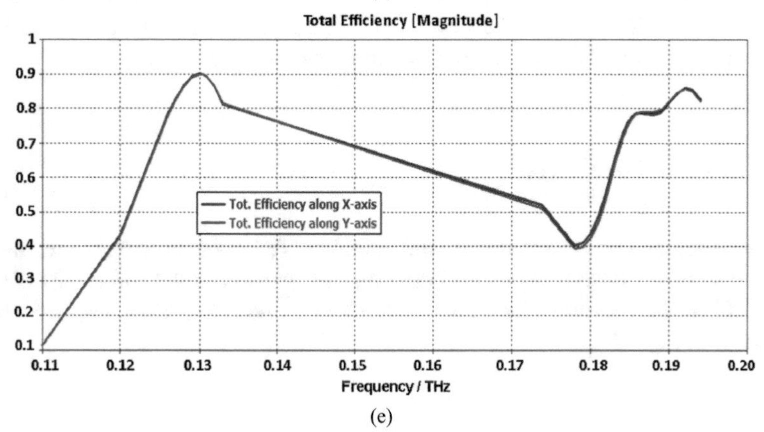

(e)

Fig. 8 (continued)

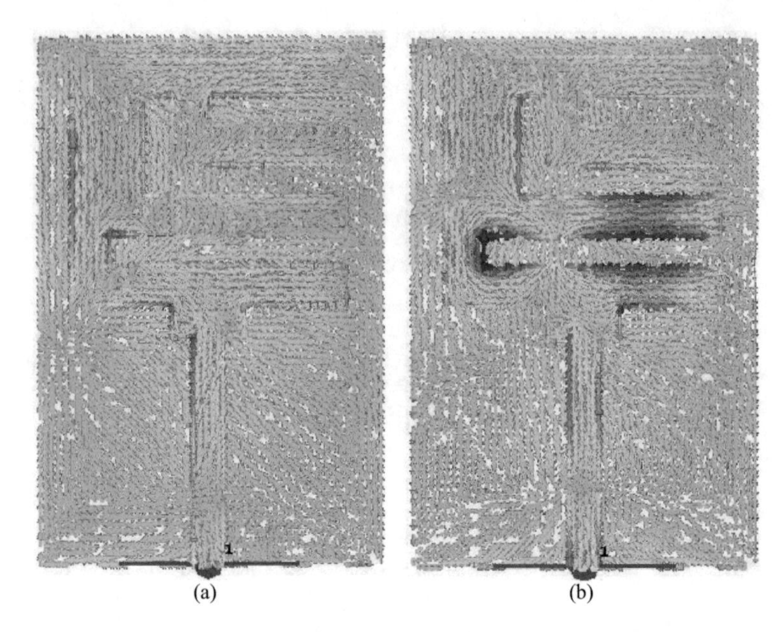

Fig. 9 Surface current density; **a** at 0.13 THz, **b** at 0.185 THz

3 Antenna Simulation on Badge

After simulating in free space, this proposed low-profile antenna has also been simulated on badge as presented in Fig. 11. The size of the Jeans box is 3 × 3 mm^2. Their properties are given as the relative permittivity of the Jeans is $\varepsilon_r = 1.78$, and loss tangent $\tan\delta = 0.085$. The return loss behavior on Badge is shown in Fig. 12, and the 3D pattern radiation is presented in Fig. 13.

3.1 Return Loss (dB)

The proposed dual-band antenna is simulated in CST MW Studio software, and the return loss of the design under the proximity of Jeans is found to be −21 dB and −43.2 dB at the operating frequency of 0.13 THz and 0.185 THz, respectively, as presented in Fig. 12.

3.2 3D Radiation Pattern

The simulated 3D radiation pattern of the antenna being simulated on the badge at 0.13 and 0.185 THz is also presented. The gain of the proposed design is found to

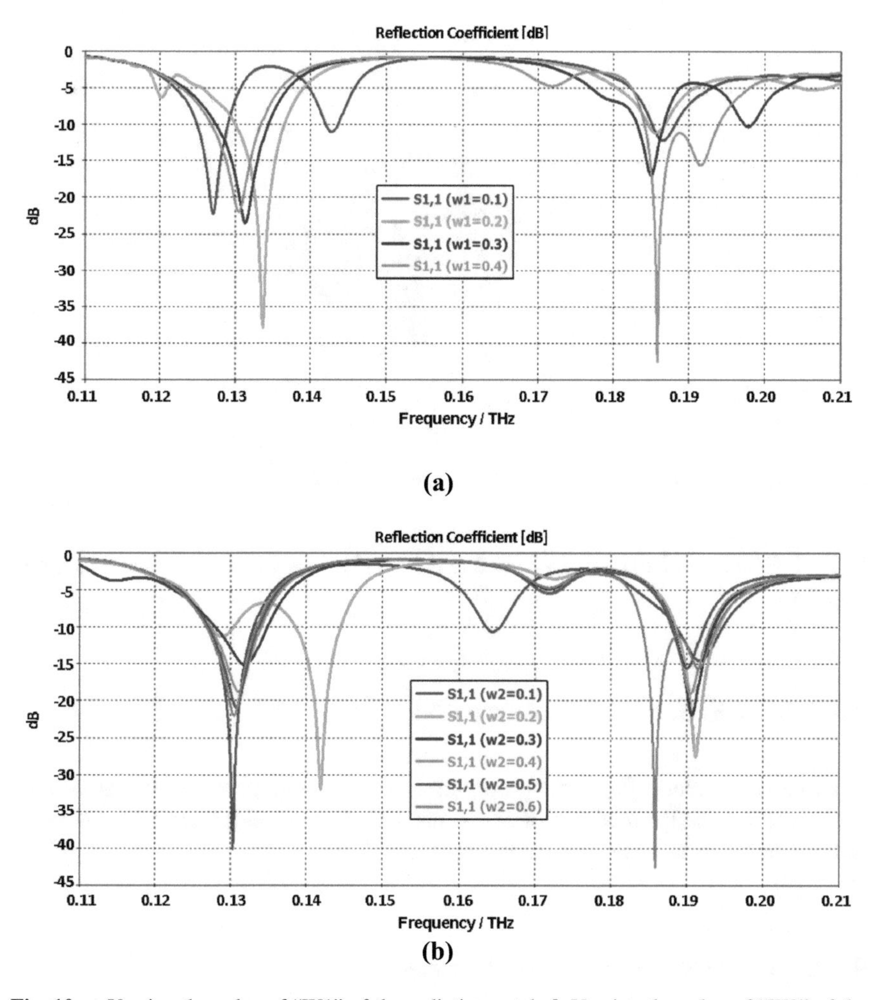

Fig. 10 **a** Varying the value of "W1" of the radiating patch. **b** Varying the value of "W2" of the radiating patch

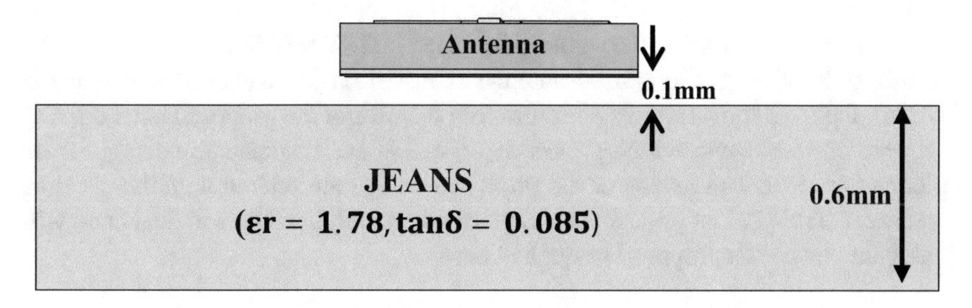

Fig. 11 Side view of the antenna embedded on badge for military applications

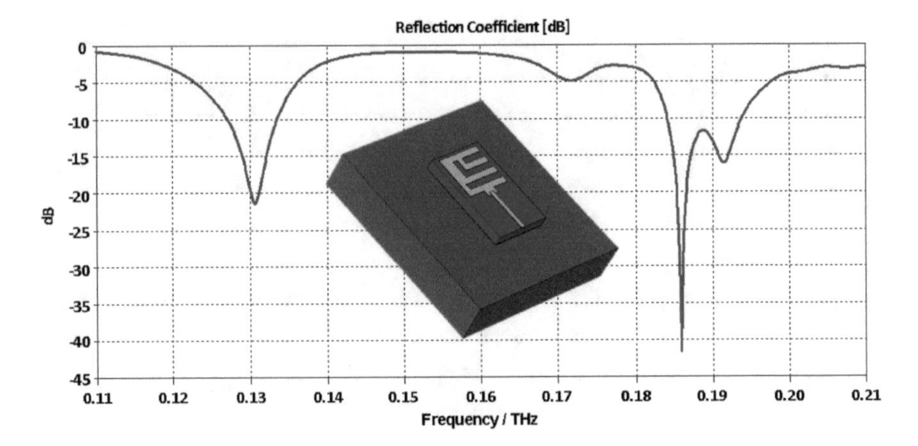

Fig. 12 S11 parameter of the antenna under the proximity of Jeans badge

be 3.61 dB and 6.42 dB at 0.13 THz and 0.185 THz, respectively, as can be seen in Fig. 13.

3.3 Total Radiation Efficiency (%)

The totally efficiency (%) of the proposed dual-band design is more than 80%. And when it is kept badge made up of Jeans material, then the efficiency of the antenna reduces only 2% as it can be seen in Fig. 14.

4 Comparison with Related Works

In reply to the high frequency (THz), the proposed technique is configured with stripline feeding and a ground plane. About 0.08 mm is the width of the feedline, which is improved based on the results performed in CST MW Studio. After finishing the design barriers, the proposed antenna is tested on the badge at a distance of 0.1 mm. Figure 2 shows simulated return loss features of the proposed antenna. This is a very precise result, which proves that the BW and resonant frequency for the antenna are clear. The results of the proposed design are related with the previous research in Table 2. The projected antenna is very small, flexible, and dual band with peak gain covers the proposed frequency bands.

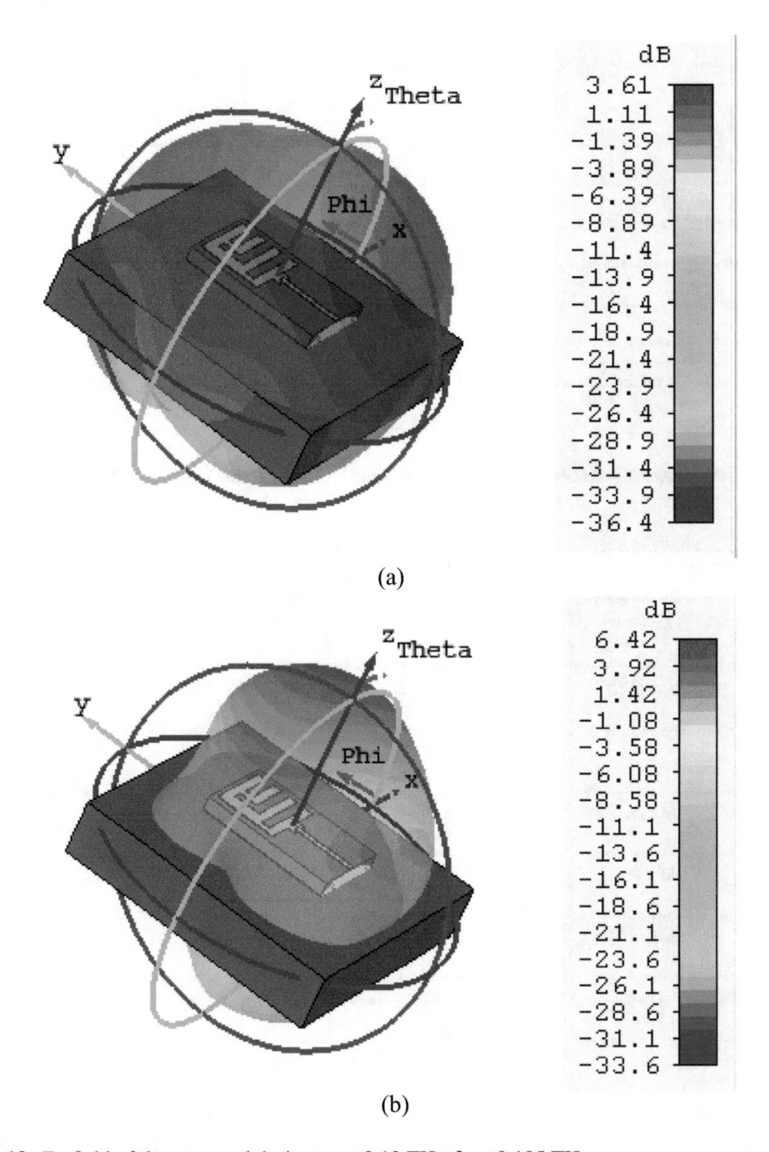

Fig. 13 Farfield of the proposed design; **a** at 0.13 THz, **b** at 0.185 THz

5 Conclusion

A compact size dual-band circularly polarized (CP) terahertz antenna for military and security applications has been proposed in this research. The proposed antenna is being operated in the range 0.126–0.133 THz and 0.184–194 THz for security and military applications. The overall size of the design is $1.5 \times 1 \times 0.15$ mm^3, and the patch is printed on a polyimide substrate with $\varepsilon_r = 3.5$, and $\tan \delta = 0.0027$. The

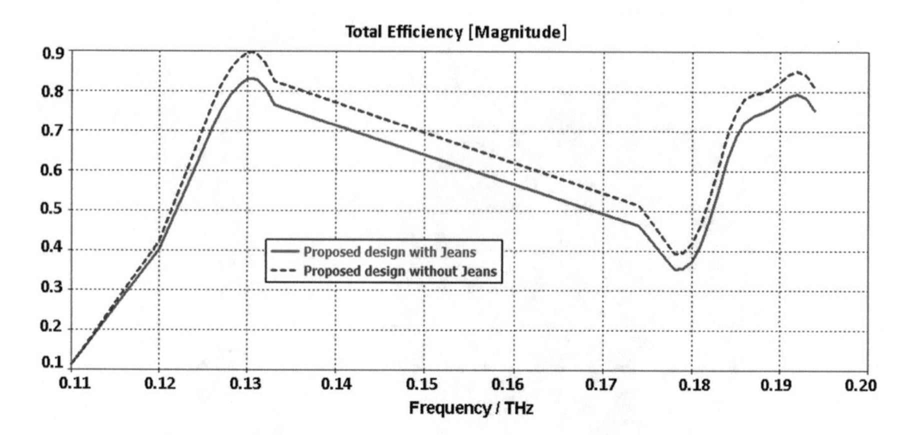

Fig. 14 Overall radiation efficiency of the antenna with and without Jeans badge

Table 2 Comparison with previous research

Ref. no	Frequency (THz)	Volume (mm^3)	Substrate material	Gain (dB)	Total efficiency (%)	Polarization
[6]	0.7, 0.85	$1 \times 1 \times 0.2$	Roger RT 6006	3.09, 3.06	<51	LP
[7]	–	$30 \times 28.4 \times 0.25$	Roger RT 5880	–	–	LP
[3]	0.6, 0.5, 2	$14 \times 14 \times 1.59$	FR-4	4.02, 7.48, 5.57	–	LP
[8]	0.694	$0.2 \times 0.42 \times 0.02$	Polyimide	5.22	–	LP
[9]	0.34–0.40	$1.8 \times 1.8 \times 0.04$	GaAs	–	73	LP
(This work)	0.130, 0.185	$1.5 \times 1 \times 0.15$	Polyimide	4.87, 5.59	>80	CP

simulated results like realized gain, total radiation efficiency, and 10 dB impedance bandwidth at terahertz frequencies are found to be 4.87 and 5.59 dB, 89.6 and 77.9%, and 0.651 and 0.979 THz. The antenna gives circularly polarized behavior at both frequencies and has axial ratio of less than 3 dB. This proposed design is a perfect candidate for security and military terahertz applications. Furthermore, we will try to increase the gain of the antenna and better impedance bandwidth.

References

1. A. Zaghloul, S.J. Weiss, W. Coburn, Antenna developments for military applications. Access J. **25**, 41–53 (2010)
2. C.S. Srinivas, D.S.S. Satyanarayana, Microstrip patch antenna design for military applications. Int. Res. J. Eng. Technol. (IRJET) **6** (2019)
3. M. Gupta, V. Mathur, A. Kumar, V. Saxena, D. Bhatnagar, Microstrip hexagonal fractal antenna for military applications. Frequenz **73**, 321–330 (2019)
4. S. Ergun, S. Sonmez, Terahertz technology for military applications. J. Manage. Inf. Sci. **3**, 13–16 (2015)
5. J. DicCbshfsE, M.K. Abd Rahim, M. Azfar Abdullah, N.A. Samsuri, F. Zubir, K. Kamardin, Design, implementation and performance of ultra-wideband textile antenna. Prog. Electromagnet. Res. **27**, 307–325 (2011)
6. A. Sharma, G. Singh, Rectangular microstirp patch antenna design at THz frequency for short distance wireless communication systems. J. Infrared Millimeter Terahertz Waves **30**, 1 (2009)
7. A.K. Patnaik, K.A. Sarika, M.S. Patrudu, P.H. Meghana, A. Suresh, Multiband reconfigurable antenna for military applications. Int. J. Eng. Res. **9** (2020)
8. A.S. Dhillon, D. Mittal, E. Sidhu, THz rectangular microstrip patch antenna employing polyimide substrate for video rate imaging and homeland defence applications. Optik **144**, 634–641 (2017)
9. N. Hussain, I. Park, Design of a wide-gain-bandwidth metasurface antenna at terahertz frequency. AIP Adv. **7**, 055313 (2017)

Tera-Bit Per Second Quantum Dot Semiconductor Optical Amplifier-Based All Optical NOT and NAND Gates

Siddhartha Dutta, Kousik Mukherjee, and Subhasish Roy

Abstract The proposed inverter based on Quantum Dot Semiconductor Optical Amplifier (QDSOA) is analyzed by calculating extinction ratio (ER), contrast ratio (CR), Quality factor (Q), amplitude modulation (AM) considering amplified spontaneous noise (ASE). For three different unsaturated gains, 10, 15 and 20 dB, these parameters are optimized with the variation of control power to the QDSOA. Input-output bit patterns and pseudoeye diagrams are included in the analysis. The present chapter investigates and compares the design and performance of NAND gates using the NOT gates and without using NOT gate or inverter.

Keywords Optical logic · Quantum dot SOA · Inverter · Quality factor · Noise

1 Introduction

Semiconductor Optical Amplifier (SOA) is a versatile gain media for all optical signal processing (AOSP) and finds applications in various optical signal processing devices for optical computation and communication [1–3]. Beside integration capability SOAs have many functional applications like wavelength conversion, optical demultiplexing, logic elements make it attractive for AOSP [4]. Moreover, SOA exhibits strong nonlinearity, low power and small size and have been used for many applications in last few decades [5–10]. Optical Switch likes Tera hertz optical asymmetric demultiplexer (TOAD) [11], Mach Zehnder Interferometers (MZI) [12] is used to design various all optical logic gates because logic gates are key element for AOSP. SOA uses bulk, quantum well (QW) and quantum dot (QD) structures [4], and quantum dot SOA (QDSOA) has several advantages over its bulk and QW counterparts. Quantum dot SOA has low noise level, enhanced output power, reduced

S. Dutta · S. Roy
Physics Department, Visva Bharati, Shantiniketan, India

K. Mukherjee (✉)
Physics Department, Banwarilal Bhalotia College, Asansol, India

Centre of Organic Spintronics and Optoelectronics Devices (COSOD), Kazi Nazrul University, Asansol, India

S. Das et al. (eds.), *Advances in Terahertz Technology and Its Applications*,
https://doi.org/10.1007/978-981-16-5731-3_5

threshold current density, increased temperature stability, high gain bandwidth, and quicker gain recovery compared to bulk or QW structures [13]. Recently, lots of proposal of optical logic devices utilize these advantages are proposed and analyzed [14–21]. These devices include optical logic gates, pseudorandom bit sequence generator and encryption-decryption systems and many more. Single QDSOA assisted by filters finds applications in designing various logic gates [14, 22, 23]. Two-photon absorption assisted performance in QDSOA [24] can boost the operating speed of the devices and turbo switch configuration based on QDSOA [25] is also very attractive. Dual rail switching mode application can enhance operating speed of the devices proposed [26]. Interferometric configurations based on QDSOA such as Mach Zehnder Interferometer are useful and find applications in different optical logic processors [24, 27–33]. However, it is difficult to maintain the balance of the interferometric device, particularly phase difference between the arms [14, 23]. Moreover, they may create problem in cascading and results in reduction of processing speed. Beside MZI structure, there are other SOA-based interferometric structures like Sagnac switch, ultrafast nonlinear interferometer (UNI), etc. exhibit interesting characteristics in terms of logic gate implementation and their applications[34–40]. However, non-interferometric structure-based logic gates using QDSOA can boost the processing speed up to 2.5 Tb/s and QDSOA can also work with optical injection in place of electrical current [41]. QDSOA-based devices also show large Quality factor (Q), extinction ratio (ER), contrast ratio (CR), with low bit error rate (BER) and is established now [42]. A recent work by one of the author Mukherjee [43] has proposed 1 Tb/s 4 bit digital to analog converter (DAC) based on QDSOA with improved dynamical range and absolute error. This establishes the superiority of QDSOA over other related devices. The present chapter uses a single QDSOA without using any filter and interferometric structure to analyze the performance of optical inverter or NOT gate. An inverter is used to design NAND gate and hence any other logic gates and processors. A NAND gate without inverter is also designed and its performance is compared with the inverter-based one. This confirms the ability of QDSOA to process high-speed data with considerable values of ER, CR, Q factor, etc. The present chapter also analyzes the performance of the NAND gates at a speed of 1 Tb/s. Moreover, the NAND gates proposed in this chapter have lower complexity compared to other similar proposals [8, 19]. Amplified spontaneous emission (ASE) noise is an important consideration and degrades performance [28, 29] and the present paper also investigates ASE noise effects. Another important fact about the proposed logic devices is the use of counter propagating pump probe signals.

2 Working Principle of the Inverter and Mathematical Modeling

The inverter is based on cross-gain modulation (XGM) in QDSOA. In XGM, an intensity-modulated pump signal modulates the gain of the QDSOA due to gain saturation effect. It means that when control or pump signal is low; the QDSOA gain

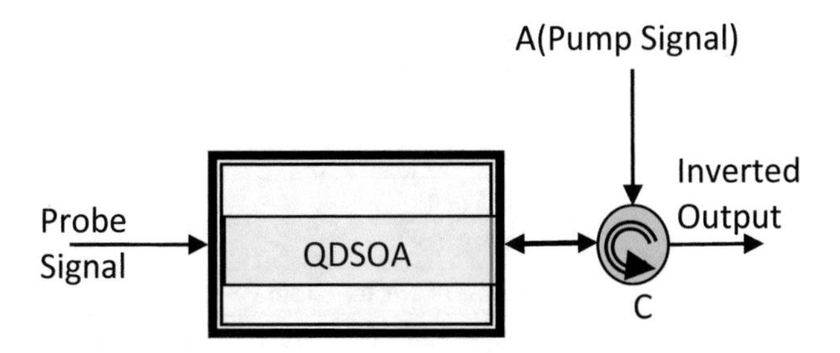

Fig. 1 All optical inverter using QDSOA

is high and vice versa. This gain variation will impose intensity variation on another weak signal called probe and output at probe wavelength gives inverted logic. For co-propagating pump probe scheme of XGM they should have different wavelength and a bandpass filter (BPF) is used to get the signal at probe wavelength [14]. To avoid use of an extra BPF, the present design uses counter propagating scheme of XGM, where pump probe signals propagate in opposite direction and may be of same wavelength [13].

Therefore, the present work uses a filter free configuration as shown in Fig. 1. In Fig. 1, A is the pump signal, which is the input of the inverter logic. P_{in} is the data input or probe. Both pump and probe are intensity-modulated signals. C is the circulator that directs pump signal toward the QDSOA, and QDSOA output toward output port indicated by the arrow. The pump signal actually modulates the gain and the time dependent gain of the QDSOA is given by $G(t) = \exp[(g_{\max}(2f - 1) - \alpha_{\text{int}})L]$, where $g_{\max}$ is the maximum modal gain of the QSDSOA, α_{int} is material absorption coefficient, f is the electron density of the ground state (GS), L is the SOA length. When A is high, the gain of the QDSOA becomes saturated or low ($G(t) = G_s$). Hence, the probe signal experiences this gain and the inverter output becomes $P_{\text{out}} = G_s\, P_{\text{in}}$, which is low. This corresponds to the operation '1' to '0'. When A is low or '0', the gain of the QSSOA is high, i.e., $G(t) = G_0$, unsaturated gain and the output of the logic inverter is $P_{\text{out}} = G_0\, P_{\text{in}}$. For calculating $G(t)$, f is calculated by numerically solving the coupled differential Eqs. (1)–(5) [14],

$$\frac{\partial N}{\partial t} = \frac{J}{eL_w} - \frac{N(1-h)}{\tau_{w2}} + \frac{N_Q h}{\tau_{2w} L_w} - \frac{N}{\tau_{wR}} \tag{1}$$

$$\frac{\partial h}{\partial t} = -\frac{h}{\tau_{2w}} - \frac{N(1-h)L_w}{\tau_{w2} N_Q} + \frac{(1-f)h}{\tau_{21}} - \frac{f(1-h)}{\tau_{12}} \tag{2}$$

$$\frac{\partial f}{\partial t} = \frac{(1-f)h}{\tau_{21}} - \frac{f(1-h)}{\tau_{12}} + \frac{f^2}{\tau_{1R}} - \frac{v_g L_w g_{\max}(2f-1)P}{N_Q A_{\text{eff}} h\nu} \tag{3}$$

$$\frac{\partial P}{\partial z} = \frac{[g_{\max}(2f-1) - \alpha_{\text{int}}]P}{A_{\text{eff}} h\nu} \tag{4}$$

Table 1 Parameters for simulation [43]

Modal gain (maximum) $g_{\max} = 1400\ \text{m}^{-1}$
Absorption coefficient (material) $\alpha_{\text{int}} = 200\ \text{m}^{-1}$
Current density, $J = 10\ \text{kA/m}^2$
$L_w = 0.25\ \mu\text{m}$
$N_Q = 5.0 \times 10^{10}$
Escape time of electron from ES to WL (τ_{2w}) = 1 ns
Spontaneous radiative lifetime [WL(τ_{wR})] = 0.2 ns
Electron relaxation time [WL to ES(τ_{w2})] = 3 ps
Electron relaxation time [ES to GS(τ_{21})] = 0.16 ps
Group velocity (V_g) = 8.3×10^7
Electron escape time [GS to ES (τ_{12})] = 1.2 ps
Spontaneous radiative lifetime [quantum dots (τ_{1R})] = 0.4 ns
$A_{\text{eff}} = 0.75\ \mu\text{m}^2$
τ, full width half maxima = 0.2 ps

Here P is the total power of control signal (P_c) and probe signal (P_{in}) and J is the injected current density. N, and h represent electron density of the wetting layer (WL), occupation probabilities of quantum dot excited state (ES) respectively. In the above equations, Quantum dot active layer surface density and effective thickness are N_Q and L_w responsively. A_{eff} is the effective cross-section of the QDSOA. Equations (1)–(4), solved numerically to analyze the performance of the optical inverter. The parameters used for simulation is shown in Table 1. The input signal A is modulated at a rate of 1 Tb/s, and its power is optimized for different parameters like extinction ratio (ER), contrast ratio (CR), Quality factor (Q value), amplitude modulation (AM) and bit error rate for three unsaturated QDSOA gains 10, 15, and 20 dB. After optimizing these parameters for control pulse power, their dependence on the amplified spontaneous noise (ASE) factor N_{sp} is investigated. The ASE noise power is given by $P_{\text{ASE}} = N_{sp}(G_0 - 1)EB_0$, E is the energy of the photons and B_0 is the bandwidth over which noise is considered. We have used expressions for ER $= 10\log(P^1_{\max}/P^0_{\min})$dB, CR $= 10\log(P^1_{\text{av}}/P^0_{\text{av}})$dB, AM $= 10\log(P^1_{\max}/P^1_{\min})$dB, $Q = (P^1_{\text{av}}/P^0_{\text{av}})/(\text{sd}_1 + \text{sd}_0)$dB, and relative eye opening (REO) $= [1 - (P^0_{\max}/P^1_{\min})] \times 100\%$, where $\left(P^1_{\max}, P^1_{\min}, P^1_{\text{av}}, \text{sd}_1\right)$ and $\left(P^0_{\max}, P^0_{\min}, P^0_{\text{av}}, \text{sd}_0\right)$ are maximum, minimum, average and standard deviations of high (1) and low (0) states respectively, for calculating these performance metrics.

3 Results and Discussions

Equations (1)–(4) are solved numerically using fourth-order RK (Runge Kutta) method with return to zero (RZ) input pulse given by [44],

$$P_{\mathrm{cp}}(t) = P_{\mathrm{peak}} \mathrm{e}^{-\frac{t^2 4 \ln 2}{\tau^2}} \tag{5}$$

where P_{peak} is the control pulse peak power and τ is the full width half maximum.

These types of short pulses with high repetition rates can be generated by mode locking as described in [44–46]. The choice of the pulse parameters is very critical for the performance of the device proposed.

Figure 2a, b shows plot of ER with control pulse energy for different unsaturated amplifications (10, 15 and 20 dB) without and with ASE noise respectively. The value of ER increases with control pulse power and becomes saturated at 0.1 mW. This is

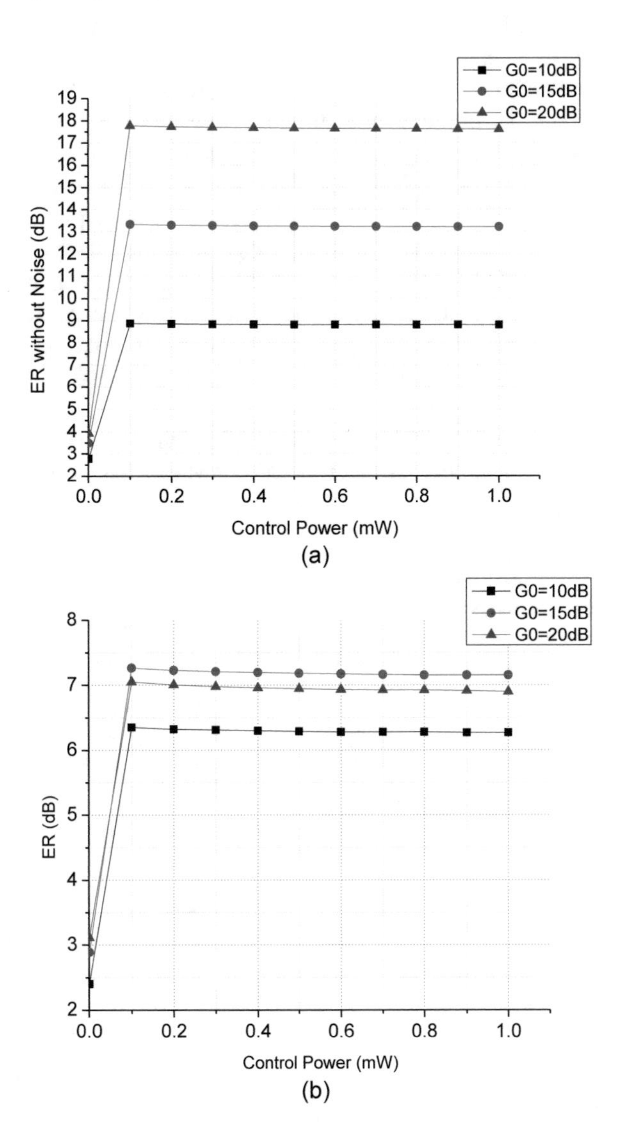

Fig. 2 **a** Dependence of ER on control pulse power without considering ASE noise. **b**. Dependence of ER on control pulse power considering ASE noise

due to sufficient gain saturation effect at low optical power (0.1 mW) for quantum dot SOA. When noise is not considered, the value of ER is higher for higher unsaturated gains. However, for higher gain (20 dB = 100), the ER value decreases compared to 15 dB gain if ASE noise is considered. This is because higher noise power affects the low bits ('0') considerably for higher gains.

Figure 3a shows dependence of CR on control pulse power with similar type of variation as ER considering noise. In this case, if we do not consider noise, CR value increases with unsaturated gain for a constant control power, but while considering noise the CR value reduces for 20 dB gain compared to 15 dB gain due to noise effect on low bits as explained in earlier paragraph.

Quality factor (Q), an important metric to analyze performance for a digital circuit, and it is related to bit error rate through complementary error function (erfc), BER = [erfc $(Q/\sqrt{2})$]/2. A $Q > 6$ corresponds to BER ~ 10^{-12} for transmission and reception of bits. In this communication, we have found Q values in the range 19–21 dB (more than 80) for the given unsaturated gains between 10 and 20 dB. The variation is shown in Fig. 4. It is clear from Fig. 4 for higher gains Q decreases with control power. Although the difference between average powers of high state('1's) and low states('0's) increases with higher gins, but due to dynamics of the gain recovery of the QDSOA (another pulse comes before complete gain recovery or gain saturation), spreading in the power levels of '1's and '0's considerably increases the standard deviations for higher gains. This causes Q value to decrease with gains. The Q values reach its maximum at control power 0.1 mW as ER and CR for three different gains.

Figure 5 shows the variation of AM with control power and it reaches its saturation value for 0.1 mW for all the gains. AM less than 1 dB is a prerequisite for efficient

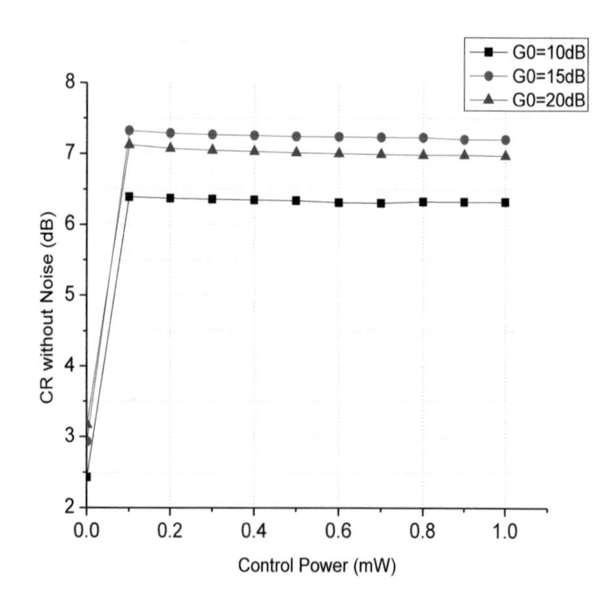

Fig. 3 Dependence of CR with control power considering ASE noise

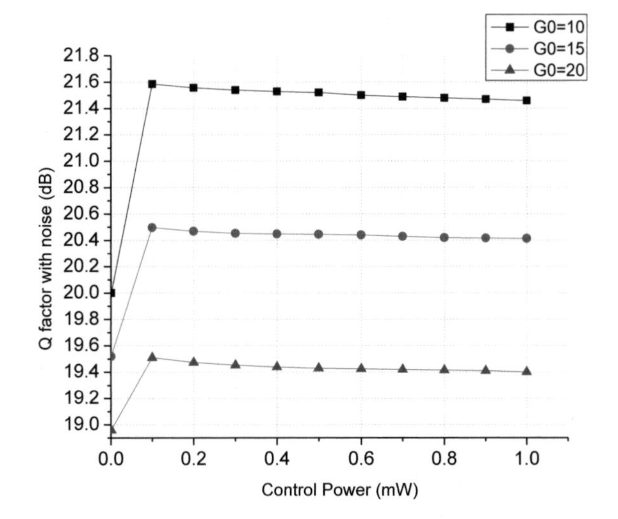

Fig. 4 Dependence of Q factor on control power

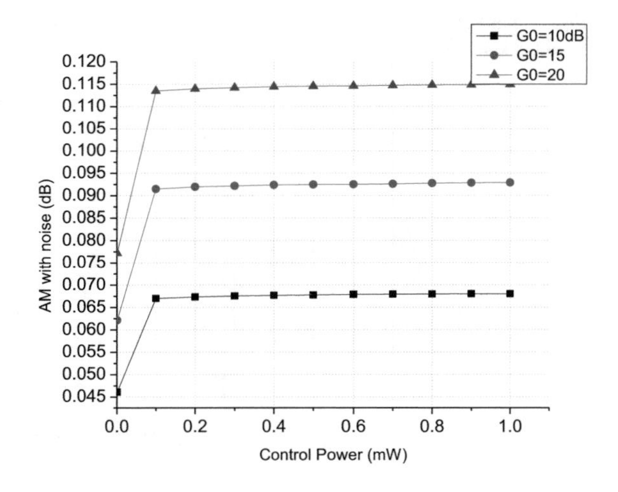

Fig. 5 Dependence of AM on control power

logic gates and processors. For 10 dB it is nearly 0.06 dB, and for 20 dB it is nearly 0.11 dB, which are much less than 1 dB ensure uniformity in the power levels of high('1') states. Now the calculations have been done with control power 0.1 mW which gives optimized values of ER, CR, and Q. Here we have considering maximum value of AM for 0.1 mW, otherwise ER, CR and Q value will be much low if we try to reduce AM. Therefore, there is a compromise that we have used maximum AM for achieving maximum values of other parameters; however, AM is still much lower than 1 dB, maximum 0.12 dB is calculated for 20 dB gain with $N_{sp} = 1$, which is significantly low.

Figures 6, 7 and 8 show the dependence of ER, CR, and AM on ASE noise factor N_{sp}. These factors decreases with increasing noise factor and for larger gains the

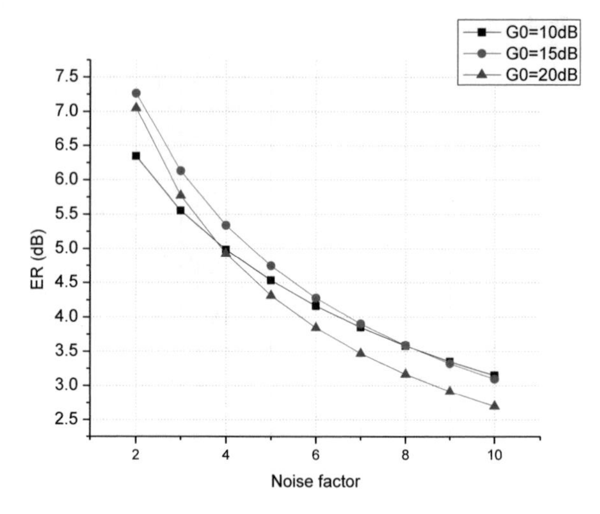

Fig. 6 Variation of ER with ASE noise factor N_{sp}

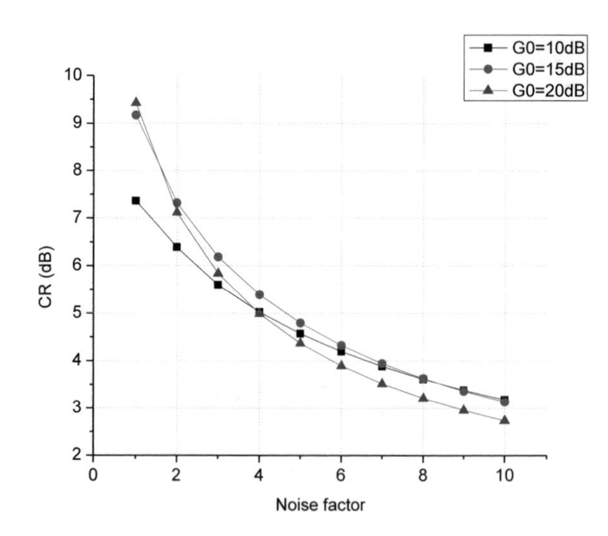

Fig. 7 Variation of CR with ASE noise factor N_{sp}

rate of change is faster due to increased effect of ASE noise. However, the Q value remains almost independent on ASE noise factor because both differences between power levels of high and low states and standard deviations are independent of noise. The former is because the noise powers cancel and for the later, it is because the noise power affects all low and all high states equally. Figure 9 shows variation of Q factor with N_{sp}. For low gain (10 dB), the noise behavior of ER, CR is not very sharp as the noise power is very low. However, for higher gains (>15 dB) the decrease is sharper.

In Fig. 10, the input-output bit pattern at 1 Tb/s is depicted. The eye diagrams for three different gains are also shown in Fig. 11. The operating speed is 1 Tb/s as clear from Fig. 11. Pseudoeye diagram (PED) is shown in Fig. 11. It shows that relative

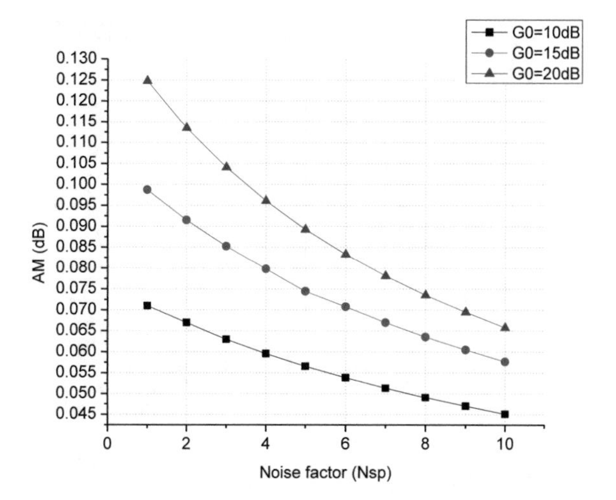

Fig. 8 Variation of AM with ASE noise factor N_{sp}

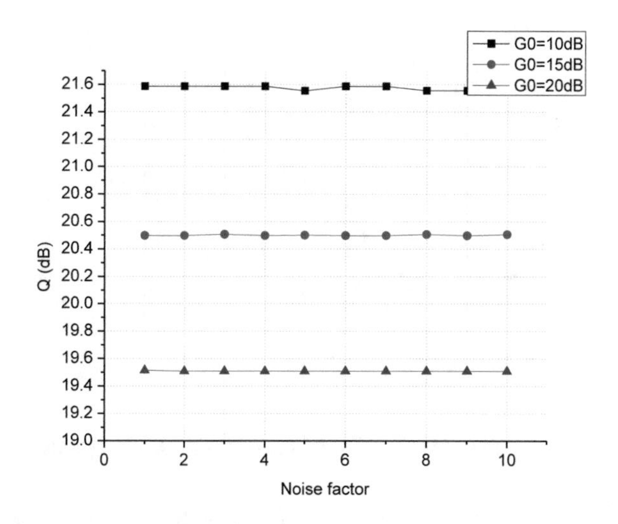

Fig. 9 Variation of Q value with Noise factor N_{sp}

eye opening (REO) increases with gain. REO of values 87.04% for 10 dB, 95.35% for 15 dB and 98.33% for 20 dB are calculated. This confirms large eye opening for these devices and large switching window.

From the eye diagram, it is clear that the relative intensity zero state becomes lower in case of higher gains. This also ensures high ER, CR and Q value for higher gains.

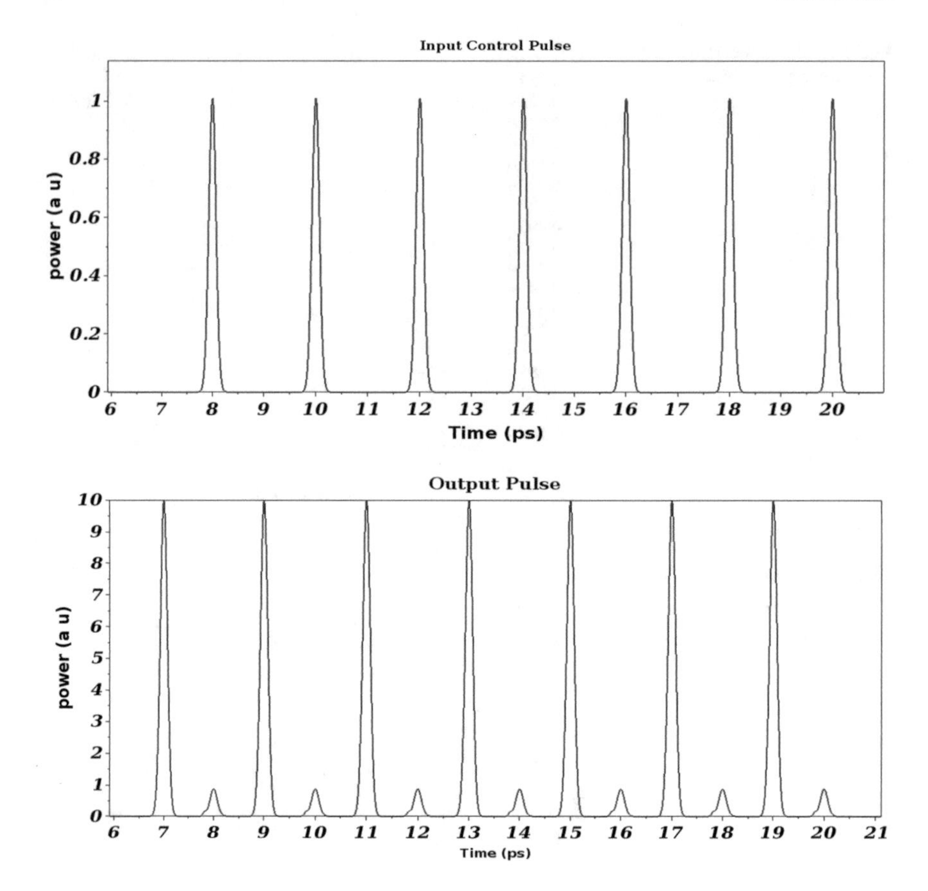

Fig. 10 Input and output pulse train

4 Application of the Inverter in Designing NAND Gate

The Inverter proposed and analyzed in the above sections is used to design the NAND gate (Fig. 12). The QDSOA1 and QDSOA2 are used as NOT gate and their outputs are combined. Therefore, the NAND gate is also non-interferometric structure. The operation is as follows:

Case 1: Controls are absent ($A = 0$, $B = 0$), no gain saturation of the QDSOAs, and gains are high (G_0). Therefore both the inputs are high giving a high ('1') output.

Case 2: Control $A = 0$, $B = 1$, QDSOA1 has high gain, but QDSOA2 experiences gain suppression, resulting in high output 1 and low output 2 giving final output high ('1').

Case 3: Control $A = 1$, $B = 0$, QDSOA1 has low gain due to gain saturation, but QDSOA2 exhibits high gain; a high output 2 and low output 1 giving final output high again as case 2('1') are achieved.

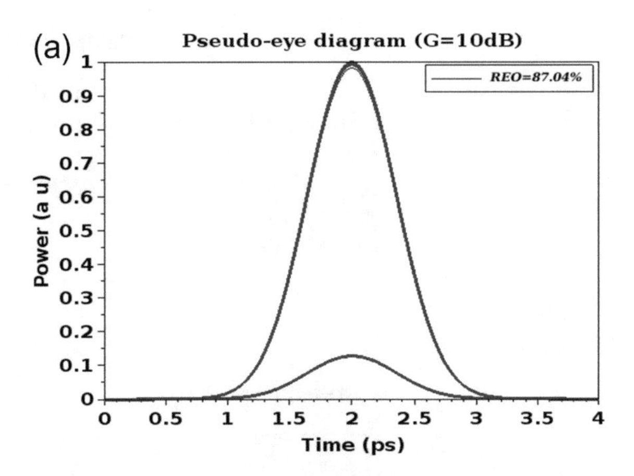

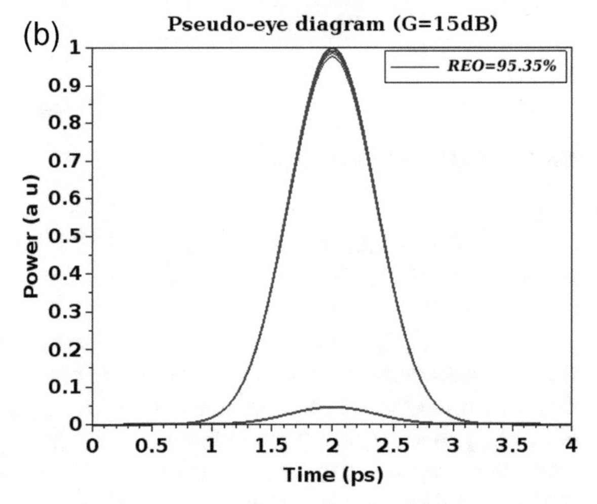

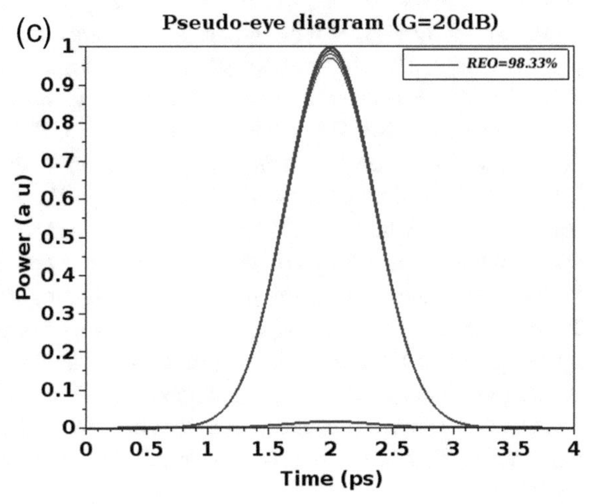

Fig. 11 Pseudoeye diagrams for different gains

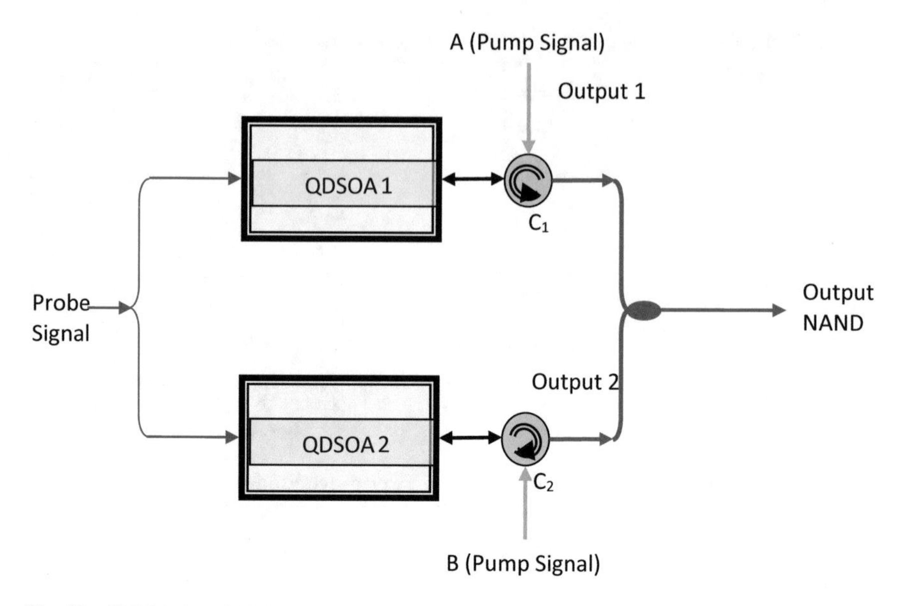

Fig. 12 QDSOA-based NAND gate

Case 4. Control inputs are present ($A = 1$, $B = 1$), gain saturation happens for both the QDSOAs, and gains are low. Therefore, both the inputs are low giving a low final output ('0').

The device in Fig. 12 works as optical NAND gate. For the analysis of the NAND gate, we have considered ER, CR, value and their dependence on Electron relaxation time for ES to GS(τ_{21}) and ASE noise factor N_{sp}. Figure 13 shows the variation of ER and CR with τ_{21} for three different gains. The values of ER and CR become maximum at $\tau_{21} = 0.16$ ps and this value of τ_{21} is used for all simulation. Figure 14 shows dependence of ER, CR and REO on N_{sp}.

Figure 14 shows that with the increase in noise factor N_{sp}, values of ER, CR and REO decrease. The effect of noise is more for higher gains than that for lower gains since ASE noise is directly proportional to the gains $G_0 > > 1$.

Bit patterns of NAND gate is shown in Fig. 15. The corresponding PED is shown in Fig. 16. Due to the spreading of high intensity levels, the REO for NAND gate is lower than that of inverter for a given gain. For $N_{sp} = 1$, it becomes less than 90% for NAND gate where as for same condition it is more than 98% for inverter. Correspondingly, the pseudoeye diagrams show lower eye opening and lower switching window. Pseudoeye diagrams show that for higher gains, relative intensity of lower states becomes less, and eye openings increase. This also confirms the fact that for higher gain REO is larger. This signifies that the NAND gate (Fig. 12) shows AM values more than 1 dB which is not accepted. Figure 16a corresponds to AM of 2.6 dB, and Fig. 16b, c represents AM values of 2.76 dB and 2.84 dB respectively. Large patterning effect in the output causes this degradation of AM. Therefore, parallel

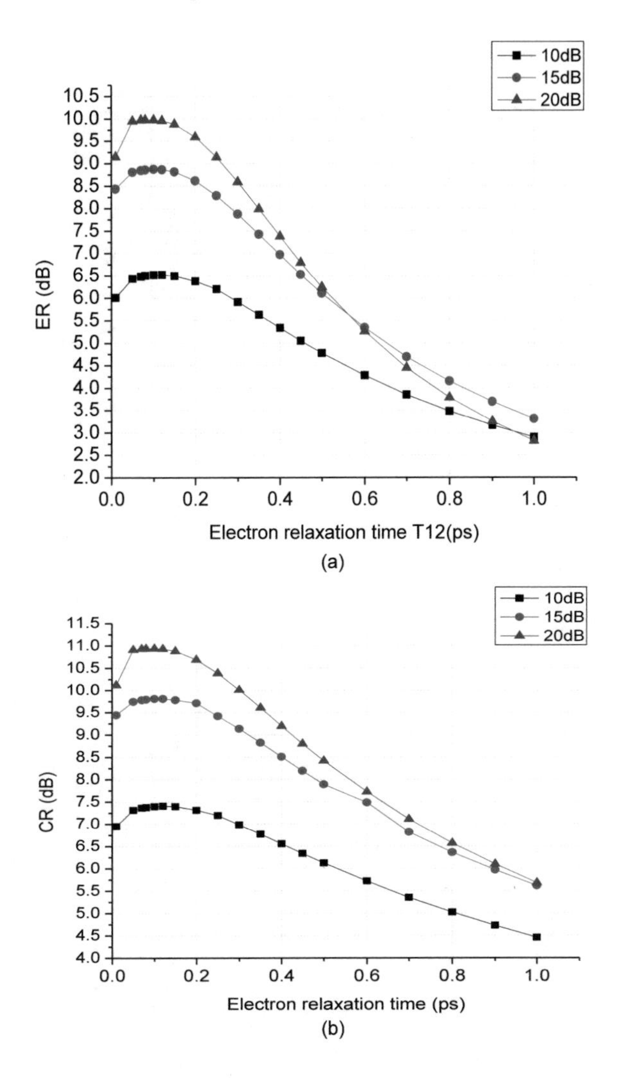

Fig. 13 Dependence of **a** ER **b** CR on electron relaxation time

addition of two NOT gates or inverters gives not so good NAND gate. To make the performance better alternative proposal of NAND gate without using the inverter is given in the section below.

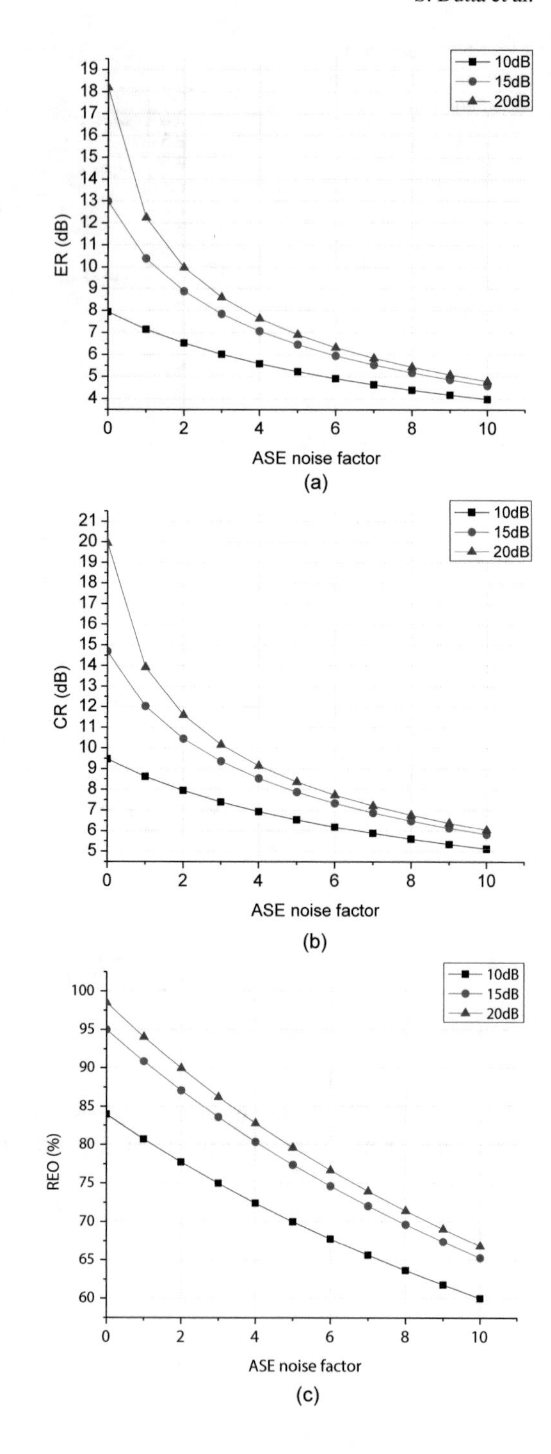

(a)

(b)

(c)

Fig. 14 Variation of **a** ER, **b** CR, and **c** REO with ASE noise factor N_{sp}

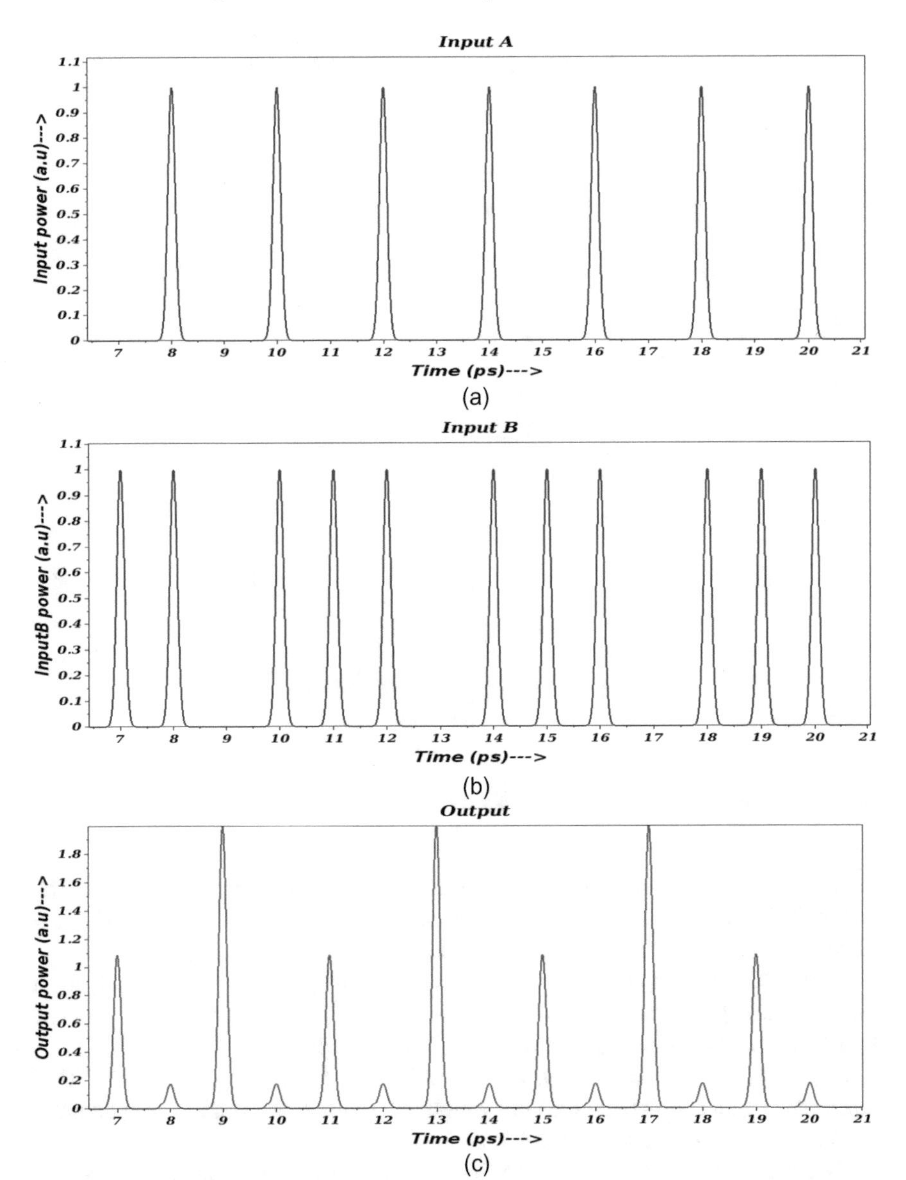

Fig. 15 Input and output bit patterns of NAND gate at a rate of 1 Tb/s

5 Implementation of NAND Gate Without Using Inverter

From Figs. 15 and 16, it is clear that inverter-based NAND gates shows large patterning effects and hence reduced amplitude modulation (AM), ER, CR, and Q factor, etc. To overcome this alternative design of NAND gate is implemented

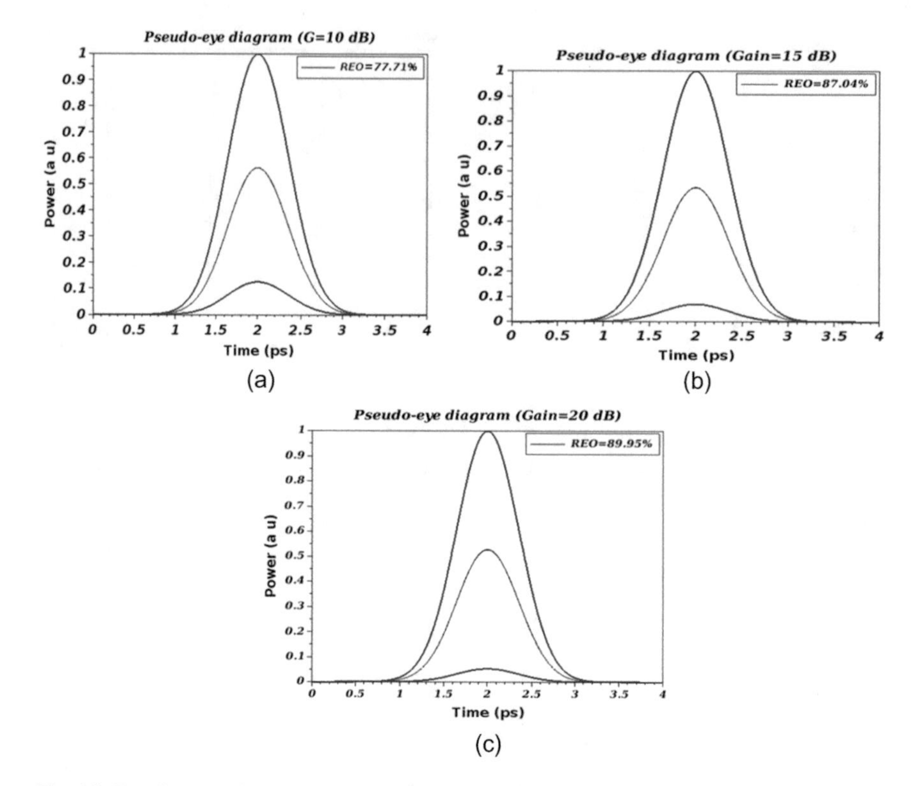

Fig. 16 Eye diagrams for **a** 10 dB, **b** 15 dB and **c** 20 dB for NAND gate

using QDSOA as shown in Fig. 17. The structure is similar to that of the NAND gate shown in Fig. 12 with modifications in the probe signals used in the QDSOAs. The QDSOA2 uses A as probe signal. The working is as follows:

Case 1: Control inputs are absent ($A = 0$, $B = 0$), no gain saturation of the QDSOAs, and gains are high (G_0). QDSOA1 gives high output. QDSOA2 receives no probe signal and hence gives no output. Therefore overall output is high.

Case 2: Control input $A = 0$, $B = 1$, QDSOA1 has high gain, but QDSOA2 experiences gain compression resulting high output1 and zero output 2(since probe A is zero) giving final output high ('1').

Case 3: Input $A = 1$, $B = 0$, QDSOA1 has low gain due to gain saturation, but QDSOA2 exhibits high gain; resulting high output2 and low output1 giving final output high again as case 2('1').

Case 4: Both inputs are absent ($A = 1$, $B = 1$), gain saturation happens for both the QDSOAs, and gains are low. Therefore, both the inputs are low giving a low final output ('0').

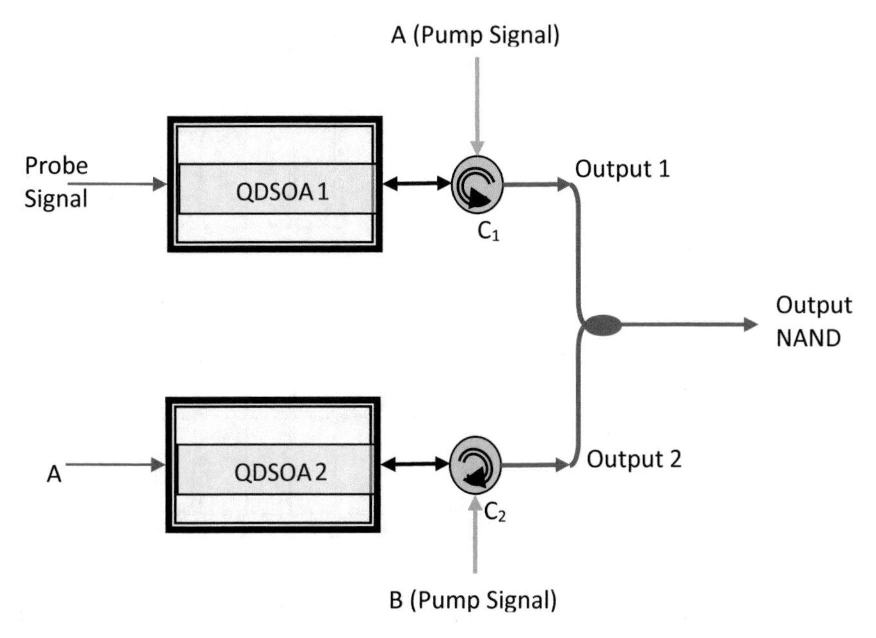

Fig. 17 QDSOA-based NAND gate

Figure 18 shows input and output bit patterns of NAND gate of Fig. 17 with $N_{sp} = 2$ for 20 dB unsaturated gain. Figure 18 clearly indicates reduced patterning effects compared to Fig. 16.

Figure 19 shows the variations of ER, and CR of the NAND gate of Fig. 17 with control power. These quantities show maximum values at control power 1mW. Their values are 16.7, and 18.45 dB show significant improvement compared to inverter-based NAND gate (corresponding values are 10 dB, 11.5 dB for $N_{sp} = 2$). Such improved performance with Tb/s operating speed is feasible with QDSOA, similar structure with bulk SOA operates at Gb/s speed only [47].

The relative eye opening is calculated to be 97.86% much larger than the corresponding value 89.95% for the inverter-based NAND gate. This improvement confirms the lower patterning effect in this NAND gate without inverter in Fig. 20.

The dependence of AM on control power is shown in Fig. 21. The minimum value is 0.11 dB much smaller than the accepted value of 1 dB. This also confirms improvement over the NAND gate using inverter.

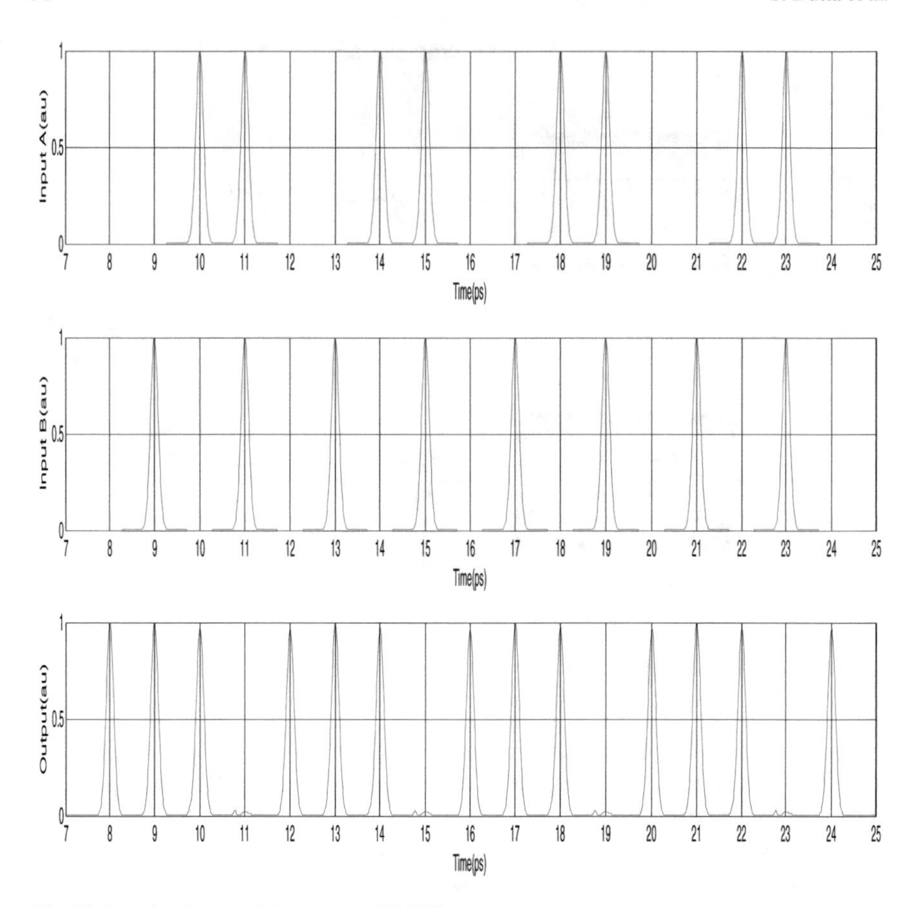

Fig. 18 Input and output bit patterns of NAND gate

6 Conclusions

We have presented all optical logic inverter and NAND gate as an application of the inverter is presented with simulation using cross-gain modulation in QDSOA. Alternative implementation of NAND gate without using the inverter is also described and shows better performance. The gates use quantum dot SOA in non-interferometric configuration due to its inherent properties, which makes QDSOA an excellent element for all optical processing. The optimized values of the parameters and analysis of different metrics show applicability of these devices for future optical computing and communication devices. Moreover, use of QDSOA makes this type of logic gates faster than similar logic gates with bulk SOA establishes superiority of QDSOA over bulk devices.

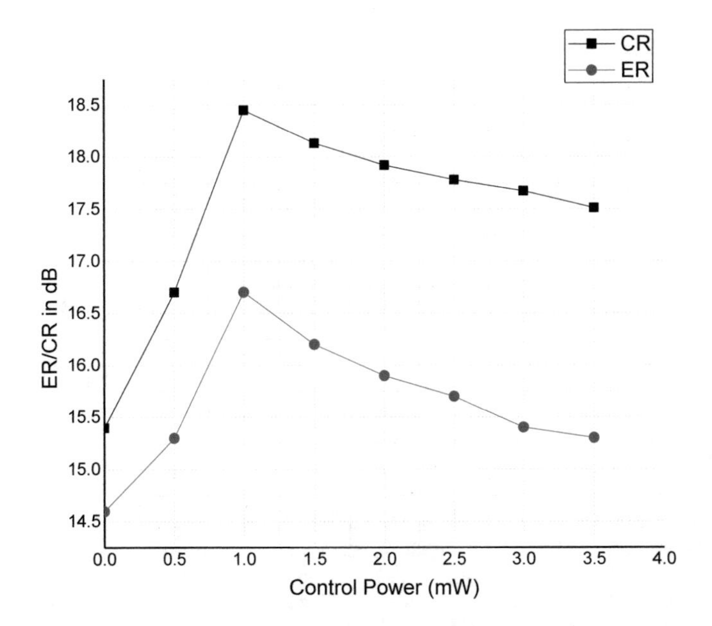

Fig. 19 Dependence of ER/CR on control power

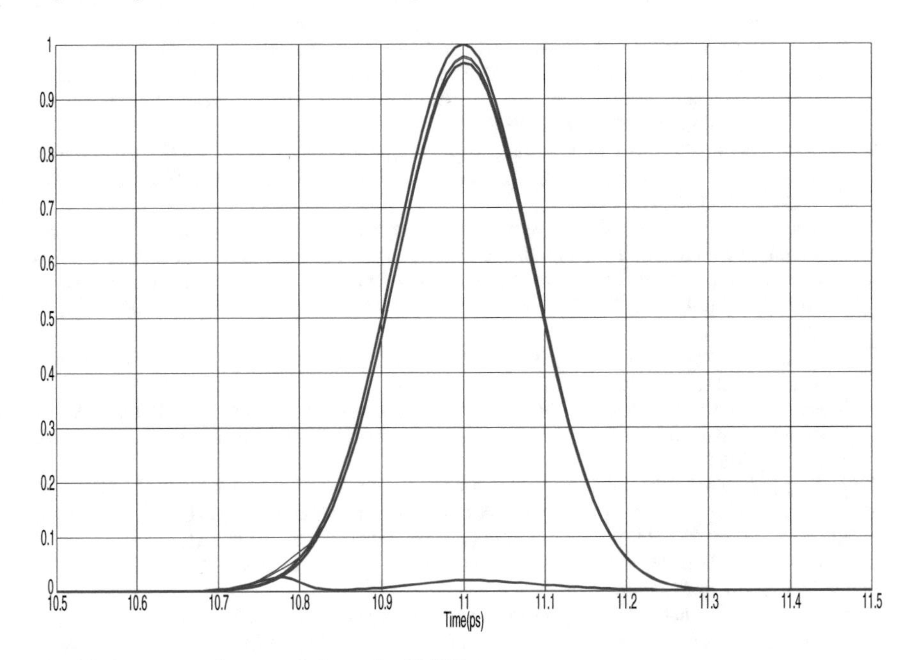

Fig. 20 Pseudoeye diagram of alternative NAND gate

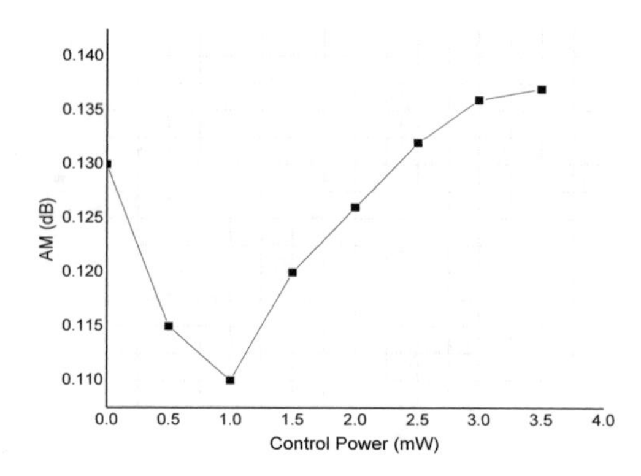

Fig. 21 Dependence of AM on control power

References

1. J. Renaudier, B. Duval, D. Lanteri, A. Boutin, L Letteron, Y. Frignac, et al., Recent advances in 100+nm ultra-wideband fiber-optic transmission systems using semiconductor optical amplifiers. J. Lightwave Technol., 1–1 (2020). https://doi.org/10.1109/jlt.2020.2966491
2. X. Zhang, S. Thapa, N. Dutta, All-optical XOR gates based on dual semiconductor optical amplifiers. Cogent Phys. **6**(1), 1660495 (2019)
3. K. Mukherjee, K. Maji, A. Raja, All optical four bit two's complement generator and single bit comparator using reflective semiconductor optical amplifier. IJNBM **9**(1/2) (2020)
4. N. K. Dutta, Q. Wang, in *Semiconductor Optical Amplifiers*, 2nd edn. (World Scientific, 2013)
5. J. Dong, X. Zhang, S. Fu, J. Xu, P. Shum, D. Huang, Ultrafast all-optical signal processing based on single semiconductor optical amplifier. IEEE J. Sel. Top. Quant. Electron. **14**(3) (May/June 2008), 770–778. ISSN 1077–260X (2008)
6. A. Kotb, C. Guo, Reflective semiconductor optical amplifiers-based all-optical NOR and XNOR logic gates at 120 Gb/s. J. Modern Opt. **67**(18), 1424–1435 (2020). https://doi.org/10.1080/09500340.2020.1862333
7. K. Mukherjee, P. Ghosh, Alternative method of implementation of frequency encoded N bit comparator exploiting four wave mixing in semiconductor optical amplifiers. Optik **123** (2012)
8. A. Kotb, C. Guo, 120 Gb/s all-optical NAND logic gate using reflective semiconductor optical amplifiers. J. Mod. Opt. **67**(12), 1138–1144 (2020). https://doi.org/10.1080/09500340.2020.1813342
9. S.K. Garai, All-optical quaternary logic gates. J. Mod. Opt. **60**(12), 993–1005 (2013). https://doi.org/10.1080/09500340.2013.826388
10. R. Sarojini, A. Sivanantha Raja, S. Selvendran, K. Esakki Muthu, Cross polarization modulation based wavelength conversion with very low pump power in SOA: an investigation. Optik **185**, 852–858 (2019)
11. K. Maji, K. Mukherjee, A. Raja, An alternative method for implementation of frequency-encoded logic gates using a terahertz optical asymmetric demultiplexer (TOAD). J. Comput. Electron. **18**, 1423–1434 (2019). https://doi.org/10.1007/s10825-019-01393-5
12. P. Singh, D.K. Tripathi, H.K. Dixit, Designs of all-optical NOR gates using SOA based MZI. Optik **125**(16), 4437–4440 (2014)
13. D.K. Gayen, T. Chattopadhyay, Designing of optimized all-optical half adder circuit using single quantum-dot semiconductor optical amplifier assisted Mach-Zehnder interferometer. J. Lightwave Technol. **31**(12), 2029–2034 (2013)
14. K. Komatsu, G. Hosoya, H. Yashima, All Optical NOR gate using a single quantum dot SOA assisted an optical filter. Opt. Quant. Electron. **50**(131), 1–18 (2018)
15. X. Zhang, N.K. Dutta, Effects of two-photon absorption on all optical logic operation based on quantum-dot semiconductor optical amplifiers. J. Mod. Opt. **65**(2), 166–173 (2018). https://doi.org/10.1080/09500340.2017.1382595

16. A. Kotb, C. Guo, All-optical OR and NOR gates using quantum-dot semiconductor optical amplifiers-assisted turbo-switched Mach-Zehnder interferometer and serially delayed interferometer at 1 Tb/s. Optik, online article (2020). https://doi.org/10.1016/j.ijleo.2020.164879
17. S. Thapa, X. Zhang, N.K. Dutta, Effects of two-photon absorption on pseudo-random bit sequence operating at high speed. J. Mod. Opt. **66**(1), 100–108 (2019). https://doi.org/10.1080/09500340.2018.1512671
18. A. Fardi, H. Baghban, M. Razaghi, Dynamics of pulse amplification in tapered-waveguide quantum-dot semiconductor optical amplifiers. Optik (2020). https://doi.org/10.1016/j.ijleo.2020.164396
19. Shu-Chao Mi, Hai-Long Wang, Shu-Yu Zhang, Qian Gong, Research of all-optical NAND gates based on quantum dot semiconductor optical amplifiers cascaded connection XGM and XPM. Optik **20** (2020). https://doi.org/10.1016/j.ijleo.2019.163551
20. F. Hakimian, M.R. Shayesteh, M.R. Moslemi, Optimization of a quantum-dot semiconductor optical amplifier (QD-SOA) design using the genetic algorithm. Opt. Quant. Electron. **52**, 48 (2020). https://doi.org/10.1007/s11082-019-2174-4
21. W. Li, H. Hu, N.K. Dutta, High speed all-optical encryption and decryption using quantum dot semiconductor optical amplifiers. J. Modern Opt. **60**(20), 1741–1749. https://doi.org/10.1080/09500340.2013.856486
22. D. Fouskidis, K.E. Zoiros, A. Hatziefremidis, Reconfigurable all-optical logic gates (AND, NOR, NOT, OR) with quantum-dot semiconductor optical amplifier and optical filter. J. Sel. Top. Quant. Electron. **27**(2, art. no. 7600915) (2021)
23. E. Dimitriadou, K.E. Zoiros, All-optical XOR gate using single quantum-dot SOA and optical filter. J. Lightwave Technol. **31**(23), 3813–3821 (2013)
24. A. Kotb, K.E. Zoiros, C. Guo, 2 Tb/s all-optical gates based on two-photon absorption in quantum dot semiconductor optical amplifiers. Opt. Laser Technol. **112**, 442–451 (2019)
25. A. Kotb, K.E. Zoiros, C. Guo, 1 Tb/s all-optical XOR and AND gates using quantum-dot semiconductor optical amplifier-based turbo-switched Mach-Zehnder interferometer. J. Comput. Electron. **18**, 628–639 (2019)
26. K.E. Zoiros, C. Demertzis, On the data rate extension of semiconductor optical amplifierbased ultrafast nonlinear interferometer in dual rail switching mode using a cascaded optical delay interferometer. Opt. Laser Technol. **43**(7), 1190–1197 (2011)
27. D. Kastritsis, K.E. Zoiros, E. Dimitriadou, Design of ultrafast all-optical pulsed-mode 2 × 2 crossbar switch using quantum-dot semiconductor optical amplifier-based Mach-Zehnder interferometer. J. Comput. Electron. **15**, 1046–1063 (2016). https://doi.org/10.1007/s10825-016-0863-9
28. A. Kotb, K.E. Zoiros, On the design of all-optical gates based on quantum-dot semiconductor optical amplifier with effect of amplified spontaneous emission. Opt. Quant. Electron. **46**(8), 977–989 (2014)
29. A. Kotb, K.E. Zoiros, Simulation of all-optical logic XNOR gate based on quantum-dot semiconductor optical amplifiers with amplified spontaneous emission. Opt. Quant. Electron. **45**(11), 1213–1221 (2013)
30. E. Dimitriadou, K.E. Zoiros, On the feasibility of 320 Gb/s all-optical AND gate using quantum-dot semiconductor optical amplifier-based Mach-Zehnder interferometer. Prog. Electromagnetics Res. B **50**, 113–140 (2013)
31. E. Dimitriadou, K.E. Zoiros, Proposal for ultrafast all-optical XNOR gate using single quantum-dot semiconductor optical amplifier-based Mach-Zehnder interferometer. Opt. Laser Technol. **45**(1), 79–88 (2013)
32. E. Dimitriadou, K.E. Zoiros, On the design of reconfigurable ultrafast all-optical NOR and NAND gates using a single quantum-dot semiconductor optical amplifier-based MachZehnder interferometer. J. Opt. **14**(10, art. no. 105401) (2012)
33. E. Dimitriadou, K.E. Zoiros, Proposal for all-optical NOR gate using single quantum-dot semiconductor optical amplifier-based Mach-Zehnder interferometer. Opt. Commun. **285**(7), 1710–1716 (2012)

34. G. Papadopoulos, K.E. Zoiros, On the design of semiconductor optical amplifier-assisted Sagnac interferometer with full data dual output switching capability. Opt. Laser Technol. **43**(3), 697–710 (2011)
35. K.E. Zoiros, P. Avramidis, C.S. Koukourlis, Performance investigation of semiconductor optical amplifier-based ultrafast nonlinear interferometer in nontrivial switching mode. Opt. Eng. **47**(11, art. no. 1150060, 1–6 (2008)
36. K.E. Zoiros, C. Botsiaris, R. Chasioti, C.S. Koukourlis, Characterization and optimization of SOA-based UNI recirculating shift register switching window. Opt. Quant. Electron. **39**(14), 1167–1181 (2007)
37. K.E. Zoiros, J. Vardakas, C.S. Koukourlis, T. Houbavlis, Analysis and design of ultrahigh speed all-optical semiconductor optical amplifier assisted Sagnac recirculating shift register with an inverter. Opt. Eng. **44**(6, art. no. 065001), 1–12 (2005)
38. T. Houbavlis, K.E. Zoiros, Numerical simulation of semiconductor optical amplifier assisted Sagnac gate and investigation of its switching characteristics. Opt. Eng. **43**(7), 1622–1627 (2004)
39. T. Houbavlis, K.E. Zoiros, SOA-assisted Sagnac switch and investigation of its roadmap from 10 to 40 GHz. Opt. Quant. Electron. **35**(13), 1175–1203 (2003)
40. T. Houbavlis, K.E. Zoiros, 10-GHz all-optical recirculating shift register with semiconductor optical amplifier (SOA)-assisted Sagnac switch and SOA feedback. Opt. Eng. **42**(9), 2483–2484 (2003)
41. S. Alavizadeh, H. Baghban, A. Rostami, Quantum-dot semiconductor optical amplifier performance management under optical injection. J. Mod. Opt. **60**(6), 509–514 (2013). https://doi.org/10.1080/09500340.2013.794392
42. A. Kotb, C. Guo, All-optical NOR and XNOR logic gates at 2 Tb/s based on two-photon absorption in quantum-dot semiconductor optical amplifiers. Opt. Quant. Electron. **52**, 30 (2020). https://doi.org/10.1007/s11082-019-2142-z
43. K. Mukherjee, Tera bit per second all optical digital to analog converter using quantum dot semiconductor optical amplifier. Accepted for publication in Journ. Computational Electronics, Springer
44. J.S. Vardakas, K.E. Zoiros, Performance investigation of all-optical clock recovery circuit based on Fabry-Pérot filter and SOA-assisted Sagnac switch. Opt. Eng. **46**(8, art. no. 085005), 1–21 (2007)
45. K. Zoiros, K. Vlachos, T. Stathopoulos, C. Bintjas, H. Avramopoulos, 40 GHz modelocked SOA fiber ring laser with 20 nm tuning range, in *Optical Fiber Communications Conference (OFC), 2000, paper TuR3*, pp. 254–256 (2000).
46. K. Zoiros, T. Stathopoulos, K. Vlachos, A. Hatziefremidis, T. Houbavlis, T. Papakyriakopoulos, H. Avramopoulos, Experimental and theoretical studies of a high repetition rate fiber laser, mode-locked by external optical modulation. Opt. Commun. **180**, 301–315 (2000)
47. S.H. Kim, J.H. Kim, B.G. Yu, Y.T. Byun, Y.M. Jeon, S. Lee, D.H. Woo, S.H. Kim, All-optical NAND gate using cross gain modulation in semiconductor optical amplifiers. Electron. Lett. **41**(18)

Performance Analysis of Oversampled OFDM for a Terahertz Wireless Communication System

Abdelmounim Hmamou, Mohammed El Ghzaoui, and Jaouad Foshi

Abstract Thanks to the development of wireless communication technology, the whole world has become a small village, especially in the last 10 years. In fact, there are several reasons for this revolution, the most important of which is the transition to higher frequencies that would allow high-speed data transmission. In order to reach this high speed, it is necessary to increase the frequency up to ten terahertz (THz). In this case, we are talking about what is called THz wireless communication, the case where the frequency is between 0.1 and 10 THz. Thus, the bandwidth used in this frequency range is wide or ultra-wide. This introduces negative impacts on this communication system, mainly the effect of selective fading in frequency (multiple trips). The appropriate selection of a modulation technique makes it possible to remove these effects. Suitable solutions to this problem include the Oversampled orthogonal frequency division multiplexing (OOFDM) technique. The objective of this work is to improve and optimize the performance of a wireless communication system in the THz frequency band using the OOFDM modulation technique.

Keywords OFDM oversampled · Terahertz band · THz wireless communication · Selective fading · Channel modeling · BER · CCDF

1 Introduction

The rapid evolution that is observed in the world is fundamentally based on the advanced developments of communications technologies including wireless communication systems that have facilitated access to data and have made it possible to increase the number of objects connected. Thus, the continuous development of high-speed wireless communication applications is emerging, while those based on traditional spectra are gradually disappearing. To obtain a good quality of wireless communication with high-speed, several research activities that were interested in

A. Hmamou (✉) · J. Foshi
TTI Team, ERTTI Laboratory, Moulay Ismail University of Meknes, Meknes, Morocco

M. El Ghzaoui
Faculty of Sciences, Sidi Mohamed Ben Abdellah University, Fes, Morocco

S. Das et al. (eds.), *Advances in Terahertz Technology and Its Applications*,
https://doi.org/10.1007/978-981-16-5731-3_6

the development of this field. Most of this work has recently focused on increasing the frequency of operation. As a result, the researchers are interested in the Terahertz (THz) frequency domain that offers the possibility of achieving higher data rates [1–6]. The advantages of the THz band over other frequency spectra currently used, e.g., the fifth generation (5G) wireless communication technology will be insufficient if considering the explosive growth of machine connectivity such as the Internet of Things (IoT) [7]. Many people will think in sixth generation (6G), but the technology of 6G is still very undefined [8].

Due to the increasing demand for broadband by users, a number of opportunities correlated with the deployment of the THz frequency band on a large scale. The THz band could therefore contribute to the realization of a resilient communication infrastructure that will eventually be established to meet the demands of users [9]. The THz frequency range is generally between 0.1 and 30 THz, but currently engineers and researchers are only interested in the low Terahertz band from 0.1 to 10 THz [1, 2, 9–12]. The waves in this band have many advantages such as the sufficiently wide frequency band, large capacity and high security [13], The THz beam being more diffracting and less attenuated by non-metallic and dry objects [14].

Several research and industrial institutes have been studying this frequency band for about ten years in order to be able to use it in wireless communication systems such as, officially the first international standard was published in 2018 called IEEE 802.15.3d standard, which offers throughput up to 100 Gbps [15]. In this standard, THz wireless communication systems operate with a carrier frequency ranging from 252 to 321 GHz, the range known as the IEEE terahertz band.

The operation of THz wireless communication systems depends mainly on the quality of the transmission channel, modulation, demodulation and transmission and reception technology [12]. In addition, the bandwidth used in the THz domain can reach very high values. Thus, the transmission of data with THz frequencies can give rise to major difficulties, particularly those related to the effect of multiple paths, more precisely the impact of selective fading in frequency. This requires finding the best solutions against these negative effects. Indeed, in this chapter, we will look for an appropriate technique that allows us to make a modulation and demodulation in the field of THz frequencies. This led to the use of the orthogonal frequency division multiplexing (OFDM) technique. On the other hand, there are several genres of this technique in the literature. Because of the limitations of conventional OFDM in the THz frequency domain, another kind of OFDM modulation with better properties, which seems to be a very promising solution. This modulation is called oversampled OFDM. The objective of this chapter will be to analyze the performance of the oversampled OFDM (OOFDM) technique of a wireless communication system in the THz frequency domain.

This chapter is divided into three parts, in the first part we recall the main elements of the communications systems in the THz frequency domain, thus highlighting the currently exploitable THz range, presents some applications of THz waves and also challenges that may limit the uses of THz radiations. The second part is focused to the efficient realization of a wireless data transmission system in the THz band by introducing an appropriate modulation technique called oversampled OFDM. Finally, in the last part of this chapter, with the simulation approach, we present and discuss the results obtained.

2 Terahertz Wireless Communication

2.1 Terahertz Frequency Domain

In the frequency bands of the electromagnetic spectrum, the terahertz frequency domain generally covers the range from 0.1 to 30 THz corresponding to wavelengths between 0.01 and 3 mm. This frequency band is included between the domain of microwave waves and infrared as shown in Fig. 1. Indeed, when we search in the literature, we find several definitions of this domain. However, most of the recent work in wireless communication systems have been exploited specifically the band 0.1 THz to 10 THz [1, 2, 9–12] due to many factors such as the lack of adequate technologies (sources, detectors, modulators, …) used in this band, the manufacturing of these inexpensive technologies and also the increasing complexity with the frequency increase. This band also named the spectral band of T-rays that located between the electronic, where the phenomena are governed mainly by the thermal energy k.T and the field of photonics, governed by the energy of photons h.ν. Comparing them with the microwave and infrared waves, we note the results illustrated in Table 1.

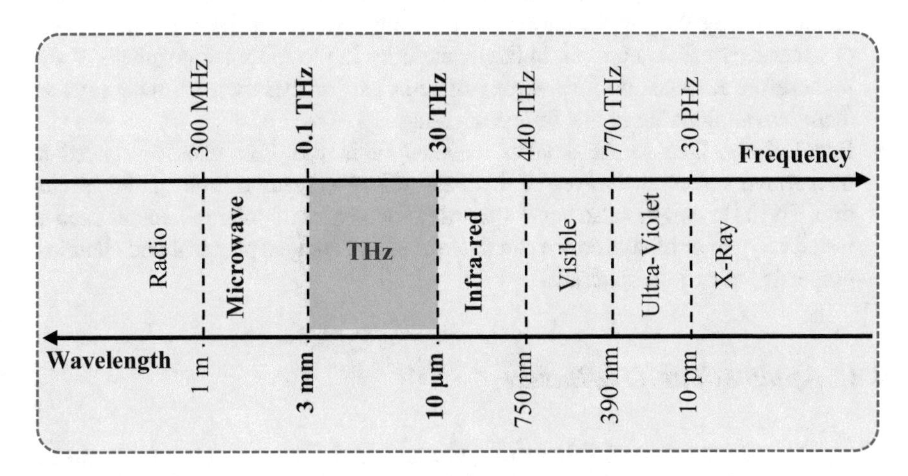

Fig. 1 Positioning of the terahertz frequency range in the electromagnetic spectrum

Table 1 Characteristics quantities of THz waves compared in the different frequency ranges

Spectral band	Microwaves	THz waves	Infrared waves
Frequency (THz)	$[3 \times 10^{-4}$–$0.1]$	[0.1–30]	[30–440]
Energy (eV)	$[1.24 \times 10^{-6}$–$4.13 \times 10^{-4}]$	$[4.13 \times 10^{-4}$–$0.124]$	[0.124–1.81]
Wavelength (mm)	$[3$–$10^{3}]$	[0.01–3]	$[7.4 \times 10^{-4}$–$10^{-2}]$

2.2 THz Band Characteristics

The THz frequency band has many characteristics, especially because of its specific physical properties. Among these characteristics are:

- **Penetration power**: Terahertz waves have a very high penetration power and are thus able to pass through many non-metallic materials such as organic materials, textiles, paper, wood, cardboard, plastics. Moreover, these waves are low-energy and non-ionizing, they have a photon energy of the order of the meV. In contrast, for example, X-rays can sometimes reach the MeV. Thus, it can be said that this energy is lower than that of most chemical bonds. These waves are therefore considered harmless to living organisms.
- **Propagation attenuation**: Although the THz band offers the possibility to increase the data rate and the bandwidth, the radiations of this band undergo a propagation attenuation which is very sensitive to the humidity and the distance transmitter–receiver. This attenuation becomes important in the presence of water molecules. For example, beyond one terahertz the attenuation could reach a value greater than 1 dB/m, with many peaks of strong absorption due to the resonances of water vapor that is present in large quantities in the earth's atmosphere. Water is therefore responsible for a strong attenuation for these waves, which makes them less exploitable in this frequency range.
- **Resolutions**: Due to the smaller wavelength in the THz band compared to microwave waves, the waves of this band allow to reach a high spatial resolution. The THz range is also very interesting for studies of polar molecules such as water, oxygen or others, so we can determine and analyze physical and chemical properties of certain materials.

2.3 Applications THz Band

The main characteristics of THz waves presented in the previous paragraph make it possible to make better and better use of the range of THz waves and making them very promising for many fields of application including spectroscopy which is one of the first THz wave applications. This technique is now becoming increasingly widespread in the field of research. It allows the study and analysis of simple and complex biological structures to identify the chemical species present in a sample as well as contributing to the study of dielectric properties and electrical conductivity of materials, THz spectroscopy also allows at the level of security, to detect illegal and dangerous substances such as explosives, drugs and toxic gases [16]. THz waves are also used in the field of medical imaging, for example, to establish medical diagnoses of diseases [1]. In the sector of monitoring and quality control, the exploitation of THz radiation is now beginning to spread more and more, such as the most popular applications is the airport body scanner that allows to detect drugs, weapons or dangerous substances, and also allows to distinguish different types of materials. Thus, THz waves have major advantages compared to other radiations. In order to achieve a higher throughput, the carrier frequency must be in the THz band, so that it can be exploited in wireless broadband communication [3, 5, 13, 17]. Briefly, we can summarize these applications under two main axes which are THz telecommunications and THz detection. The main axis targeted in this work is communications in the THz domain, which will be detailed in the following paragraph.

2.4 Technologies of THz Wireless Communication System

The active components enabling THz radiation to be generated and/or detected are the fundamental elements in the emergence and evolution of these technologies. These elements consist mainly of transmitters and detectors, which can be classified according to their sensitivity, output power, frequency and tunability in several categories. In the following, we will present the main transmitters and detectors that have been developed to date.

- **THz waves Transmitters**

Transmitters play a central role in the technology of wireless communications systems in the THz band where they can be classified according to the nature of the physical phenomenon at the origin of these waves, in electronics, optics, optoelectronics and mechanics.

- Electronic THz transmitters can be classified into 4 families:

  Negative differential resistance oscillators (IMPATT diodes, RTD diodes, GUNN structures, etc.).
  Active transistor circuits: two types of transistor have been used in these circuits, HEMT (High Electron Mobility Transistor) and HBT (Heterojunction Bipolar Transistor).
  Frequency multipliers: the main elements of this type of transmitter are Schottky diodes.
  Vacuum electronic sources: there are several types such as klystron, gyrotron, progressive wave tube and BWO (Backward Wave Oscillator).

- Optical transmitters: currently this type of transmitter remains unusable in the THz band due to high thermal noise.

- Optoelectronic transmitters: also called quantum cascade lasers there are many such as, photodiodes (PN; PIN; TTR; UTC) are widely used at THz frequencies, photoconductors and photogenerators.
- Mechanical transmitters

Traditional solid semiconductor sources are limited by the parasitic effects and transit time of carriers, which reduces their performance at high frequencies, particularly in the THz range.

- **THz waves Detectors**

Today, there are many THz detectors on the market, and others in the process of testing. According to the measurement method, they can be classified into two categories:

- Coherent detectors: they depend on the method of measuring the electromagnetic field and which gives information on the amplitude and phase of the signal received. Three types can be distinguished:

  The Schottky barrier diodes: they formed for a long time. Their principle is to transport electrons through a potential barrier of a semiconductor metal junction. They are used as detectors in THz wireless communications.
  Transistors: High electron mobility transistors (HEMTs) and field-effect transistors (FETs) are available. HEMT is more important than FET because they operate at room temperature and their expensive manufacturing.
  Mixers: their principle is to decrease the frequency of a detected signal, they can cover a frequency band from 0.05 to 0.5 THz and they can also detect a frequency modulation up to 500 MHz.

- Non-coherent detectors: they have two types:

  Pyroelectric detectors: they are mainly based on the properties of ferroelectric materials. Their manufacture is based on crystals, mainly $LiTaO_3$ tantalum lithium.
  Bolometers: They are based on the absorption of incident photons and the transformation of their energy into thermal energy. But they cannot detect high-throughput signals because of their sensitivities to ambient thermal energy. These detectors detect wide band THz waves.

2.5 THz Wireless Communication

In recent years, the speed of a wireless communications network has evolved more rapidly, it was 100 Mbps in 2009 and currently reached 10.5Gbps (IEEE 802.11ax 2021 standard). Despite this significant evolution, the increase of users, the emergence of new uses such as video streaming in high definition (4 K) and ultra-high definition (8 K), autonomous vehicles, Online games impose additional constraints on this communication network. The frequency bands used today are still unable to meet all these requirements.

In order to efficiently meet these requirements and the increasing demand for data rate, the increase of the carrier frequencies to the terahertz band represents the best solutions. In recent years, several research teams have shown that it is possible to achieve ultra-high data rates by exploiting the THz band [18].

High-speed THz wireless communications can be distinguished according to the distance between the transmitter and the receiver into three types: Short range THz wireless communication, on distances less than 1 m, can be used to transmit or receive information such as high definition videos or high quality photos; medium range, on distances between 1 and 100 m, can be applied in a building to create a high-speed local network; long-range THz wireless communication, on distances greater than 100 m, this type is reserved for interconnection between the core network and different entities or also the exchange of data between military ships.

Compared to microwave and infrared systems, THz wireless communication offers several advantages. As the possibility of increasing the data rate and bandwidth, can also be used to transmit multiple channels simultaneously in high definition, although THz waves are deeply absorbed by water vapor, there are a number of atmospheric windows transmitting these waves in the air with low attenuation and having very directional beams.

Indeed, it is not easy to transmit waves with frequencies above 1 THz, due to several factors such as absorption peaks due to the presence of water molecules in the air and significant propagation attenuation for long-range communication, and other technical and technological difficulties that we will address in the next paragraph.

2.6 *Challenges of THz Wireless Communication Systems*

The efficient and practical realization of a wireless communication network in the THz band requires finding solutions compatible with this network. Devices (transmitters, receivers and antennas) and their quality, as well as the capacity of the channels used, play an important role in the development of a wireless communications system. They also pose major challenges, especially when moving to higher frequency bands such as the THz band. In addition, technical difficulties can also pose constraints and limitations on the performance of this communication system.

The bandwidth used in the THz range will be wide or ultra-wide, so the conventional modulation and demodulation technique cannot be suitable for both transmit and receive systems. This requires finding a modulation technique that allows to take full advantage of the properties of the THz band.

In addition to these challenges that we have mentioned, there are other equally important difficulties, such as coding (to characterize the sources of error at THz band frequencies), synchronization, dynamic massive MIMO and traditional challenges (attenuation, attenuation, absorption) [19].

3 Modulation for THz Wireless Communication System

3.1 Justification for the Choice of Modulation Technique

The choice of the modulation technique to transmit signals in a communication system must satisfy the easy implementation, bandwidth efficiency and BER (Bit Error Rate) binary error rate must be minimum. Among the most widely used transmission techniques, which meet the conditions mentioned above, is the OFDM (Orthogonal Frequency Division Multiplexing) technique, which consists of dividing the available frequency spectrum into several independent symbol subcarriers and due to their robustness against frequency selectivity [20]. To optimize the spectral occupancy of the OFDM signal and to minimize the interference between subcarriers, we choose a minimal spacing between subcarriers ($\Delta f = \frac{1}{T_S}$). His operation called orthogonality, which means that the maximums of the subcarriers all align with the zeros of the other subcarriers [21]. The principle of this method is to distribute simultaneously the data of the very short T_d period on several frequency channels that correspond to N subcarriers, each of these subcarriers has a T_S duration such as $T_S = N \times T_d$. The modulation of data symbols is achieved by an Inverse Fast Fourier Transform (IFFT) which also generates the signal in the time domain at the transmitter level. While at the reception level, the Fast Fourier Transform (FFT) transposes the OFDM symbol from the time domain to the frequency domain.

So, the OFDM technique allows to convert a frequency selective channel into N non-selective sub-channels, this class of modulation is called classical OFDM. Despite this, as we mentioned earlier, this technique has strengths against multi-path channels. This technique has disadvantages such as, it is less robust against the selective channel especially in the case duration of impulse response of the channel is greater than the symbol time, also it uses only the rectangular waveform. In addition to the appearance of undesirable lobes in the power spectral density (PSD). Consequently, these limitations make this technique useless, especially in the case of high frequencies such as in the THz band.

The idea is to find a modulation technique that allows to use a non-rectangular waveform, to minimize the out-of-band energy, to optimize the time–frequency localization, and thus will be an efficient solution for a THz wireless communication system. However, Oversampled Orthogonal Frequency Division Multiplexing (OOFDM) modulation is a suitable solution that overcomes all the constraints we mentioned above. The working principle of this modulation will be presented in the next section.

3.2 Oversampled OFDM Modulation in THz Band

To write the expression of the oversampled OFDM signal we distinguish two cases, continuous time and discrete time modulation.

3.2.1 Continuous Time

Continuous time writing does not usually generate systems that are directly usable in practice. This is the reason why the study of this case was brief.

The expression of the oversampled OFDM signal is written in the following form:

$$s(t) = \sum_{p=0}^{P-1} \sum_{q=-\infty}^{+\infty} c_{p,q} h(1 - qT_0) e^{j2\pi p F_0} \quad (1)$$

where:

$c_{p,q}$ complex symbols from a given constellation
P number of carriers
h any waveform
T_0 the symbol time and F_0 the inter-carrier gap where $F_0 > \frac{1}{T_0}$.

3.2.2 Discrete Time

Digital transmission systems are the most frequent, so it is necessary to focus more on the discrete time behavior.

- **Transmitter**:

The purpose of the multi-carrier transmitter is to distribute the data in time and frequency. This transmission technique makes it possible to isolate areas of the time–frequency plane that are affected differently by the transmission channel. This is particularly useful in the case of dynamic power allocation, the choice of the number of bits per symbol, or the configuration of the error correction encoder. The system structure of the oversampled OFDM modulation can be illustrated in Fig. 2.

On this transmission chain, each binary data represents a certain number of bits. Each data can be associated with a complex number $c_{m,n} = \left[c_{0,n}, c_{1,n}, \ldots, c_{1,M-1}\right] \in C_B$, called Symbol. The set C_B constitutes a constellation. Assuming an appropriate coding process, the symbols are considered independent and identically distributed. Thus, we represent the oversampled OFDM transmitter in Fig. 3.

The mathematical expression of the oversampled OFDM signal in discrete time is written:

$$s(\alpha) = \sum_{x=0}^{P-1} \sum_{y} c_{x,y} \varphi(\alpha - yM) e^{j2\pi \frac{x}{P}\left(\beta - \frac{E}{2}\right)} \quad (2)$$

$c_{x,y}$ are the symbols from the constellation, for example, M-QAM, M-PSK, ...
φ any waveform
M number of symbols per block
v strictly positive integer.

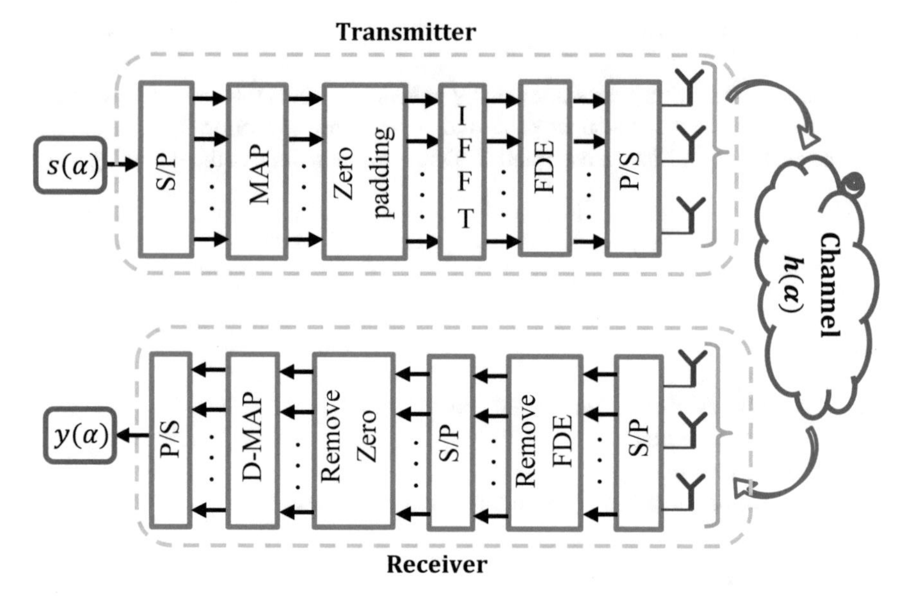

Fig. 2 Structure of oversampled OFDM modulation system

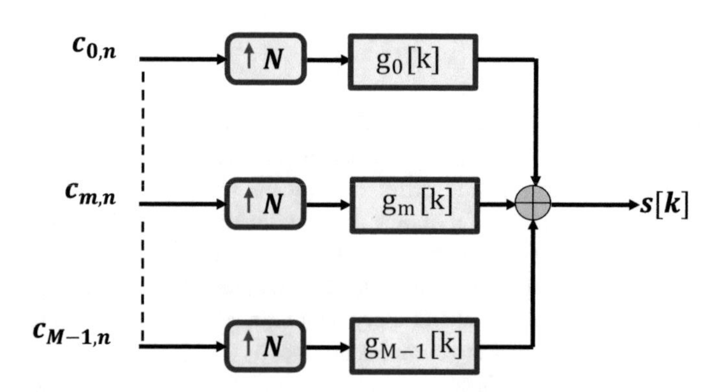

Fig. 3 Discrete time oversampled OFDM transmitter

The demodulation equations are realized thanks to the complex scalar product in discrete:

$$\hat{c}_{x,y} = \varphi_{x,y}, s = \sum_{\alpha=-\infty}^{+\infty} \varphi^*_{x,y}(\alpha) s(x) \tag{3}$$

where $\varphi_{x,y}(\alpha) = \varphi(\alpha - yM) e^{j2\pi \frac{x}{P}\left(\beta - \frac{E}{2}\right)}$.

The frequency selectivity of the transmission channel may cause interference between inter-symbol Interference (ISI) symbols, resulting in an increase in the binary error rate. To minimize this loss of data, a frequency equalization is introduced at reception. Using the Frequency Domain Equalizer (FDE) or the Cyclic Prefix (CP) to simplify equalization. In our communication system based on the OOFDM technique we use the FDE, because of their simplicity and good performance.

The simplification of the effect of the transmission channel is done by an Inverse Discrete Fourier Transform (IDFT), followed by a multiplication vector $c(x)$, followed by a Discrete Fourier Transform (DFT) defined by the equation [20]:

$$F_P = \frac{1}{\sqrt{P}} \mathrm{e}^{\left(-j2\pi\alpha\frac{x}{\mathrm{P}}\right)} \tag{4}$$

Using linear equalization techniques such as Zero Forcing (ZF) and Minimum Mean Square Error (MMSE), such as ZF minimizes interference between symbols and MMSE minimizes the mean square error between the transmitted signal and its estimate.

- **Receiver**:

Generally, after the processing done in the transmitter, the signal $s(\alpha)$ is transmitted via a transmission channel with a noise. At the receiver, the signal is received. After the processing which is detailed in the diagram in Fig. 2, the information sent is recovered.

The block diagram of the oversampled OFDM receiver is shown in Fig. 4.

In the case of a discrete signal, the effect of oversampling appears clearly. However, a risk of interference is always possible. It is thus possible to explain more clearly why we are talking about oversampled OFDM because the term oversampled OFDM is also used for OFDM systems where the modulated signal is obtained for $N = N'$, such as N is number of symbol and N' number of sample. It is thus possible to explain more clearly why we are talking about oversampled OFDM. Indeed, we observe that the sampling period is less than the critical sampling period. As a result,

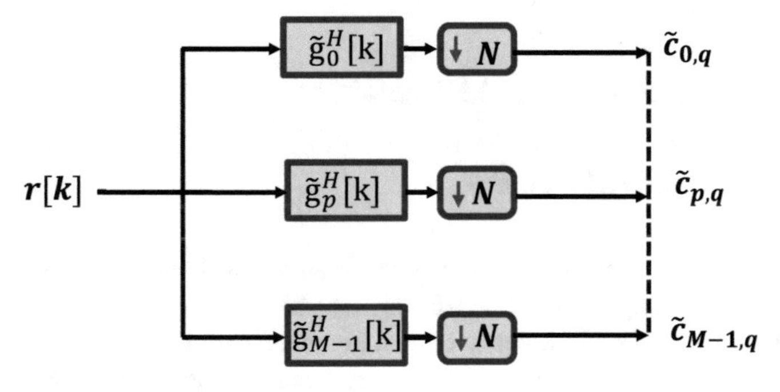

Fig. 4 Discrete time oversampled OFDM receiver

signal samples are sent by symbol time (whereas in the case of OFDM it is only transmitted samples). This signal is therefore oversampled relative to the conventional OFDM signal. But it is important to note that the signal is not an OFDM signal that has been oversampled with an oversampling ratio or by an integer.

- **Time–Frequency Localization**:

In the case of a discrete x signal, we use the discrete time–frequency localization, noted η and given by the following expression:

$$\eta(x)=\frac{1}{\sqrt{4m(x)M(x)}}. \tag{5}$$

where $m(x)$ and $M(x)$ are the time and frequency moments respectively.

- **Power spectral density of the oversampled OFDM signal**:

The oversampled OFDM modulation emission filter has interesting spectral properties (good frequency localization for example). Therefore, an emission spectrum with significant out-of-band attenuations that has significant benefits for transmissions in the terahertz frequency band. The Square Root of Raised Cosine (SRRC) is a widely used reference filter for digital communications systems. For transmission at the rate $F_0 = \frac{1}{T_B}$, the frequency expression of the SRRC function is given by.

$$R_c(\upsilon)=\begin{cases} \frac{1}{\sqrt{F_0}} & |\upsilon| \leq (1-\alpha)\frac{F_0}{2}, \\ \frac{1}{\sqrt{F_0}}\cos\left(\frac{\pi}{2\alpha}\left(\frac{|\upsilon|}{F_0}-\frac{1-\alpha}{2}\right)\right) & (1-\alpha)\frac{F_0}{2} \leq |\upsilon| \leq (1+\alpha)\frac{F_0}{2} \\ 0 & (1+\alpha)\frac{F_0}{2} < |\upsilon| \end{cases} \tag{6}$$

where α is the roll-off factor ($0 \leq \alpha \leq 1$).

4 Simulation Results and Discussions

4.1 Power Spectral Density

To avoid interference phenomena between different applications, current standards provide for very strong constraints on the signal emission spectra. So, Power Spectral Density must be studied in order to draw a big and clear picture on the modulation technique over the desired system. In this context, signals having a well-confined emission spectrum and having significant out-of-band attenuations then possess interesting assets. The study of Power Spectral Density (PSD) then becomes a key point for this type of transmission like THz wireless transmissions. The prototype filter is the central point of these waveform modulations because it controls the PSD of a modulation technique. Indeed, these modulations were mainly introduced to be able to use waveforms other than rectangular because the latter has not very interesting spectral properties (poor frequency localization for example). For oversampled OFDM, the expression of the PSD is:

$$\Phi_{\mathrm{OOFDM}}(f) = \frac{\sigma_c^2}{T_B} \sum_{k=1}^{N} \left| G(f - f_k) \right|^2 \tag{7}$$

In Fig. 5, we depicted the PSD of the oversampled OFDM using an SRRC filter with a factor of (roll-off) equal to 0.5 (the choice of roll-off of the SRRC filter is also directly related to the time–frequency structure of the modulation), for 512 carriers and of length 2048. It will be noted first of all that the phenomena of power fluctuation in the band of the signal do not exist in oversampled OFDM (this is not like the case of OFDM with guard interval where the PSD has slight fainting of power in the useful band of the signal), which makes it possible to erase one of the major drawbacks of OFDM with IG. Also, it is observed that the results obtained by oversampled OFDM modulation give much better results than OFDM. DSPs also have the same characteristics as the prototype filters used. Based on this simulation we can noted a noticeable improvement in spectral characteristics compared to OFDM.

Table 2 shows the points where the characteristics of the oversampled OFDM and OFDM signals diverge by comparing between them.

In this table, we find that even if the signal structures appear close, there are many points where the signals diverge. We call "gap between 2 waveforms (F. O.)" the offset

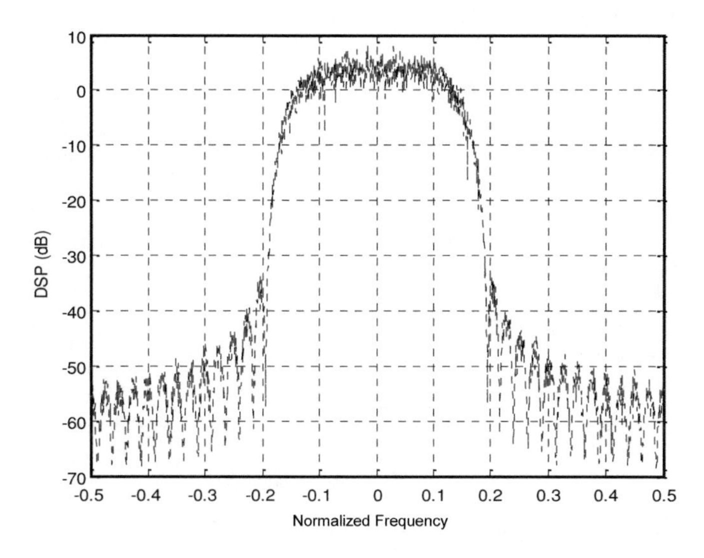

Fig. 5 PSD of oversampled OFDM for 512 subcarriers

Table 2 Differences between the characteristics of the OOFDM and OFDM signals

	OOFDM	OFDM
Symbols	Complexes	Complexes
T_0F_0	$\eta = N'/N$	1
Phase	Arbitrary (null)	Arbitrary (null)
Sample/T_s	N'	N
Subcarriers	$N' = \eta N$	N

in the number of samples existing between 2 waveforms. Thus, for oversampled OFDM, it is N' while it is N for OFDM. This different between them make the oversampled OFDM suitable for wireless communication by improving the power spectral density.

4.2 Cumulative Distribution Function

The results in Fig. 6 show the variation in the Cumulative Distribution Function (CDF) as a function of the delay spread for a number of paths equal to 3, 6 and 10.

From these curves, we can deduce that the delay spread is increasing for an increasing number of trips. Thus, we find that the number of trips, and there is a large difference between the curves (at 3 trips, 6 trips and 10 trips). On the other hand, between the curves with 3 and 6 trips, we have a small deviation associated with a minor error. The CDF are well separated with an increasing path order, which shows that the number of paths is detrimental to the total power of a signal transmitted on this type of channel.

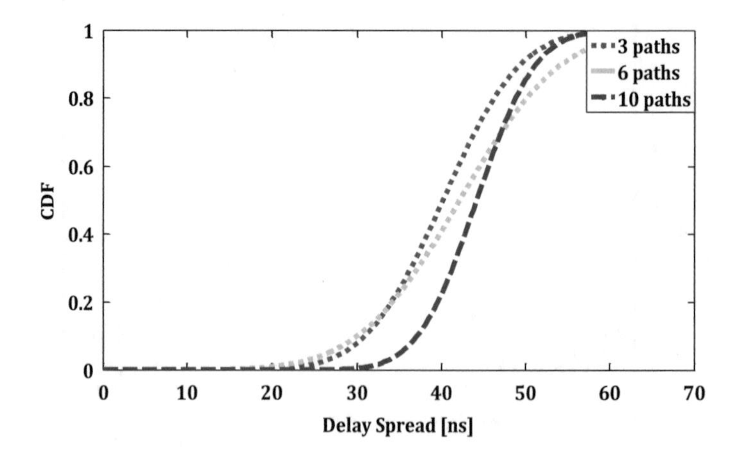

Fig. 6 CDF versus delay spread for 3, 6 and 10 paths

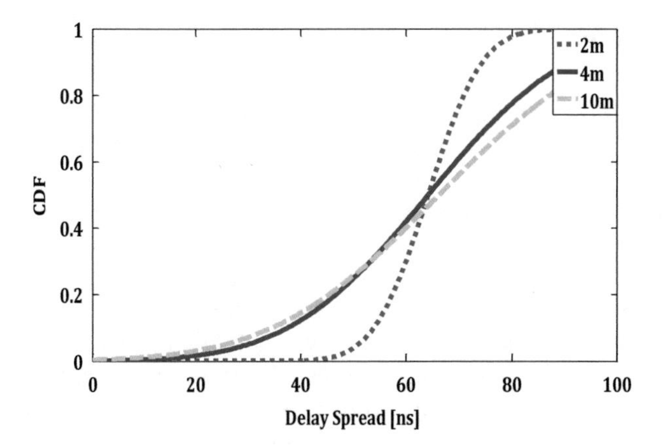

Fig. 7 CDF versus delay spread for 2, 4, 10 m

The same conclusion can be drawn from Fig. 7. FDCs are well separated with an increasing number of paths. Thus, as the distance increases, the delay spread also increases. From Figs. 6 and 7, it can be noted that the simulated delay spread matches extremely well with the normal distribution.

Figure 8 depicted the CDF of the delay spread for 0.1, 0.24 and 0.3 THz. It can be noted that the small value of delay spread can be found for 300 GHz and higher for others frequencies.

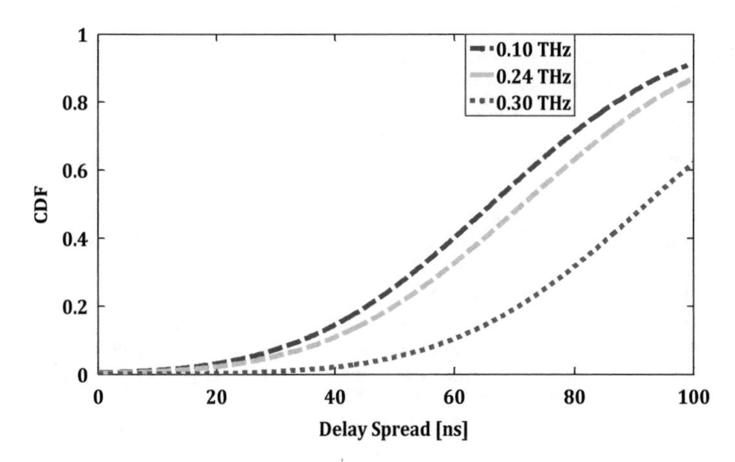

Fig. 8 CDF versus delay spread for 0.1, 0.24 and 0.3 THz

5 Conclusion

The principle is to use the THz frequency band to obtain a good quality of wireless communications and to achieve transmission rates much higher than those currently obtained.

In this chapter, the THz frequency domain, some properties of the waves of this domain and the devices (transmitters, detectors and antennas) have been highlighted. We have seen that the field of applications of THz waves is growing day by day. These applications cover many fields such as medical imaging, industry, spectroscopy, security and telecommunications. Using the oversampled OFDM modulation technique, we performed the wide and ultra-wide modulation of the Terahertz transmission signals. In comparison with the performance of the classical OFDM technique, the results of this work show clearly the best performance of the OOFDM technique on a THz wireless communication system.

References

1. H. Sarieddeen, N. Saeed, T.Y. Al-Naffouri, M.-S. Alouini, in *Next Generation Terahertz Communications: A Rendezvous of Sensing, Imaging, and Localization*. arXiv:1909.10462 [eess]. (2020).
2. S.A. Busari, K.M.S. Huq, S. Mumtaz, J. Rodriguez, Terahertz massive MIMO for beyond-5G wireless communication, in *ICC 2019—2019 IEEE International Conference on Communications (ICC)*. pp. 1–6 (IEEE, Shanghai, China, 2019). https://doi.org/10.1109/ICC.2019.8761371
3. I. Dan, G. Ducournau, S. Hisatake, P. Szriftgiser, R.-P. Braun, I. Kallfass, A terahertz wireless communication link using a superheterodyne approach. IEEE Trans. Terahertz Sci. Technol. **10**, 32–43 (2020). https://doi.org/10.1109/TTHZ.2019.2953647
4. Y. Chi, N. Kuang, Y. Tang, W. Chen, X. Ma, Z. Li, Q. Wen, Z. Chen, Modulator design for THz communication based on vanadium dioxide metasurface, in *2019 IEEE/CIC International Conference on Communications Workshops in China (ICCC Workshops)*, pp. 142–146 (IEEE, Changchun, China, 2019). https://doi.org/10.1109/ICCChinaW.2019.8849947
5. J.F. Federici, L. Moeller, K. Su, Terahertz wireless communications, in *Handbook of Terahertz Technology for Imaging, Sensing and Communications*, pp. 156–214 (Elsevier, 2013). https://doi.org/10.1533/9780857096494.1.156
6. T. Yilmaz, O.B. Akan, Utilizing terahertz band for local and personal area wireless communication systems, in *2014 IEEE 19th International Workshop on Computer Aided Modeling and Design of Communication Links and Networks (CAMAD)*, pp. 330–334 (IEEE, Athens, Greece, 2014). https://doi.org/10.1109/CAMAD.2014.7033260
7. M. El Ghzaoui, A. Hmamou, J. Foshi, J. Mestoui, Compensation of non-linear distortion effects in MIMO-OFDM systems using constant envelope OFDM for 5G applications. J. Circ. Syst. Comput. **29**, 2050257 (2020). https://doi.org/10.1142/S0218126620502576
8. J.F. O'Hara, S. Ekin, W. Choi, I. Song, A perspective on terahertz next-generation wireless communications. Technologies **7**, 43 (2019). https://doi.org/10.3390/technologies7020043
9. H. Elayan, O. Amin, R.M. Shubair, M.-S. Alouini, Terahertz communication: The opportunities of wireless technology beyond 5G, in *2018 International Conference on Advanced Communication Technologies and Networking (CommNet)*, pp. 1–5 (IEEE, Marrakech, 2018). https://doi.org/10.1109/COMMNET.2018.8360286

10. S. Bashir, M.H. Alsharif, I. Khan, M.A. Albreem, A. Sali, B. Mohd Ali, W. Noh, MIMO-terahertz in 6G nano-communications: channel modeling and analysis. Comput. Mater. Continua. **66**, 263–274 (2020). https://doi.org/10.32604/cmc.2020.012404
11. Z. Hossain, J.M. Jornet, Hierarchical bandwidth modulation for ultra-broadband terahertz communications, in *ICC 2019—2019 IEEE International Conference on Communications (ICC)*, pp. 1–7 (IEEE, Shanghai, China, 2019). https://doi.org/10.1109/ICC.2019.8761547
12. Z.T. Ma, Z.X. Geng, Z.Y. Fan, J. Liu, J., H.D. Chen, Modulators for terahertz communication: the current state of the art. Research **2019**, 1–22 (2019). https://doi.org/10.34133/2019/6482975
13. Y. Zhao, Y. Huang, H. Wu, Y. Zhang, X. Xu, M. Yang, Z. Liu, S. Ding, THz wireless communication system based on wavelet transform. Int. J. Commun. Syst. **33**, e4496 (2020). https://doi.org/10.1002/dac.4496.https://doi.org/10.1002/dac.4496
14. A. Siemion, Terahertz diffractive optics—smart control over radiation. J. Infrared Millimeter Terahertz Waves **40**, 477–499 (2019). https://doi.org/10.1007/s10762-019-00581-5
15. M. Shehata, K. Wang (Desmond), J. Webber, M. Fujita, T. Nagatsuma, W. Withayachumnankul, in *IEEE 802.15.3d-Compliant Waveforms for Terahertz Wireless Communications* (2021). https://doi.org/10.36227/techrxiv.14393093.v1
16. L. Juery, in *Communication térahertz sans fil à haut débit avec un transistor à haute mobilité électronique comme détecteur*, 146.
17. H.-J. Song, K. Ajito, A. Wakatsuki, Y. Muramoto, N. Kukutsu, Y. Kado, T. Nagatsuma, Terahertz wireless communication link at 300 GHz, in *2010 IEEE International Topical Meeting on Microwave Photonics*, pp. 42–45 (IEEE, Montreal, QC, Canada, 2010). https://doi.org/10.1109/MWP.2010.5664230
18. B. Aghoutane, M. El Ghzaoui, H. El Faylali, Spatial characterization of propagation channels for terahertz band. SN Appl. Sci. **3** (2021). https://doi.org/10.1007/s42452-021-04262-8
19. I.F. Akyildiz, J.M. Jornet, C. Han, Terahertz band: next frontier for wireless communications. Phys. Commun. **12**, 16–32 (2014). https://doi.org/10.1016/j.phycom.2014.01.006
20. A. Rahmati, K. Raahemifar, T.A. Tsiftsis, A. Anpalagan, P. Azmi, OFDM signal recovery in deep faded erasure channel. IEEE Access. **7**, 38798–38812 (2019). https://doi.org/10.1109/ACCESS.2018.2876646
21. J. Mestoui, M.E. Ghzaoui, A. Hmamou, J. Foshi, BER performance improvement in CE-OFDM-CPM system using equalization techniques over frequency-selective channel. Procedia Comput. Sci. **151**, 1016–1021 (2019). https://doi.org/10.1016/j.procs.2019.04.143

Terahertz Antenna: Fundamentals, Types, Fabrication, and Future Scope

Sunil Lavadiya and Vishal Sorathiya

Abstract Terahertz technology has grown in popularity in recent years due to the rapid development of wireless communication applications. To begin, the evolution of Terahertz antennas is briefly studied, and the fundamental concepts of THz antennas are employed. THz antennas are then classified as dielectric antennas, metallic antennas, recent novel material antennas. Following that, the most recent scientific advances in THz horn antennas, photoconductive antennas, on-chip antenna, microstrip antennas, lens antennas, on-chip antenna, graphene sheet-based THz antenna will be discussed. The technological challenges like the smaller size and relatively high loss for the developing THz antennas are addressed, along with promising methods. This chapter also discusses THz antenna designing technology and the critical problems and potential study directions for THz antennas. THz technologies open the new door for the application like radio astronomy, radar imaging, remote sensing, graphene-based plasmonic resonator, broadband communication, high data rate, high switching RF components, and fast-pulse optical time-domain spectroscopic techniques.

Keywords Terahertz antennas · Quantum cascade laser · Photoconductive antenna · Technological challenges · Graphene

1 Introduction

The popularity of wireless communication gadgets and the huge data traffic have reached a new era of accelerated growth [1]. A vast number of systems are now moving from PCs to cellular devices like mobiles, which are easier to bring and run in real-time, however, these conditions often results in a quick rise in data usage

S. Lavadiya (✉) · V. Sorathiya
Department of Information and Communication Technology, Marwadi University, Rajkot, Gujarat 360002, India
e-mail: sunil.lavadiya@marwadieducation.edu.in

V. Sorathiya
e-mail: vishal.sorathiya@marwadieducation.edu.in

S. Das et al. (eds.), *Advances in Terahertz Technology and Its Applications*,
https://doi.org/10.1007/978-981-16-5731-3_7

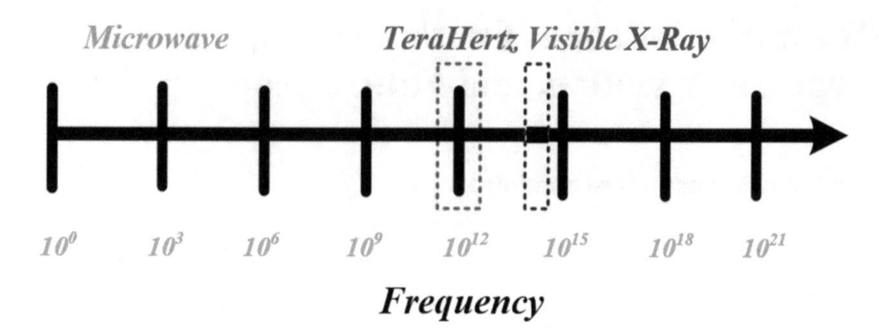

Fig. 1 The spectrum of electromagnetic waves

and a shortage of the high bandwidth (BW) [2]. As per the survey, the industry data requirements in the next decade will be expected to exceed Gbps or even more [3]. To utilize the new frequency bands is a viable alternative solution, that is THz em wave and its null field between microwaves and infrared light [4]. Terahertz waves have a wavelength of 0.03 mm to 3 mm and a frequency range of 0.1 THz to 10 THz [4]. The entire electromagnetic spectrum is represented in Fig. 1.

Terahertz (THz) wave has the following outstanding properties: (a). Lower harmful: THz waves have smaller single-photon energy than X-rays, with around one component per million. As a result, THz wave applications in the biomedical industry, screening the body for disease detection and management and less harmful to animals [5]. (b). Object Scanning: THz waves better penetrate certain non-metallic or non-polar surfaces due to lower wavelength. THz signals may be invisible to transparent opaque surfaces, which helps for the HD photographs. As a result, the THz wave is commonly used for sensing devices, such as full-body scanners at airports [6]. (c). Exceptional spectral resolution: The THz band dominates the spectrum of most big molecules. Radiation of the THz wave is critical for detecting hazardous products such as bombs, pathogens, chemicals, and so on [7]. (d). High data rate: In electronics, the THz wave enables the broader electromagnetic spectrum that helps for the broader rate of information transmission, which may hit new heights also at TBPs.

2 Development of Terahertz Antennas

THz research started in the nineteenth century, although it was not deliberate as a separate discipline at the time. The majority of THz research is in the far-infrared field. During the 1980s, the development of THz radiation sources allowed THz waves to be used in realistic structures [8]. Since the starting of the twenty-first century, wireless communication and networking have advanced exponentially, therefore its need for information has imposed increasingly as well strict conditions on the transfer rate of communication data. Many consumers utilize multiple-input and multiple-output (MIMO) technologies to advance spectral efficiency and device capability through spatial multiplexing to address the issue of spectrum congestion. With the progress of 5G networks, each user's Internet link level would reach Gbps, and base station data traffic will also accelerate. Infrared contact has a low transmitting gap and a set location for its communication equipment. As a result, the THz wave was used to construct a high-speed communication device [9]. THz waves have a far larger transmission bandwidth than microwaves since their frequency range is around a thousand times that of mobile communication. IEEE 802.15.3d-2017 was the first THz wireless networking protocol to be announced [10], which describes data transfer between the point-to-point at the lower THz frequency (252–325 GHz).

THz networking systems have a huge range and a fast data transfer rate that millimeter waves cannot match. It is mostly used in short-distance terrestrial communication, and space networking. Due to the high propagation rate and reasonable anonymity will easily satisfy current needs. As a result, the establishment of the THz contact mechanism has piqued the interest of all countries around the world, and several studies have been conducted. THz antennas have been increasingly established as an essential component of the THz networking framework. For the first time, in 2004 the 0.12 THz communication was developed in 2013 the 0.3 THz communication rate was achieved (Table 1).

3 Basic Terahertz Antennas

THz antennas come in a variety of shapes and sizes, including pyramidal cavities with dipoles, angle reflector arrays, bow-tie dipoles, photoconductive antennas, THz horn antennas, dielectric lens planar antennas, graphene-based antennas [18, 22]. THz antennas are divided into three types depending on their processing content: dielectric antennas, metallic antennas, and modern material antennas.

3.1 Metallic Antennas (Horn Antenna)

Horn antenna is considered under the metallic antenna. Horn antennas are a kind of metallic antenna that is configured to operate in the THz. Generally, the conical horn is used as a receiver. Many advantages exist for dual-mode antennas, and corrugated including a rotationally symmetric radiation design with a gain of up to 30 dBi, and cross-polarization up to the −30 dB, and the coupling performance of 97–98%. The BW of horn antennas is approximately 30–40% [23]. Since the THz wave's frequency is so large, the horn antenna's size is so limited that processing the horn's tip end is difficult, particularly when designing antenna arrays. Since the THz wave's frequency is so large, the horn antenna's size is so tiny. This makes processing the horn's tip end challenging, particularly in the design of antenna elements, and the

Table 1 Summaries THz development from 2004 to 2021

References	Year	Fre. (THz)	Research results
[11]	2004	0.12	Data transmission rates of up to 10 Gbps are possible
[12]	2006	0.12	Photon technology combined with a receiving capacity of less than −30 dBm
[13]	2007	0.30	Set of 4 × 4 microstrip antennas
[14]	2009	0.12	Direct fabrication of a modulator and demodulator (BPSK) on the MMIC
[15]	2011	~0.22	For the first time, data transmission rates of up to 25 Gbit/s were achieved. The maximum transmission rate is 30 Gbps. Bringing together photonics and electronics
[16]	2011	1–15	Tera hertz antenna based on graphene
[14]	2012	0.30	The UTC-PD transmit capacity is less than 200 microwatts on both sides of the detector and emitter. The antenna gain is 40 dB achieved
[17]	2013	0.30	The data transfer capacity of the detector may be increased to 50 Gbps or 100 Gbps by improving the detector's baseband circuit bandwidth
[18]	2016	0.24	A 0.13 μm SiGe HBT circularly polarized radar
[19]	2019	0.35, 0.50	Tera hertz dielectric rod antenna arrays with more than 20 dBi gain and a relative bandwidth of 28%. The proposed GaN/AlGaN detectors' broadband detection capability at frequencies ranging from 0.2 to 1.2 THz
[20]	2020	0.35	The photonic crystal waveguide track's defect-row configuration is designed to eliminate the Bragg-mirror effects thus providing a six-fold increase in bandwidth
[21]	2021	0.20	High-isolation antenna array using SIW and realized with a graphene layer for sub-terahertz wireless applications

process technique's difficulty contributes to high costs and restricted output. Since the bottom of a complicated horn configuration is difficult to manufacture, a basic horn antenna in the shape of a conical/tapered is commonly used, which reduces the expense and complexity while maintaining the antenna's efficiency.

3.2 Traveling-Wave Corner Cube Antenna

A traveling-wave corner cube antenna, as seen in Fig. 2, consists of a traveling-wave antenna installed on the dielectric film (1.2 mm) and positioned in the cavity formed on a silicon wafer, and the design works with the Schottky diodes. Its comparatively easy layout and low manufacturing specifications make it suitable for the usage of more than 0.6THz. The antenna's cross-polarization levels and sidelobe, are higher, therefore its coupling performance is poor (about 50%) [24].

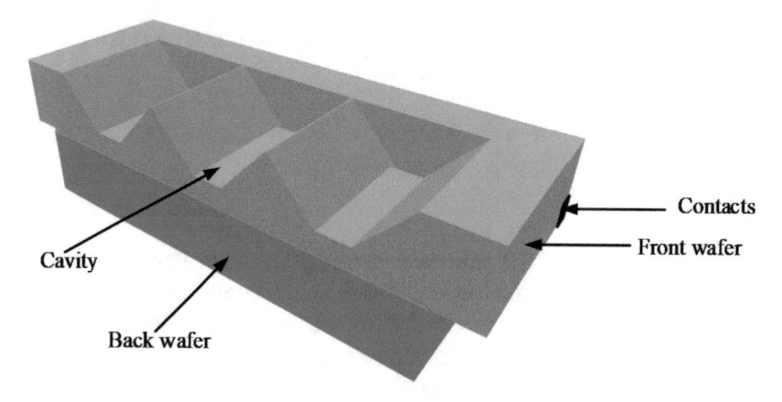

Fig. 2 Traveling-wave corner cube antenna. **a** three-dimensional design structure. **b** side view

3.3 Dielectric Antenna

A dielectric antenna is made up of a dielectric material and an antenna radiator. The proper design helps to achieve better impedance matching, and they have the benefits of being quick to use, fast to integrate, and low in cost. Dual U-shaped antennas, Butterfly antennas, log-periodic sinusoidal antennas, and logarithmic periodic antennas are examples of edge-emitting antennas (narrowband and wideband), which can match with the low-impedance detectors for the dielectric antennas. Furthermore, a genetic algorithm may be used to build more complicated bent-wire antenna geometries [25]. As the frequency approaches the THz band, however, the surface wave effect is produced because the dielectric substrate and antennas are coupled. This fatal flaw results in a substantial loss of energy that results in lower efficiency of the antenna. As the antenna radiation angle exceeds the cut-off angle shown in Fig. 3 [23]. The more thickness results in more modes. This couples the

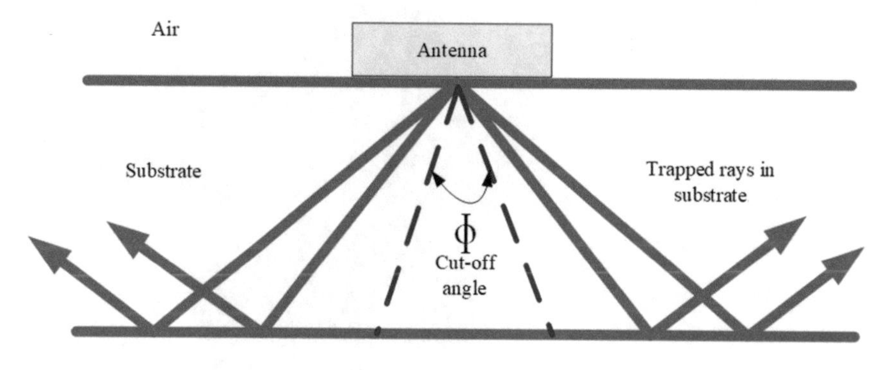

Fig. 3 The surface wave influence of an antenna is depicted schematically

antenna radiation. The binding efficiency raises as the substrate medium increases as the number of high-order modes increases.

3.4 Novel Antenna

Researchers have projected a dipole antenna using the carbon nanotube in 2006. The dipole, as seen in Fig. 4a, is constructed of carbon nanotubes rather than metal investigates the optical and infrared properties of carbon nanotube dipole antennas. The finite length carbon nanotube dipoles are addressed, including current distribution, input impedance, gain, and radiation mode. The input impedance vs frequency curve of a carbon nanotube dipole antenna. It has been analyzed that input impedance has several null points at a higher frequency range, and the antennas can achieve several resonance points at various frequencies. The carbon nanotube resonances within a lower bounded frequency band, and heavily oscillate apart from this range [26].

In 2012, a bundle of carbon nanotubes in the dielectric layers was developed as shown in Fig. 5. The outer dielectric layer is made of metamaterial, and the inner use

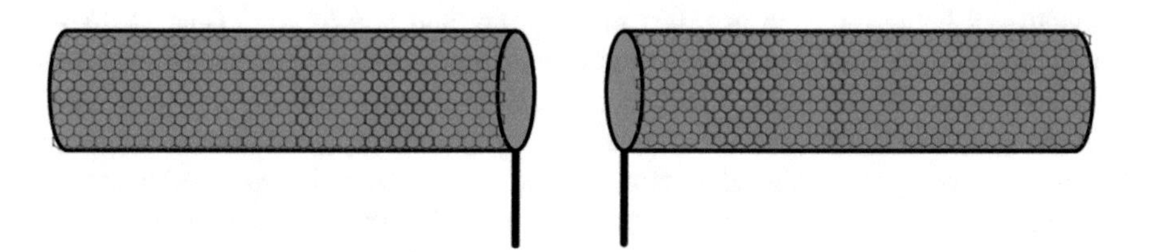

Fig. 4 Dipole antenna based on carbon nanotube

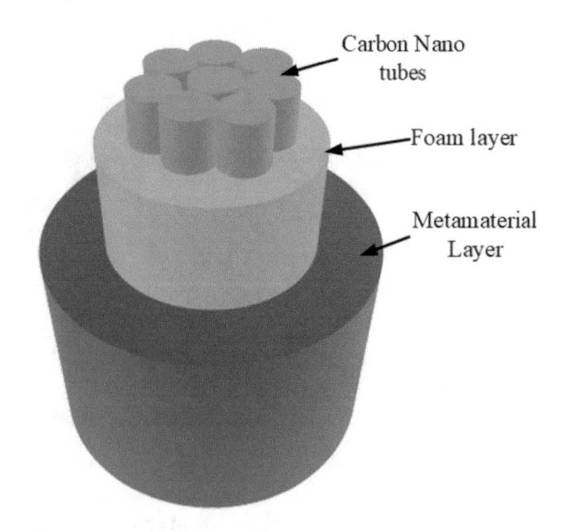

Fig. 5 The geometry of new carbon nanotube antenna

dielectric foam. The radiation efficiency of the antennas is improved by research as opposed to single-walled carbon nanotubes [27, 28].

The above-mentioned modern material THz antennas are often three-dimensional. The planar graphene antenna structure is becoming increasingly common for increasing bandwidth and fabricating conformal antennas. By changing the bias voltage and producing surface plasmons, graphene has outstanding dynamic continuous control characteristics. Surface plasmons occur at the interface of positive dielectric constant substrates, and negative dielectric constant substrates [29]. Conductors, such as precious metals and graphene, have a significant amount of "free electrons." These free electrons are often referred to as plasmas. Because of the conductor's internal potential area, these plasmas are in a steady condition with no outer disruption. Plasma deviates due to the effect of em waves. the em mode generates a wave in the form of a TM mode, which propagates at the interface. According to the Drude model's concept, the metal cannot naturally pair with electromagnetic waves in free space and turn electricity. To inspire the surface plasmon wave, additional components must be used [30]. When a metal electrode conducts electricity parallel to the axis, the surface plasmon wave suppresses quickly in the vertical orientation of the wire-substrate interface, resulting in a skin reaction [31]. Due to the skin effect and limited dimensions antenna deteriorates at the high-frequency band. On the opposite, Graphene is capable of absorbing and modulating light in a broad range of ways. At the THz frequency range, graphene's in-band transition dominates, and the reciprocal fluctuation of plasma makes graphene superior surface plasmon material properties. The surface plasma of graphene not only has a higher attachment and lower leakage rate, but it also has a lower leakage rate, electrical tunability, and higher attachment [22, 32]. In the THz spectrum, graphene often has complex conductivity. As a result, slow-wave propagation is compatible with the plasma mode at THz. These properties show that graphene can be used to replace metal in the THz frequency spectrum [33–35].

3.5 *Terahertz Photoconductive Antennas*

This antenna is used in the THz wave production and recognition. Invention and advancement of photoconductive antennas affect the THz communication mechanism and related fields. When a photoconductive switch (InP, GaAs) is energized by the laser beam, an electron–hole pair is created. If an external field applies to the gap of a photoconductive switch the current (DC) is created. If the laser signal is adequately brief, around 100 fs, the induced photoconductive current generates the THz signal. The PCA is shown schematically in Fig. 6. The PCA antenna model consists primarily of an antenna void photoconductive substrate and an electrode. The antenna distance is the point at which the laser beam irradiates photoconductive substances directly. The laser signal is centered on electrode distance and absorbed by the photoconductive substrate. To enhance the directivity lens is adjust on the PCA substance. PCA substrates that are widely used are ZnTe, GaP, and GaAs.

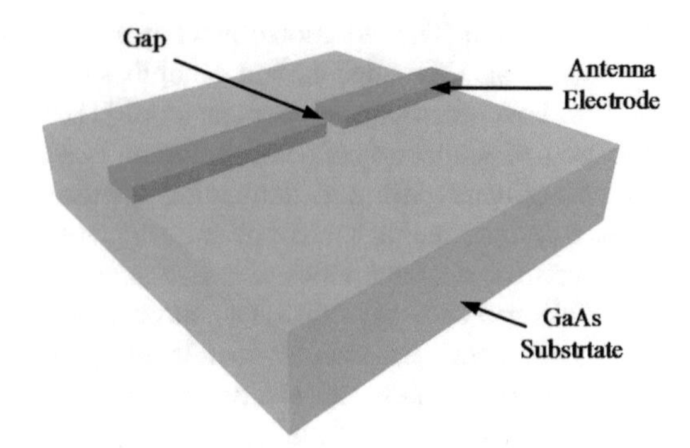

Fig. 6 PCA schematic

3.6 *Reflector Antenna*

A reflector antenna is used in a variety of uses, including microwave networking to optical telescopes. Reflector antennas are used in the majority of satellite transmission networks. Radioastronomy may not have progressed that much, as wavelength begins to shrink, resulting in a reduction in the efficient reception field. Reflector antennas are typically electrically massive, with diameters ranging from few wavelengths to tens of hundreds of wavelengths. Such electric sizes are typically associated with high gain or a large effective reception field. In general, an electrically wide antenna produces a very small beam, typically a couple of degrees or much narrower than the 3 dB beam diameter. Since the energy is so condensed in such a small beam, it produces a strong gain, which is important for the applications of remote sensing [36].

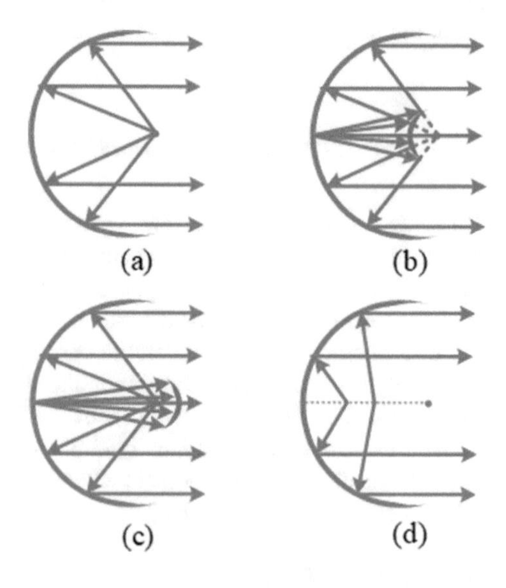

Fig. 7 Terahertz reflectors. **a** paraboloidal reflector **b** Cassegrain reflector **c** Gregorian reflector **d** spherical reflector

As seen in Fig. 7, the exact shape of a reflector antenna may be paraboloid, spheroid, ellipsoid, or hyperboloid. The most common is the paraboloidal reflector. Optical rays launched from the focal point to the paraboloidal reflector become parallel collied after being mirrored by the reflector. This property implies is output aperture plane has an equivalent phase. A Cassegrain antenna is an enhanced structure that employs a hyperboloidal sub-reflector to maximize its efficient focal range. An astronomy telescope, in the shape of a spherical reflector, may also transmit signals from various incident angles.

3.7 *Terahertz Lens Antennas*

The lens has imaging and focusing capabilities that help improve the efficiency of THz antennas by lowering sidelobe and cross-polarization ratios, achieving, high gain, and high directivity. Accelerating lenses and delaying antennas are the two most popular forms of lenses. They are distinguished by varying the electrical duration of the electromagnetic wave direction [37, 38]. The simultaneous configuration of metal plates creates the metal plate prism. As if it is broadcast in a waveguide, the electromagnetic wave travels through a metal plate prism. The metal plate lens is high-frequency sensitive due to its close contact with the arrangement of the workpiece, rendering it unfit for THz antenna design. Metal plate lenses require high accuracy work, they are seldom used in the THz band. However, a metal lens that satisfies the design criteria of THz antennas has recently been created as an artificial lens. The low loss-based dielectric material is required for the dielectric lens with the property like center arranged and then around the lens and has focusing and processing capabilities. Dielectric lenses may be made in a variety of sizes, including hemispherical, ellipsoidal over-hemispherical, and stretched hemispherically. In addition, the energy can radiate further into the dielectric layer as the breadth of the substrate rises [39]. The Fresnel zone plate of the lens consists of so-called full-wave zones with a radius b_w calculated using the following Eq. (1) [37].

$$b_w = \sqrt{2w\lambda_0 F + (w\lambda_0)^2} \tag{1}$$

$$b_s = \sqrt{\frac{2s\lambda_0 F}{p} + \left(\frac{s\lambda_0}{p}\right)^2} \tag{2}$$

$$t_c = \frac{F}{n+1}\left[\sqrt{1 + \left(\frac{D}{2F}\right)^2 \frac{n+1}{n-1}} - 1\right] \tag{3}$$

$$t_1 = \frac{\lambda_0}{p(n-1)} \tag{4}$$

$$t_p = \frac{(p-1)\lambda_0}{p(n-1)} \quad (5)$$

$$T = t_c + t_s \quad (6)$$

$$T = t_p + t_s \quad (7)$$

where w ¼ 1, 2,…, W is a current integer and W is the number of full-wave zones in the FZP. The full-wave zones are equi-phase because the phase from one to another full-wave zone is altered by 2π. The FZP lens's full-wave Fresnel zone is separated into $p = 2, 4, 8,$ subzones. The phase-in each subzone s is modified by $2\pi/p$ about the phase in the next subzone. The b_s of the outer radius of the sth subzone is provided by Eq. (2). where $s = 1, 2,.., S$, $s = w_p$, and S ¼ W_p. The size of the plane-hyperbolic dielectric lens is calculated using Eq. (3). Where aperture diameter is D and central thickness (depth) is t_c. In each grooved full-wave zone, the corrugations form a stair with equal steps of height t_1 determined by the total depth of the FZP lens is t_p. The entire thickness of the ordinary or FZP lens is given by Eq. (6). The refraction lens is represented by Eq. (7).

3.8 *Terahertz Microstrip Antennas*

A thin dielectric layer with a metal patch is used to build a microstrip antenna. The microstrip antenna is compact in scale, light in weight, easy to produce, and wearable, making it ideal for mass manufacturing. Many different styles of microstrip antennas have been produced in recent years, slotted, T-type, clustered, dual-band, and single-band. Since the substrate of microstrip antennas is frequency sensitive, very thin, existing research on THz microstrip antennas is focused on the lower frequency band (0.1–1 THz) [40].

In 2017, researchers created a microstrip antenna with a broader broadband (26.4 GHz) and a defective ground structure using a photonic bandgap substrate. Few researchers created a compact patch antenna, then for the substrate PBG structure is used, that enhance the performance, directed ground concept implemented in the ground area, metamaterial property is enabled using rings for enhancing the performance [41, 42]. The benefit and bandwidth were increased by optimizations. Figure 8 depicts a prototype of the antennas.

3.9 *THz On-Chip Antennas*

The THz high signal on the chip would be significantly attenuated due to lengthy communication links and higher atmospheric losses. Meanwhile, having good

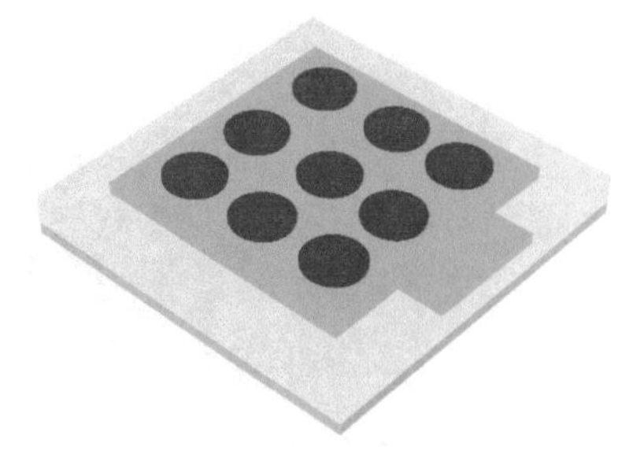

Fig. 8 Slotted microstrip patch antenna

impedance matching between various parts is difficult in such a long connection. As a result, integrating THz antennas on-chip is both feasible and essential. The rapid advancement of packaging technologies, like SiGe, and CMOS packaging technology, facilitates the realization of on-chip antennas. The most popular on-chip antenna arrangement is the rectangular patch antenna, which would be not only easy but also fast to meet CMOS technology design requirements. For example, to realize the frequency detection feature patch antenna need to be integrated on the same chip, and the operating frequencies of the antennas were 1.6, 2, 2.6, 3, 3.4, 4.1 THz. However, the feeder of on-chip antennas is either very long or very short area. Furthermore, an antenna's region is comparatively high, bandwidth is too short, the beam is not centered, and gain is minimal [43].

4 Terahertz Sources and Detectors

Even though there are lower attenuation windows below 1 THz, the signal attenuation in this band is higher than that of a conventional microwave link. To address this constraint, numerous wireless communication connection software and hardware parameters must be configured. There is a requirement for innovation in the areas of detectors, high power sources, lower loss interconnects, high gain antennas. The creation of lightweight, low-profile, low-cost terahertz wireless communication systems will result from research in these areas [44–46]. Many THz sources have been identified as research and technology progresses, which opens a new door for wireless communication. The optical photo-mixing method enables the emission of above 1 THz. The backward wave oscillator method is used for the generation of lower THz signals. However, owing to the need for a strong magnetic field, the scale of this system is excessively huge. Aside from that, several other instruments can produce low-power THz signals. Because of the lightweight terahertz system's requirements, semiconductor systems must be streamlined because they are essentially low-power devices. Previously, numerous semiconductor devices such as Tunnel Diode, IMPATT, Gunn Diode, and Schottky Diode were commonly used to produce lower/moderate power at mm and micro range wavelength. Different forms of sources have been studied at the lower THz frequency (>1 THz), however,

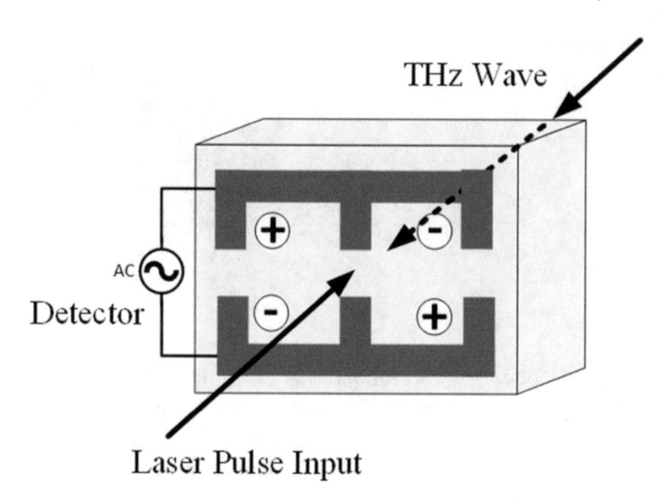

Fig. 9 THz detector using a photoconductive antenna

the output power declines as the operating frequency increases [47]. Figure 9 shows the photoconductive antenna arrangement. The laser beam would arrive at the with the change of Δt compared to the applied wave. The induced current calculated using the galvanometer.

The THz wave is conspired by the delay (Δt) between two signals. Fourier transform may be used to gain spectral details such as absorption. The electrical to optical (EO) sampling method is used for calculating the current signal. The birefringence effect occurs when the refractive index of an EO crystal varies around various axes. The THz field shifts the diffraction of the EO crystal in the EO sampling system. A linearly polarized THz wave is transformed to a slightly elliptical polarized one. The THz wave can be divided into horizontally and vertically polarized components after passing through a Wollaston prism and quarter-wave plate, balanced photodetector used for the detection. Figure 10 depicts the EO crystal identification system. Many methods for achieving high-resolution imaging have been developed. These techniques are aimed at achieving subwavelength resolution. The distance between the two electrodes is just 1.5 mm, resulting in a resolution of 6 mm over a bandwidth of 0.1–3 THz. In principle, such a probe may also be used for far-field measurements. The near-field spreading over a given area is calculated. The near-field distribution over the spherical surface, planar surface, and cylindrical surface are first detected using a probe in this process. And, to achieve far-field radiation properties.

Planar scanning is well designed for high gain reflector antennas, and electrically passive. Raster scanning is typically used to acquire data over such a planar scanning, as seen in Fig. 11. The Nyquist sampling rate limits the maximum scanning time for Δy and Δx to the half wavelength. Although the near-field approach is effective to measure the radiation pattern of the THz antenna, its limitations are evident in many ways. For electrically massive antennas, the first consideration is data acquisition time. According to the Nyquist sampling rule, the maximum sampling time is half

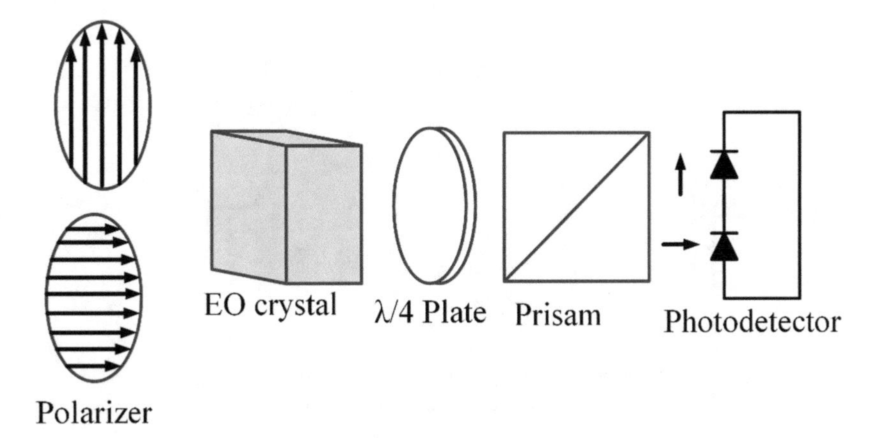

Fig. 10 EO crystal-based Balanced detection method

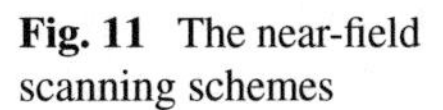

Fig. 11 The near-field scanning schemes

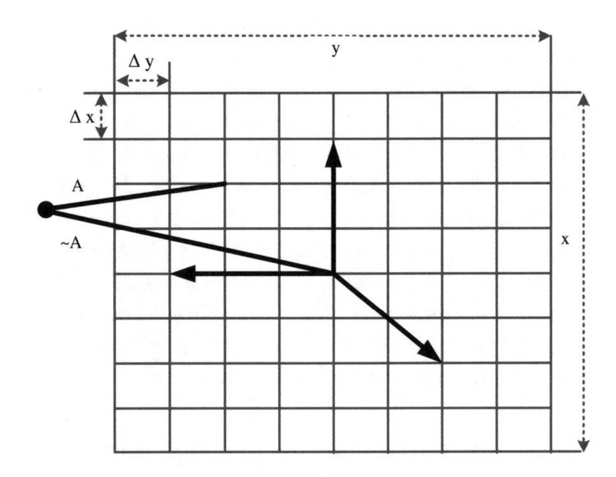

a wavelength. The input wave to the antenna should ideally be a plane wave of standardized phase and amplitude. The antenna is placing at a sufficient far distance for the measuring field pattern. The front phase shift for diameter D must be lower than 22.5° (~λ/16), to ensure precise antenna position is illustrated in Fig. 12.

5 Proposed Terahertz Lecky Wave Antenna

5.1 Introduction

Due to outstanding electrical and mechanical characteristics, graphene is piquing the curiosity of researchers. The use of graphene in antennas is significant because

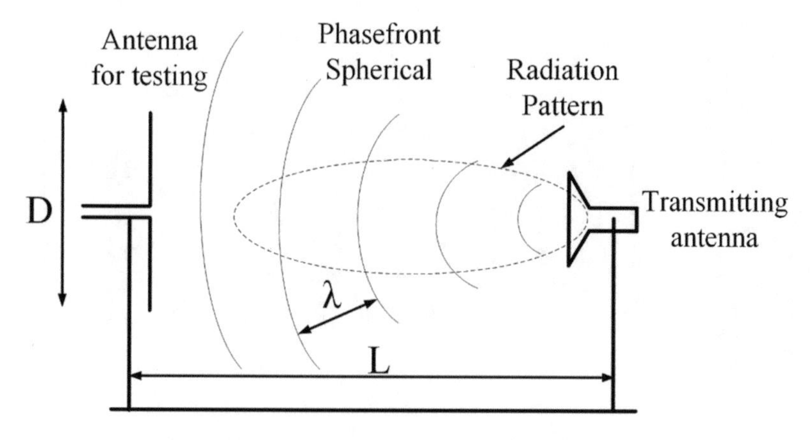

Fig. 12 A setup for the far-field measurement

graphene allows for the tailoring of a frequency spectrum as well as the modification of radiation properties. Graphene also has mechanical flexibility, making it suitable for usage in future electrical gadgets. With the substantial current enhancement, nanoantennas coated in graphene layers effectively convert photons to electrons. When compared to an antenna-less graphene device, the current improvement is eight times greater [48]. Graphene metasurfaces are crucial in improving performance at THz frequencies. Graphene can sustain surface plasmon polariton waves in the THz frequency range, making it ideal for tiny electronics. Tera hertz leaky-wave antenna with periodic silicon (Si) disturbances allows for radiation pattern tweaking through electronics is shown in Fig. 13. To modify the radiation response, silicon corrugations are generated in dielectric Si nitride waveguides. Because of these Si corrugations, a significant amount of power is lost.

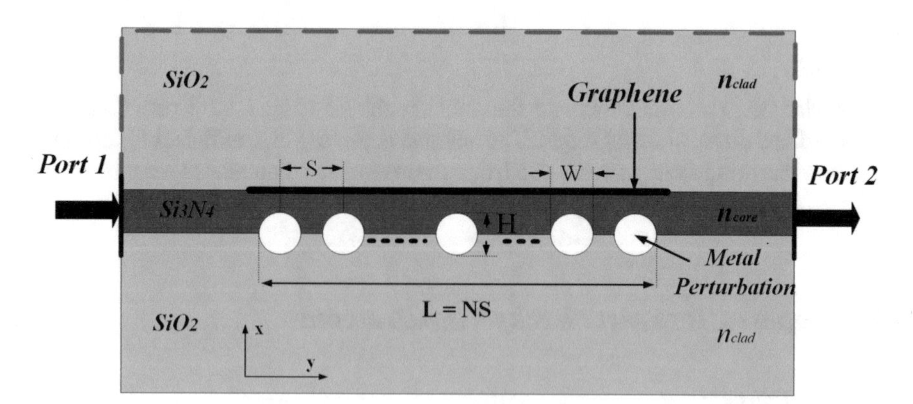

Fig. 13 Proposed Terahertz Lecky wave antenna structure

5.2 Terahertz Antenna Design

We have proposed a Lecky wave antenna for the 1–3 THz. Tera hertz Lecky wave antenna is made up using Si_3N_4 with the circular shape perturbation. The performance is observed in terms of the transmission coefficient, reflectance response, and radiation pattern. The graphene structure is used for tunability. The overall length of the structure is 6 mm, The height of the structure is 100 μm, the width of the structure is 0.5 mm, The waveguide height is 0.1 mm. The diameter of perturbation (H) is 16 μm, the gap between two perturbations (S) is 50 micrometers, the number of perturbations 60, The dimension of the graphene sheet is calculate using $L = N$ x S. The Si_3N_4 waveguide is kept on the Sio_2 substrate.

$$\sigma(\omega) = \frac{2e^2\omega_T}{4\hbar}\frac{i}{\omega + i\tau^{-1}}\log\left[2\cosh\left(\frac{\omega_F}{2\omega_T}\right)\right] + \frac{e^2}{2\hbar}\left[H\left(\frac{\omega}{2}\right) + \frac{i2\omega}{\pi}\int_0^{\infty}\frac{H\left(\frac{\omega'}{2}\right) - H\left(\frac{\omega}{2}\right)}{\omega^2 - \omega'^2}d\omega'\right] \quad (8)$$

$$H(\omega) = \sinh\left(\frac{\omega}{\omega_T}\right)/\left[\cosh\left(\frac{\omega_F}{\omega_T}\right) + \cosh\left(\frac{\omega}{\omega_T}\right)\right] \quad (9)$$

$$\omega_F = E_F/\hbar \quad (10)$$

$$\omega_T = k_B T/\hbar \quad (11)$$

where electron charge is e, frequency is ω, Temperature is T, Boltzmann constant is k_B, The drude relaxation rate is τ, The free-carrier response of graphene (intraband) is described by the first term in Eq. (8). The reaction to interband transition is represented by the second term.

5.3 Discussion on Result

Simulation is carried out using the COMSOL Tool. The performance is analyzed in terms of directivity, transmission coefficient, and reflectance response. The transmission coefficient value is less than −20 dB for the entire band of 1 to 3 THz and reflectance response also less than −10 dB for the entire bands and the resonance observed at the 2.38 THz and 2.45 THz. S_{11} and S_{21} plot for the frequency range of 1 to 3 THz are presented in the Fig. 14.

The radiation pattern is shown in Fig. 15. The broadband directivity is observed with almost more than 16 dB. The directivity behavior also changes as per the applied

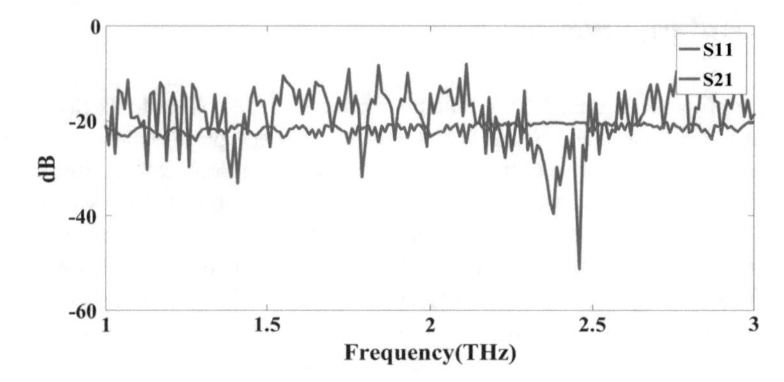

Fig. 14 S_{11} and S_{21} plot for the frequency range of 1–3 THz

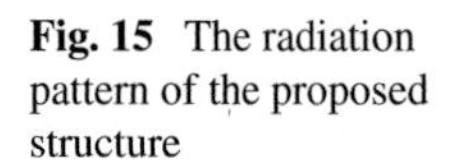

Fig. 15 The radiation pattern of the proposed structure

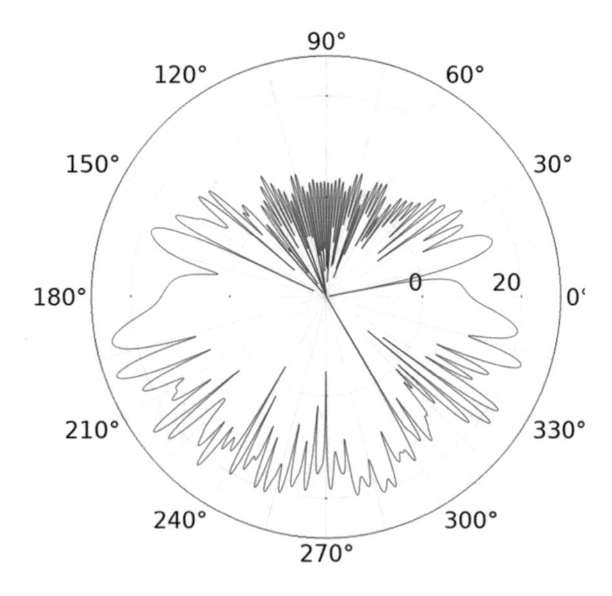

fermi voltage to the graphene sheet. The proposed design opens the new possible way of tunable terahertz communication.

6 Terahertz Antenna Process Technology

The wavelengths of THz are very shorter, as a result, it's illogical to represent that the surface of antennas is flat. The metal antenna efficiency is lower due to the rough surfaces. Using proper machining precision can be improved. The advancement of THz antennas is inextricably linked to the advancement of process technology. Most common process technologies, which include centered ion beam (FIB) technology, and 3D printing. The lithography technology is solve using the focused ion beam (FIB) method, which can be used for the development of spiral antennas, and 3D printing is most often used for printing horn antennas/waveguide, THz lenses, due to the benefits of fast prototyping at a low expense, high accuracy, and miniaturization [49–51]. Traditional machining technology is used to create the micromechanical THz process technology. In 1979 the micromechanical technology was first used in THz circuits. This technology has precise 2D and 3D power structural control, illustrating realistic methods for manufacturing a variety of good THz components. Micromachining technology, which includes lithography, laser milling, and mold reproduction, is focused on silicon technology. Researchers, for example, developed and installed a lower expensive industrial milling method for better gain. It operates from 0.32 to 0.5 THz.

Furthermore, Using Deep Reactive Ion Etching (DRIE) Fabrication of bulk silicon wafers is a reasonably well-established manufacturing technique capable of etching features with high aspect ratios. Figure 16 depicts a schematic representation of the compression alignment pin, as well as an image of the pin when compressed with tweezers to fit into the alignment pocket. When the pin is removed, it expands to fill the pocket. the advancement of silicon micromachining technologies allows for the development of silicon-based microlens resonating at 1.9 THz, and beam scanning at 0.55 THz. It benefits antenna miniaturization and improves stability and integration performance, which is used for the THz planner array. Micromachining technology is classified into three phases of development: metal micromachining, surface micromachining, and bulk micromachining. To transform silicon materials into the necessary components, bulk micromachining primarily employs etching technology. Surface micromachining was first used in integrated circuit production in 1980. Metal micromachining is used in applications like the X-ray to treat products including plastics, metals, and ceramics.

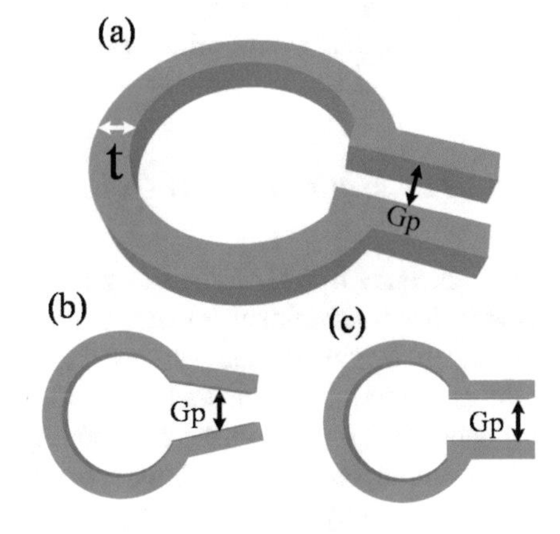

Fig. 16 **a** A schematic diagram of the silicon compression pin. **b** compression pin during assembly squeezed. **c** Photograph of the compression pin released into the alignment pocket

While micromachining technology is capable of achieving high precision, process accuracy in lower wavelength desires to be enhanced, Therefore the THz technology emerges in this era with the dense photoresist, milling, discharge, electroforming. Electroforming is the process of depositing a target substance model and then removing it from the main mold to produce a final object. In the discharge method electrical energy is used to convert a soft metal into a hard one with the sharpening. The milling method involves repairing the initial mold and rotating it at high speeds. The mold is machined using a knife to carve out the perfect product shape. This procedure is reasonably inexpensive. It is a kind of cold metal process that is widely used in today's THz antenna process. Thick photoresist negative photoresist is a chemically amplified that is used in THz antenna lithography. Finishing and roughing may be accomplished parallelly, depending on the scale of the milling cutter, and the operation generates no chemical waste. According to the above study of THz antennas, industrial milling technology is used in the majority of current THz antenna production due to its low expense, high performance, and high precision. Since the high THz band defines the limited size of the THz antennas, the THz antennas are primarily processed by micromachining. THz antenna fabrication is a relatively complex method that is dependent not just on antenna structure but also on the integration of circuits and antenna. There is currently no standardized packaging standard method. THz antenna design method would need to be both precise and low-cost.

7 Future Research on Terahertz Antenna

The specifications of THz antennas: strong mechanical properties, temperature tolerance, alkali, and acidic opposition, compact scale, relatively broad operating bandwidth, high gain, high radiation, lower cost. In response, potential research directions for THz antennas would primarily focus on the following aspects.

7.1 The High Value of Gain

The efficiency of THz antennas, as an essential part of a wireless communication device, has a significant impact on the overall system's communication output. In general, gain and efficiency are used to assess energy conversion and radiation capabilities. Directivity and main lobe width are used to assess the radiation behavior, and the bandwidth represents the accessibility of the antenna's available frequency bandwidth, among other things. Atmospheric attenuation is the most significant barrier to THz connectivity. Since free space route failure is practically unavoidable, therefore the solution is possible by enhancing the gain [52]. As a result of the greater operational antenna and without much affecting environmental loss For better usage the antennas must be broadband, high gain, and effective. Given the wideband output of

THz waves, developing high efficiency with high gain antennas is the main research focus. Because of the silicon substrate's strong dielectric constant, the majority of the energy is bound in the form of surface waves. However, due to silicon substrate's poor resistivity. the majority of this energy is lost as heat. As a consequence, silicon-based on-chip antennas typically have poor radiation performance. Since high gain THz antennas may work in the worst environments that degrade the effect of transmitting signal failure. THz antennas ideal for high gain service are mostly lens antennas and THz horn antennas at the moment. The construction of an array-based THz antenna is the most complex. Since THz antennas are so tiny, installing the THz antenna array necessitates high-quality materials with precise technology. The existing process technologies cannot fulfill the design criteria, therefore the THz antenna gain must be increased from both the process technology and the materials perspectives [53].

7.2 Size Reduction

THz antennas have a moderately very short wavelength, which accounts for the narrow antenna capacity. When the number of smart devices rises, so does the need for wearable mobile antennas. The size reduction is an essential future research path. At the moment, miniaturization based on the integration of the different antennas is more appropriate. The lens radius may be chosen as possible as a minimum. However, we must exercise caution to prevent the degradation of antenna performance. The microstrip antenna is the shortest and has the lowest return loss, gain, and directivity, which are smaller than those of other kinds of antennas. An antenna may be worn as a handheld antenna with a tiny scale, lower weight, and reproducibility feature [54].

7.3 The High Degree of Incorporation

Many studies have looked into THz antenna integration due to the high consumer demand and advancements in integration technology based on silicon semiconductors. Integration of an antenna can be improved by integrating many antennas on-chip. One of the most difficult challenges in THz packaging schemes is electromagnetic compatibility (EMC). In reality, high gain antennas are realized using a compact large antenna array. There is reciprocal interference between the array's components. The addition of filter arrangement increases the overall size of the structure. It needs to trade-off between EMC reduction, and the THz antenna miniaturization process. The size of the chip is more than the wavelength, therefore the highly precise packaging and process technology need to be developed. It is common knowledge that the advancement of fully embedded applications in recent years may be traced to CMOS technology. Because of this aspect, CMOS technology is suitable for THz wave circuit design. THz wave circuit architecture has been significantly influenced by CMOS and SiGe developments, which also extended the implementation areas of integrated circuits in THz devices. To address the complex actions of electromagnetic waves in embedded structures, modern architecture and modeling principles should be created for integration and packaging technologies.

8 Conclusion

This chapter has discussed the terahertz operating antennas and their accompanying measurements, with an emphasis on the fundamental ideas and operating principles. An antenna's design in the THz range often has far more to deliberate than its lower frequency, such as sources and manufacturing problems. Despite the photoconductive antenna is an integrated signal generator for THz. Several ways have been investigated to improve its directivity and radiation output. Due to their short wavelength, other high gain antennas such as reflector, and horn antennas have manufacturing challenges in the THz region. Another issue comes as a result of the measurement techniques of the THz antenna. In the frequency range up to 1 THz, the near-field and CATR measuring methods are feasible but not without issues. However, no practical RF measuring technology exists in the frequency range over 1 THz. THz antenna design and its measurements will continue to be a technological challenge shortly. In the long term, it is believed that the development of high power THz sources and better manufacturing processes would solve this difficulty.

References

1. F. Xu et al., Big data driven mobile traffic understanding and forecasting: a time series approach. IEEE Trans. Serv. Comput. **9**(5), 796–805 (2016). https://doi.org/10.1109/TSC.2016.2599878
2. S. Mumtaz, J.M. Jornet, J. Aulin, W.H. Gerstacker, X. Dong, B. Ai, Terahertz communication for vehicular networks. IEEE Trans. Veh. Technol. **66**(7), 5617–5625 (2017). https://doi.org/10.1109/TVT.2017.2712878
3. Z. Chen et al., A survey on terahertz communications. China Commun. **16**(2), 1–35 (2019). https://doi.org/10.12676/j.cc.2019.02.001
4. K. Guan et al., On millimeter wave and THz mobile radio channel for smart rail mobility. IEEE Trans. Veh. Technol. **66**(7), 5658–5674 (2017). https://doi.org/10.1109/TVT.2016.2624504
5. Y. Peng, C. Shi, Y. Zhu, M. Gu, S. Zhuang, Terahertz spectroscopy in biomedical field: a review on signal-to-noise ratio improvement. PhotoniX **1**(1), 12 (2020). https://doi.org/10.1186/s43074-020-00011-z
6. J. Grade, P. Haydon, D. Van Der Weide, Electronic terahertz antennas and probes for spectroscopic detection and diagnostics. Proc. IEEE **95**(8), 1583–1591 (2007). https://doi.org/10.1109/JPROC.2007.898900
7. N.V. Petrov, M.S. Kulya, A.N. Tsypkin, V.G. Bespalov, A. Gorodetsky, Application of terahertz pulse time-domain holography for phase imaging. IEEE Trans. Terahertz Sci. Technol. **6**(3), 464–472 (2016). https://doi.org/10.1109/TTHZ.2016.2530938
8. D. Grischkowsky, S. Keiding, M. van Exter, C. Fattinger, Far-infrared time-domain spectroscopy with terahertz beams of dielectrics and semiconductors. J. Opt. Soc. Am. B **7**(10), 2006 (1990). https://doi.org/10.1364/josab.7.002006

9. K.M.S. Huq, J.M. Jornet, W.H. Gerstacker, A. Al-Dulaimi, Z. Zhou, J. Aulin, THz communications for mobile heterogeneous networks. IEEE Commun. Mag. **56**(6), 94–95. https://doi.org/10.1109/MCOM.2018.8387209
10. IEEE, *IEEE Standard for High Data Rate Wireless Multi-Media Networks--Amendment 2: 100 Gb/s Wireless Switched Point-to-Point Physical Layer*, vol. 2017 (2017)
11. T. Kosugi, M. Tokumitsu, T. Enoki, M. Muraguchi, A. Hirata, T. Nagatsuma, 120-GHz Tx/Rx chipset for 10-Gbit/s wireless applications using 0.1-μm-gate InP HEMTs, in *Technical Digest - IEEE Compound Semiconductor Integrated Circuit Symposium, CSIC*, pp. 171–174 (2004). https://doi.org/10.1109/CSICS.2004.1392524
12. A. Hirata et al., 120-GHz-band millimeter-wave photonic wireless link for 10-Gb/s data transmission. IEEE Trans. Microw. Theory Tech. **54**(5), 1937–1942 (2006). https://doi.org/10.1109/TMTT.2006.872798
13. R. Piesiewicz, T. Kleine-Ostmann, N. Krumbholz, D. Mittleman, M. Koch, J. Schoebel, T. Kurner, Short-range ultra-broadband terahertz communications: concepts and perspectives. ieeexplore.ieee.org
14. H. Takahashi, T. Kosugi, A. Hirata, K. Murata, N. Kukutsu, 120-GHz-band BPSK modulator and demodulator for 10-Gbit/s data transmission, in *IEEE MTT-S International Microwave Symposium Digest*, pp. 557–560 (2009). https://doi.org/10.1109/MWSYM.2009.5165757
15. I. Kallfass et al., All active MMIC-based wireless communication at 220 GHz. IEEE Trans. Terahertz Sci. Technol. **1**(2), 477–487 (2011). https://doi.org/10.1109/TTHZ.2011.2160021
16. M. Dragoman, A.A. Muller, D. Dragoman, F. Coccetti, R. Plana, Terahertz antenna based on graphene. J. Appl. Phys. **107**(10), 104313 (2010). https://doi.org/10.1063/1.3427536
17. T. Nagatsuma et al., Terahertz wireless communications based on photonics technologies. Opt. Express **21**(20), 23736 (2013). https://doi.org/10.1364/oe.21.023736
18. T. Kleine-Ostmann, T. Nagatsuma, A review on terahertz communications research. J. Infrared Millimeter Terahertz Waves **32**(2), 143–171. https://doi.org/10.1007/s10762-010-9758-1
19. W. Withayachumnankul, M. Fujita, T. Nagatsuma, Polarization responses of terahertz dielectric rod antenna arrays, in *2019 International Conference on Microwave and Millimeter Wave Technology, ICMMT 2019 - Proceedings*, May 2019, pp. 1–3. https://doi.org/10.1109/ICMMT45702.2019.8992313
20. D. Headland, M. Fujita, T. Nagatsuma, Bragg-mirror suppression for enhanced bandwidth in terahertz photonic crystal waveguides. IEEE J. Sel. Top. Quantum Electron. **26**(2), 1–9 (2020). https://doi.org/10.1109/JSTQE.2019.2932025
21. M. Alibakhshikenari et al., High-isolation antenna array using SIW and realized with a graphene layer for sub-terahertz wireless applications. Sci. Rep. **11**(1), 10218 (2021). https://doi.org/10.1038/s41598-021-87712-y
22. A. Pandya, V. Sorathiya, S. Lavadiya, Graphene-based nanophotonic devices, in *Recent Advances in Nanophotonics - Fundamentals and Applications* (IntechOpen, 2020)
23. G.M. Rebeiz, Millimeter-wave and terahertz integrated circuit antennas. Proc. IEEE **80**(11), 1748–1770 (1992). https://doi.org/10.1109/5.175253
24. S.S. Gearhart, C.C. Ling, G.M. Rebeiz, H. Davée, G. Chin, Integrated 119-μm linear corner-cube array. IEEE Microw. Guid. Wave Lett. **1**(7), 155–157 (1991). https://doi.org/10.1109/75.84567
25. O. Markish, Y. Leviatan, Analysis and optimization of terahertz bolometer antennas. IEEE Trans. Antennas Propag. **64**(8), 3302–3309 (2016). https://doi.org/10.1109/TAP.2016.2573861
26. J. Hao, G.W. Hanson, Infrared and optical properties of carbon nanotube dipole antennas. IEEE Trans. Nanotechnol. **5**(6), 766–775 (2006). https://doi.org/10.1109/TNANO.2006.883475
27. S. Jaydeep, L. Sunil, An investigation on recent trends in metamaterial types and its applications, *i-manager's*. J. Mater. Sci. **5**(4), 55 (2018). https://doi.org/10.26634/jms.5.4.13974
28. S.F. Mahmoud, A.R. AlAjmi, Characteristics of a new carbon nanotube antenna structure with enhanced radiation in the sub-terahertz range. IEEE Trans. Nanotechnol. **11**(3), 640–646 (2012). https://doi.org/10.1109/TNANO.2012.2190752
29. M. Yan, M. Qiu, Analysis of surface plasmon polariton using anisotropic finite elements. IEEE Photonics Technol. Lett. **19**(22), 1804–1806 (2007). https://doi.org/10.1109/LPT.2007.906832

30. Y. Wang et al., Manipulating surface plasmon polaritons in a 2-D T-shaped metalinsulatormetal plasmonic waveguide with a joint cavity. IEEE Photonics Technol. Lett. **22**(17), 1309–1311 (2010). https://doi.org/10.1109/LPT.2010.2053531
31. N.N. Feng, M.L. Brongersma, L. Dal Negro, Metal-dielectric slot-waveguide structures for the propagation of surface plasmon polaritons at 1.55 μ/m. IEEE J. Quantum Electron. **43**(6), 479–485 (2007). https://doi.org/10.1109/JQE.2007.897913
32. H. Lu et al., Graphene-based active slow surface plasmon polaritons. Sci. Rep. **5** (2015). https://doi.org/10.1038/srep08443
33. S. Ke, B. Wang, P. Lu, Plasmonic absorption enhancement in periodic cross-shaped graphene arrays, in *2015 IEEE MTT-S International Microwave Workshop Series on Advanced Materials and Processes RF THz Applications IEEE MTT-S IMWS-AMP 2015 - Proceeedings*, vol. 23, no. 7, pp. 4810–4817 (2015). https://doi.org/10.1109/IMWS-AMP.2015.7325015
34. S.K. Patel, V. Sorathiya, Z. Sbeah, S. Lavadiya, T.K. Nguyen, V. Dhasarathan, Graphene-based tunable infrared multi band absorber. Opt. Commun. **474**, 126109 (2020). https://doi.org/10.1016/j.optcom.2020.126109
35. Y. He, Y. Chen, L. Zhang, S.W. Wong, Z.N. Chen, An overview of terahertz antennas. China Commun. **17**(7), 124–165 (2020). https://doi.org/10.23919/J.CC.2020.07.011
36. K. Sumathi, S. Lavadiya, P.Z. Yin, J. Parmar, S.K. Patel, High gain multiband and frequency reconfigurable metamaterial superstrate microstrip patch antenna for C/X/Ku-band wireless network applications. Wirel. Networks **27**(3), 2131–2146 (2021). https://doi.org/10.1007/s11276-021-02567-5
37. H.D. Hristov, J.M. Rodriguez, W. Grote, The grooved-dielectric Fresnel zone plate: An effective terahertz lens and antenna. Microw. Opt. Technol. Lett. **54**(6), 1343–1348 (2012). https://doi.org/10.1002/mop.26812
38. S.K. Patel, V. Sorathiya, S. Lavadiya, Y. Luo, T.K. Nguyen, V. Dhasarathan, Numerical analysis of polarization-insensitive squared spiral-shaped graphene metasurface with negative refractive index. Appl. Phys. B Lasers Opt. **126**(5), 80 (2020). https://doi.org/10.1007/s00340-020-07435-2
39. S.K. Patel, V. Sorathiya, S. Lavadiya, T.K. Nguyen, V. Dhasarathan, Polarization insensitive graphene-based tunable frequency selective surface for far-infrared frequency spectrum. Phys. E Low-Dimensional Syst. Nanostruct. **120**, 114049 (2020). https://doi.org/10.1016/j.physe.2020.114049
40. G. Zhang, S. Pu, X.Y. Xu, C. Tao, J.Y. Dun, Optimized design of THz microstrip antenna based-on dual-surfaced multiple split-ring resonators, in *IEEE Antennas and Propagation Society is an International Symposium Proceedings*, vol. 2017-January, pp. 1755–1756, 2017. https://doi.org/10.1109/APUSNCURSINRSM.2017.8072920
41. L.C. Paul, M.M. Islam, Proposal of wide bandwidth and very miniaturized having dimension of μm range slotted patch THz microstrip antenna using PBG substrate and DGS, in *20th International Conference of Computer and Information Technology, ICCIT 2017*, vol. 2018-January, pp. 1–6, 2018. https://doi.org/10.1109/ICCITECHN.2017.8281766
42. S.K. Patel, S. Lavadiya, Y.P. Kosta, M. Kosta, T.K. Nguyen, V. Dhasarathan, Numerical investigation of liquid metamaterial-based superstrate microstrip radiating structure. Phys. B Condens. Matter **585**, 412095 (2020). https://doi.org/10.1016/j.physb.2020.412095
43. S. Boppel et al., Monolithically-integrated antenna-coupled field-effect transistors for detection above 2 THz, in *2015 9th European Conference on Antennas Propagation, EuCAP 2015* (2015)
44. S. Matsuura, M. Tani, K. Sakai, Generation of coherent terahertz radiation by photomixing in dipole photoconductive antennas. Appl. Phys. Lett. **70**(5), 559–561 (1997). https://doi.org/10.1063/1.118337
45. S. Verghese, K.A. McIntosh, E.R. Brown, Optical and terahertz power limits in the low-temperature-grown GaAs photomixers. Appl. Phys. Lett. **71**(19), 2743–2745 (1997). https://doi.org/10.1063/1.120445
46. *Terahertz Sources and Systems* (Springer Netherlands, 2001)
47. P. Vidhi, J.D. Ramesh, L. Sunil, New approach of high power testing of spacecraft passive components using diplexer. i-manager's J. Commun. Eng. Syst. **9**(1), 1 (2020). https://doi.org/10.26634/jcs.9.1.17395

48. Z. Fang, Z. Liu, Y. Wang, P.M. Ajayan, P. Nordlander, N.J. Halas, Graphene-antenna sandwich photodetector. Nano Lett. **12**(7), 3808–3813 (2012). https://doi.org/10.1021/nl301774e
49. A. Jagannathan et al., Effect of periodic roughness and surface defects on the terahertz scattering behavior of cylindrical objects . Terahertz Physics, Devices Syst. IV Adv. Appl. Ind. Def. **7671**, 76710E (2010). https://doi.org/10.1117/12.852909
50. B. Zhang, Y.-X. Guo, H. Zirath, Y.P. Zhang, Investigation on 3-D-Printing Technologies for Millimeter- Wave and Terahertz Applications. Proc. IEEE **105**(4), 723–736 (2017). https://doi.org/10.1109/jproc.2016.2639520
51. L. Guo, H. Meng, L. Zhang, J. Ge, Design of MEMS on-chip helical antenna for THz application. ieeexplore.ieee.org, pp. 1–4, (2016). https://doi.org/10.1109/imws-amp.2016.7588385
52. S.P. Lavadiya, S.K. Patel, M. Rayisyan, High gain and frequency reconfigurable copper and liquid metamaterial tooth based microstrip patch antenna. AEU - Int. J. Electron. Commun. **137**, 153799 (2021). https://doi.org/10.1016/j.aeue.2021.153799
53. M.S. Rabbani, H. Ghafouri-Shiraz, Liquid crystalline polymer substrate-based THz microstrip antenna arrays for medical applications. IEEE Antennas Wirel. Propag. Lett. **16**, 1533–1536 (2017). https://doi.org/10.1109/LAWP.2017.2647825
54. Prince, P. Kalra, E. Sidhu, Rectangular TeraHertz microstrip patch antenna design for riboflavin detection applications, in *Proceedings of the 2017 International Conference on Big Data Analytics, Computational Intelligence, ICBDACI 2017*, pp. 303–306, 2017. https://doi.org/10.1109/ICBDACI.2017.8070853

1D Periodic Nonlinear Model and Using It to Design All-Optical Parity Generator Cum Checker Circuit

Tanay Chattopadhyay

Abstract All-optical parity bit generator cum checker circuit has been proposed using one-dimensional (1D) periodic nonlinear material model in this paper. This structure consists of alternating layers of different nonlinear materials. The design is reconfigurable in nature, i.e., we can use this circuit as parity generator / checker without changing the circuit design. Numerical simulation results have been performed to check the result with the theoretical pre-calculated values.

Keywords All-optical signal processing · Optical logic · Nonlinear optics

1 Introduction

Correct data transmission and receiving is very important in data communication. n-bit data can be coded with 2^n possible combinations. An error detection code is generally added to n-bit data to make the word $(n + 1)$ bit to correct and detection of error in data transmission. Parity bit has been used as error detection code today. In this code, an extra bit is inserted with the binary message. This bit is parity bit, depending on which the total number of 1's in the massage become either even or odd. In a coded data, if the total number of 1's is made even (odd) then it is called even (odd) party encoded. A parity bit generates party bit at transmitter end and party checker checks the parity at the receiver end. Recently signal processing with optics has received considerable attention due to its high speed and high bandwidth [1–3]. Due to increase of data traffic ultra-high-speed photonic communication network is essential. In network nodes, the problem of optical-to-electronic-to-optical conversion will be avoided by all-optical circuit in future. Some optical circuit designs of all-optical parity and parity checking have already been proposed in different literatures. All-optical parity checker was demonstrated by Poustie et al. with concatenation of two TOADs (terahertz optical asymmetric demultiplexer) [4], where amplified

T. Chattopadhyay (✉)
Mechanical Operation (Stage-II), WBPDCL, Sagardighi Thermal Power project, Manigram, Murshidabad, West Bengal 742237, India

S. Das et al. (eds.), *Advances in Terahertz Technology and Its Applications*,
https://doi.org/10.1007/978-981-16-5731-3_8

signal feedback to the first TOAD as control pulse hence every time one-bit differential delay is required as feedback. This bit delay is act as 145 bit memory counter circuit. Chowdhury et al. have proposed isotropic linear-nonlinear material slab (LM + NLM) based parity generating and parity checking circuit [5], which is based on Snell's law. Samanta and Mukhopadhyay proposed polarization encoded parity generator and checker circuit by isotropic nonlinear material [6]. Also Pahari [7] proposed odd–even parity bit circuit by OPNLM-based device. 4-bit parity generator circuit using XGM in SOA has been proposed by Srivastava and Priye [8]. Also some design on party circuit has been proposed ring resonator (MMR) [9, 10], SOA-MZI [11], $LiNbO_3$-NLMZI [12, 13], quantum dot Mach–Zehnder interferometer [14] and plasmonic waveguides (PWG) [15]. $LiNbO_3$ based Kumar et al. [13] design requires optical-to-electrical conversion in intermediate circuit. In Nair et al. [16] design, optical tree net architecture is used to design 4-bit parity generator-checker circuit, here $2^4–1 = 15$-TOAD-based switches are required. TOAD-based parallel parity bit generator circuit is also proposed in our previous literature [17]. In TOAD-based switch, two counters propagating pulses reach to the input coupler so propagation time is very important for switching operation, hence cascading of many number of TOAD-based switches is more challenging and difficult to design in practical for high speed operation. Also SOA-based switches requires electrical biasing to active the device, also gain dynamics is very important for switching of this device. As a result at high frequency, slow gain recovery may create unwanted patterns at the output [18, 19].

In this paper, one-dimensional (1D) periodic nonlinear system is proposed and described. In case of nonlinear material, the refractive index of the material varies with intensity of light [20–24]. This 1D periodic model can be used to design XOR gate, which can be used in parity generator and checker circuit. The proposed design is parallel in nature and also reconfigurable, i.e., we can use this circuit as parity generator and also checker without changing the circuit design. Numerical simulation has been done to compare the result with the theoretical predetermined results.

2 1D Periodic Nonlinear Model

Refractive index (n) of a nonlinear material is incident light intensity (I) dependent as [20, 21, 25, 26],

$$n = n_0 + n_{\mathrm{nl}} I \tag{1}$$

where n_0 and n_{nl} are the linear refractive index the coefficient of nonlinear refractive index (Kerr coefficient), respectively. One-dimensional periodic nonlinear system was proposed by Brzozowski and Sargent [22, 27]. These structures consist of alternating layers nonlinear materials with two different linear refractive index (n_{01} and n_{02}) and Kerr coefficients ($n_{\mathrm{nl}1}$ and $n_{\mathrm{nl}2}$) as shown in Fig. 1. Their thickness (d_1 and

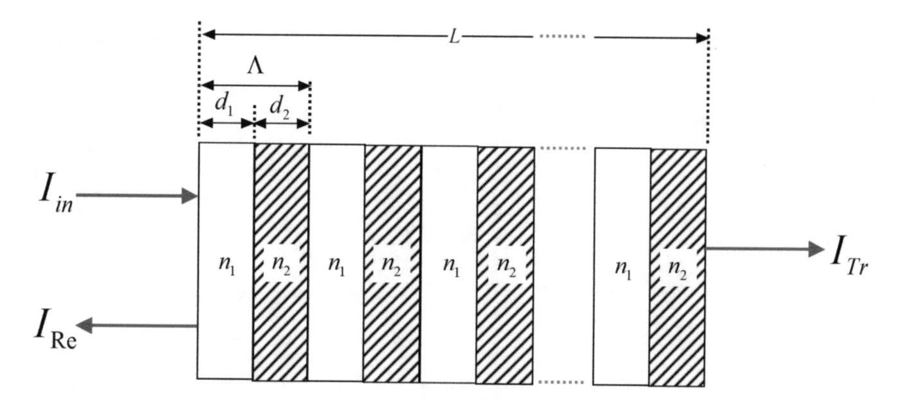

Fig. 1 One-dimensional (1D) periodic nonlinear system. These structures consist of alternating layers nonlinear materials with different linear refractive index (n_{01} and n_{02}) and Kerr coefficients (n_{nl1} and n_{nl2}), $n_1 = (n_{01} + n_{\mathrm{nl1}} I_{\mathrm{in}})$ and $n_2 = (n_{02} + n_{\mathrm{nl2}} I_{\mathrm{in}})$

d_2) are maintained such that they maintain a relation with the wavelength of the light as [27],

$$d_1 = \frac{\lambda}{2\left(n_{01} - n_{02}\frac{n_{\mathrm{nl1}}}{n_{\mathrm{nl2}}}\right)}$$
$$d_2 = \frac{\lambda}{2\left(n_{02} - n_{01}\frac{n_{\mathrm{nl2}}}{n_{\mathrm{nl1}}}\right)} \tag{2}$$

Hence this device behaves like gating having period of the grating $= \Lambda = (d_1 + d_2)$. When a light beam of intensity falls on the material layer, one portion is transmitted and other is reflected. Transmitted and reflected intensity also depends on the parameter 'a'. Where 'a' can be defined as [22, 27],

$$a = \frac{n_{01} - n_{02}}{|n_{\mathrm{nl1}}| + |n_{\mathrm{nl2}}|} \tag{3}$$

For N $(= L/\Lambda$, is the total length of the material) layers of such periodic refractive index system the transmitted light intensity can be expressed as [27],

$$I_{\mathrm{Tr}} = \begin{cases} 0 \quad \text{for} \quad I_{\mathrm{in}} < a\left(1 - \frac{1}{2^N}\right) \\ a\left[2^N(I_{\mathrm{in}} - 1) + 1\right] \quad \text{for} \quad a\left(1 - \frac{1}{2^N}\right) < I_{\mathrm{in}} < a \\ a \quad \text{for} \quad I_{\mathrm{in}} > a \end{cases} \tag{4}$$

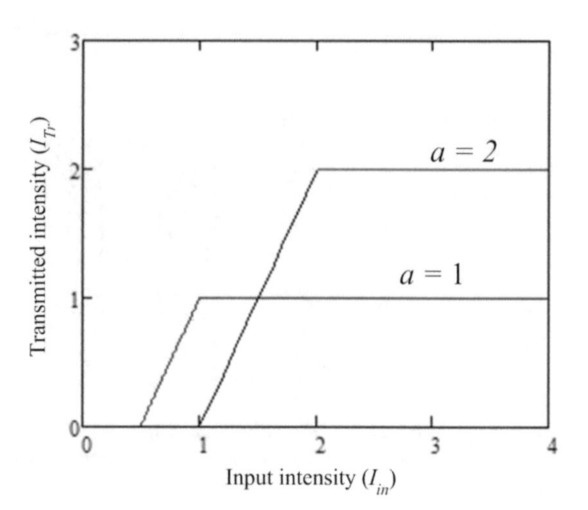

Fig. 2 The variation of transmitted intensity (I_{Tr}) with incident intensity (I_{in}) for $a = 1$ and 2 at very higher values of N

We can choose $a = 1, 2$, etc. using different nonlinear materials. As an example, $a = 2$ can be obtained by choosing $n_{01} = 1.5$, $n_{\mathrm{nl1}} = +\,0.01$ and $n_{02} = 1.54$, $n_{\mathrm{nl2}} = -0.01$ [28]. Equation (4) shows that the transmitted light intensity becomes a step function for large values on N. For very large N, I_{Tr} is also plotted in Fig. 2 for $a = 1$ and 2 respectively. From these graphs, we see that when $I_{\mathrm{in}} \geq a$, we find the transmitted light intensity will clamp to a. Brzozowski and Sargent showed that all light intensities which are less than a is reflected [22, 27]. This operation is archived for input light frequency 2.8×10^{14} Hz to 3.2×10^{14} Hz [22]. The required input beam intensity for switching of a 1 mm—long device is nearly 1×10^{6} W/cm^{2} [27]. This design can be easily fabricated on the SOI ridge waveguide by etching techniques described in reference [29–32]. The simulated refractive index profile (both linear and nonlinear refractive index) is also shown in Fig. 3. From the simulation, we find with grating period $\Lambda = 0.765$ μm.

XOR gate with this 1D periodic nonlinear system (for $a = 2$) can be easily designed [28, 33], which is shown in Fig. 4a. Here two beams 'A' and 'B' are combined with a beam combiner. The combined beam 'X' is the input of the 1D periodic nonlinear system. When the intensity level of input A and $B = 1$, then the combined beam (X) intensity level = 2. According to the property described before, we can say that in this case, light is transmitted (because I_{in} = light intensity level of the combined beam 'X' $= 2 = a$). In other cases the input intensity $= 1 < a$, so light is reflected. In this way, XOR logical operation ($A \oplus B$) is possible at the output O_{Re} shown in Fig. 4a. The input–output intensity table is shown in Fig. 4b. The transfer matrix of the XOR logic unit shown in Fig. 4 can be expressed as,

$$[T] = \begin{pmatrix} AB & AB \\ A + B - 2AB & A + B - 2AB \end{pmatrix} \tag{5}$$

And the operation is,

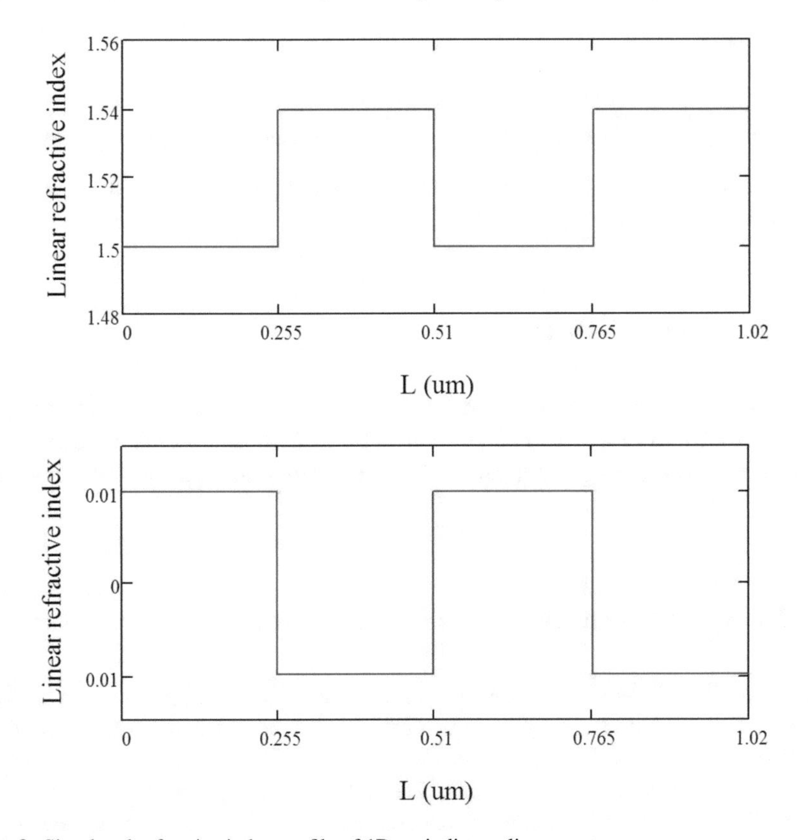

Fig. 3 Simulated refractive index profile of 1D periodic nonlinear system

Fig. 4 1D periodic nonlinear system-based XOR gate **a** all-optical circuit, BC: beam combiner, **b** Input–Output intensity table

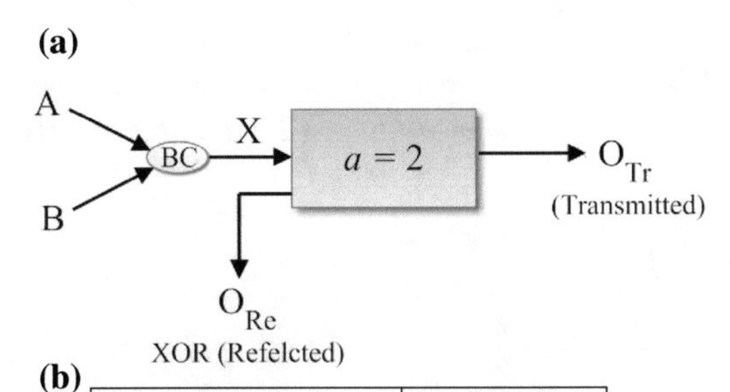

(b)

Input Intensities			**Output Intensities**	
A	B	X	O_{Re}	O_{Tr}
0	0	0	0	0
0	1	1	1	0
1	0	1	1	0
1	1	2	0	2

$$\mathbf{O} = [T]\mathbf{I} \tag{6}$$

where $\mathbf{I} = \binom{A}{B}$ and $\mathbf{O} = \begin{pmatrix} O_{\mathrm{Tr}} \\ O_{\mathrm{Re}} \end{pmatrix}$ are the input and output matrices respectively.

Numerical simulation has been done to compare the result with the theoretical predetermined results.

3 All-Optical XOR Gate

Cascading the above 1D periodic nonlinear model we can construct 8-bit XOR operation as shown in Fig. 5. The output (Out) of the circuit exhibit the relation Out = $(A1 \oplus A2 \oplus A3 \oplus A4 \oplus A5 \oplus A6 \oplus A7 \oplus A8)$. This circuit can be easily extended to n-bit easily.

4 Parity Bit Generator Cum Checker Circuit

Using the above-proposed XOR gate, we can design parity bit generator cum checker circuit. The 4-bit parity bit generator cum checker circuit is shown in Fig. 6. It is designed with four (I, II, III and IV) all-optical XOR gates shown in Fig. 6. Here A, B, C and D are four inputs. The output P' gives the logical expression $(A \oplus B \oplus C \oplus D)$. Other input P'' is combined with P'. The combined beam X' is fed to a XOR gate. The final output is P gives the logical expression, $P = \{P'' \oplus (A \oplus B \oplus C \oplus D)\}$. Here the input P'' is called reconfigurable input. By selecting input values the circuit behaves like parity checker and some time generator. The proposed 4-bit parity generator cum checker size is nearly the size of (two parallel OC + two parallel switching module (I) and (II) + one OC + (III)rd switching

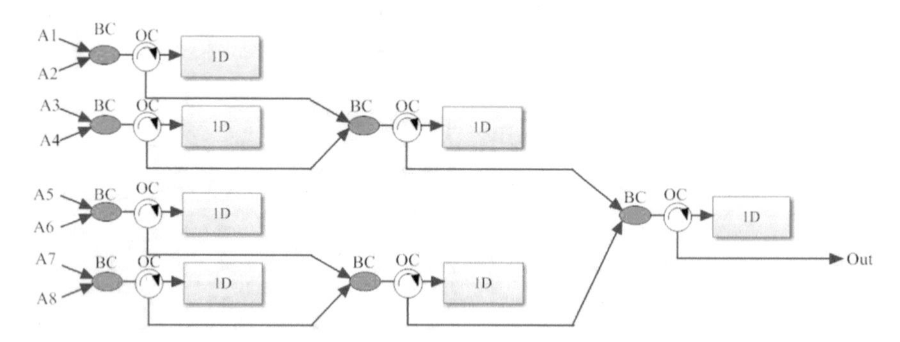

Fig. 5 All-optical 8-bit XOR operation using 1D—periodic nonlinear system. OC: optical circulator, BC: beam combiner

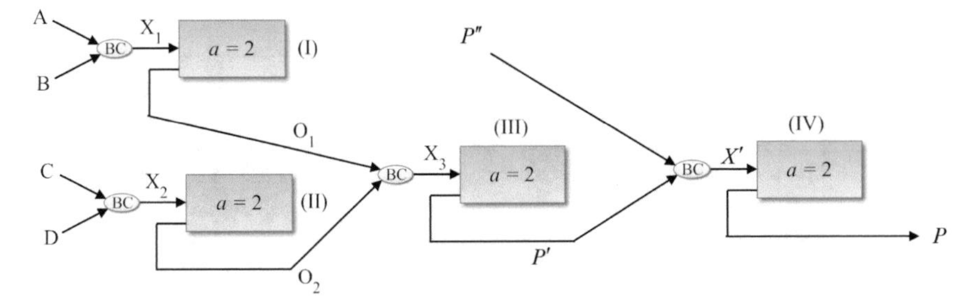

Fig. 6 4-bit Parity generator cum checker circuit using 1D periodic nonlinear system

module + one OC + (IV)th switching module + some spaces for beam combiner, i.e., Y coupler) = $(3.7 \times 3 + 14 \times 3 + 1.7 \times 3) = 58.2$ mm. Actual size will be slightly greater than this size. The output intensity at port P can be calculated from the following equations using Eq. (6) as,

$$\begin{pmatrix} O_{\text{Tr-I}} \\ O_1 \end{pmatrix} = [T_{\text{I}}] \begin{pmatrix} A \\ B \end{pmatrix} \tag{7}$$

$$\begin{pmatrix} O_{\text{Tr-II}} \\ O_2 \end{pmatrix} = [T_{\text{II}}] \begin{pmatrix} C \\ D \end{pmatrix} \tag{8}$$

$$\begin{pmatrix} O_{\text{Tr-III}} \\ P' \end{pmatrix} = [T_{\text{III}}] \begin{pmatrix} O_1 \\ O_2 \end{pmatrix} \tag{9}$$

$$\begin{pmatrix} O_{\text{Tr-}IV} \\ P \end{pmatrix} = [T_{IV}] \begin{pmatrix} P'' \\ P' \end{pmatrix} \tag{10}$$

For $P'' = 0$, $P = (A \oplus B \oplus C \oplus D)$, i.e., the circuit behaves like parity generator. The simulated waveform of the parity bit generator for different combination of inputs A, B, C and D using Eq. (6) are shown in Fig. 7, which satisfy the theoretical pre-calculated results shown in Table 1.

At the receiving end, we take these data and can check the parity. To do this we have to put $P'' = P$ (previously calculated or simulated parity bit). If the final output '$P = 0$' then we can say that there is no error in the data, i.e., parity is conserved. But if we find '$P = 1$' then we can say that during transmission from a position to another an error has occurred. For getting the theoretical truth table, we introduce some bit errors in the data at the receiving end to understand the circuit operation (denoted by '$\underline{1}$' or '$\underline{0}$' marks in the truth Table 2). '$\underline{1}$' and '$\underline{0}$' occurs when 'state-0' is transformed to 'state-1' and 'state-1' is changed to '0 state' due to error. The simulated waveform is shown in Fig. 8, which satisfy the truth Table 2.

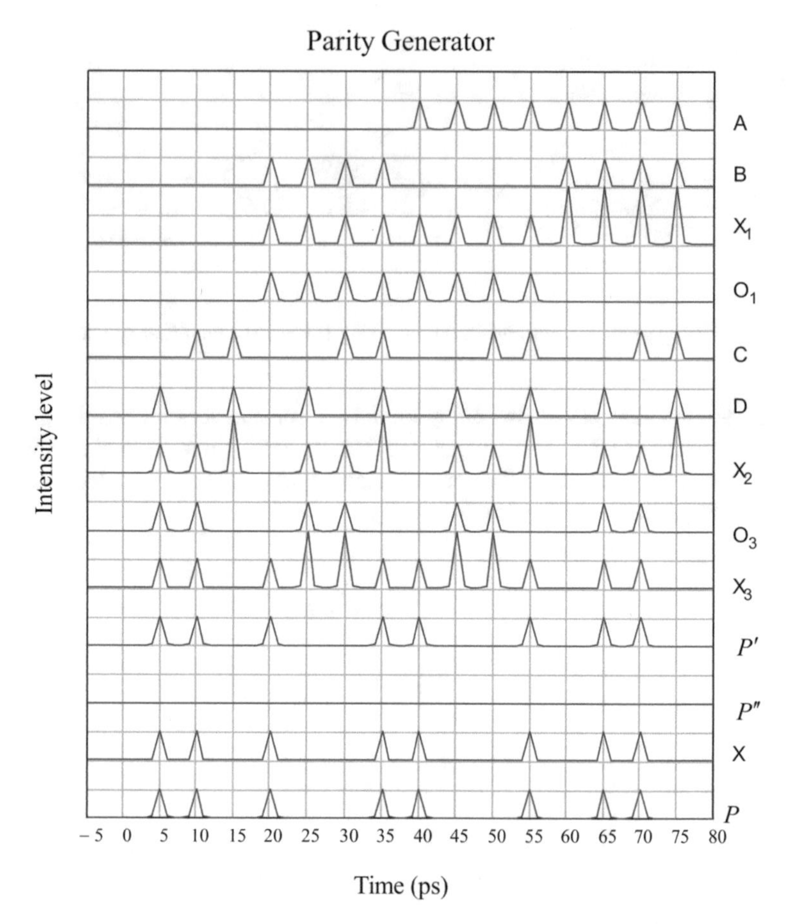

Fig. 7 Simulated waveforms of parity generator circuit for $P'' = 0$, Y-axis having the value 1×10^6 W/cm^2 per division

5 Discussion

A comparative study can be made on the basis of number of switching modules used in the circuit and identical circuit for parity checker and generator, which is shown in Table 3.

For n—bit parity generator/checker circuit n—input XOR gate is required. To design n—input XOR gate we need $(2^{\lceil \log_2 n \rceil} - 1)$ numbers of two—input XOR gates. The output of this n—input XOR gate (P') is fed to another XOR gate to obtain the final output 'P' as shown in Fig. 9a. Hence to design n—bit parity generator/checker circuit, we need total $2^{\lceil \log_2 n \rceil}$ two—input XOR gates, shown graphically in Fig. 9b. Also n—bit delay line is required in P'' input line of the last XOR gate. The switching scheme depends upon the intensity level of the incoming signal. Due to optical loss at the coupler and the nonlinear material, the intensity level can be

Table 1 Truth table of parity generator

A	B	C	D	*P*
0	0	0	0	0
0	0	0	1	1
0	0	1	0	1
0	0	1	1	0
0	1	0	0	1
0	1	0	1	0
0	1	1	0	0
0	1	1	1	1
1	0	0	0	1
1	0	0	1	0
1	0	1	0	0
1	0	1	1	1
1	1	0	0	0
1	1	0	1	1
1	1	1	0	1
1	1	1	1	0

Table 2 Parity checker truth table. Some bit errors inside the circuit is created to understand the circuit operation (denoted by '$\underline{1}$' and '$\underline{0}$')

A	B	C	D	P'	*P*	Remarks
0	0	$\underline{1}$	0	0	1	Error
0	0	0	1	1	0	
0	0	1	0	1	0	
$\underline{1}$	0	1	1	0	1	Error
0	1	0	0	1	0	
0	1	0	1	0	0	
0	1	$\underline{0}$	0	0	1	Error
0	1	1	1	1	0	
1	0	0	$\underline{1}$	1	1	Error
1	0	$\underline{1}$	1	0	1	Error
1	0	1	0	0	0	
1	0	1	1	1	0	
$\underline{0}$	1	0	0	0	1	Error
1	1	$\underline{1}$	1	1	1	Error
1	1	1	0	1	0	
1	1	$\underline{0}$	1	0	1	Error

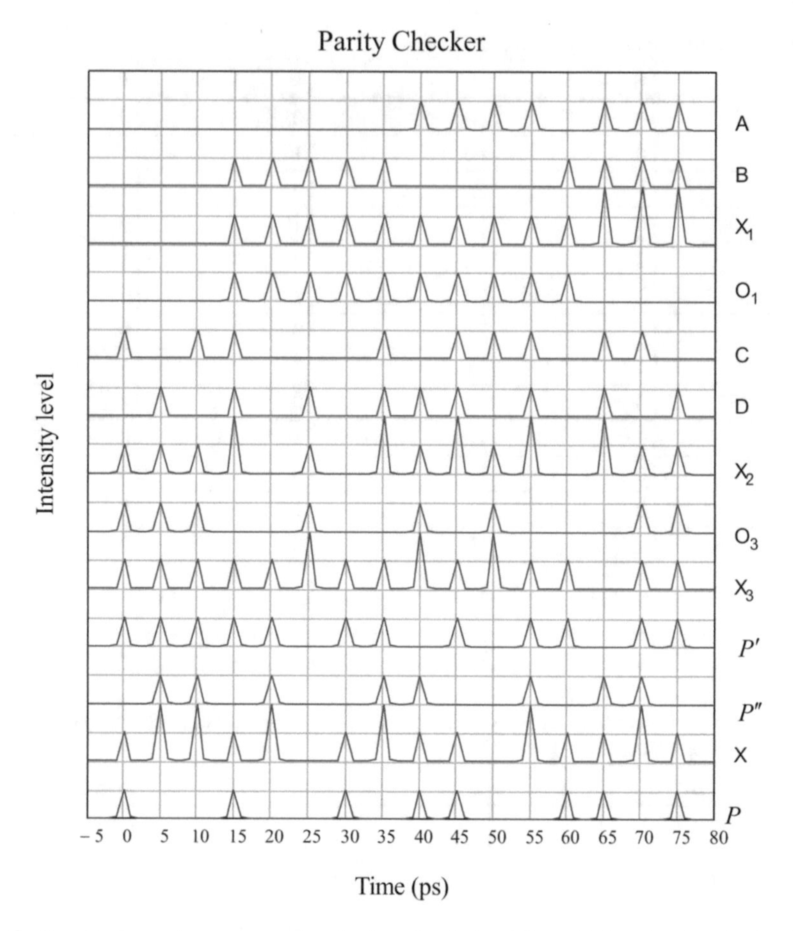

Fig. 8 Simulated waveforms of parity checker circuit for $P'' = P$ (previously calculated or simulated parity bit), Y-axis having the value 1×10^6 W/cm^2 per division

decreased after certain stages which may disturb the switching behavior. But fixed intensity can be maintained with optical isotropic nonlinear material-based design proposed by Samanta and Mukhopadhyay in their manuscript [34].

6 Conclusion

The structure of the proposed parity bit generator cum checker is all-optical in nature with application in photonic networks. We describe the circuit operation using analytical expressions. Numerical simulation results have been done to check the result with the theoretical pre-calculated values. The circuit operates in picoseconds range and can be used in future optical supercomputing and communication.

Table 3 Comparative study of proposed parity bit generator/checker circuit in different papers

References	Nature of circuit			Number of bits in their proposed circuit	Name of switching module by which the circuit proposed	Number of switching modules by which the circuit designed	Switching speed/response time
	Parity generator	Parity checker	Both circuit in combine				
Paustie et al. [4]		YES		32	TOAD	2	>40 GB/s
Chowdhury et al. [5]			YES	4	LM + NLM	5	
Samanta and Mukhopadhyay [6]	YES			4	LM + NLM	12	Response time 10^{-9}
Pahari [7]	YES	YES		3	OPNLM	3 for checker and 4 for generator	Order of GHz
Srivastava and Priye [8]	YES			4	XGM in SOA	6	
Rakshit et al. [9]			YES	4	MMR	4	
Law et al. [10]	YES			2	MMR	1	10 GB/s
Kaur and Sukla [11]	YES	YES		3	SOA-MZI	3 for checker and 2 for generator	10 GB/s
Singh et al. [12]	YES			3	NLMI	6 for even parity generator and 7 for odd parity generator	Response time in the order of μm
Kumar et al. [13]		YES		4	LiNbO3-MZI	8	Response time ~ 5 μm
Dimitriadou et al. [14]			YES	4	QD-SOA-MZI	4	160 Gb/s

(continued)

Table 3 (continued)

References	Nature of circuit			Number of bits in their proposed circuit	Name of switching module by which the circuit proposed	Number of switching modules by which the circuit designed	Switching speed/response time
	Parity generator	Parity checker	Both circuit in combine				
Wang et al. [15]		YES		4	U-shaped plasmonic waveguide	5	
Nair et al. [16]			YES	4	TOAD	15	10 Gb/s
Bhattacharyya et al. [17]	YES			4	TOAD	6	1.25 Gb/s
Proposed in this paper			YES	4	1D photonic nonlinear model	4	Response time in the order of ps

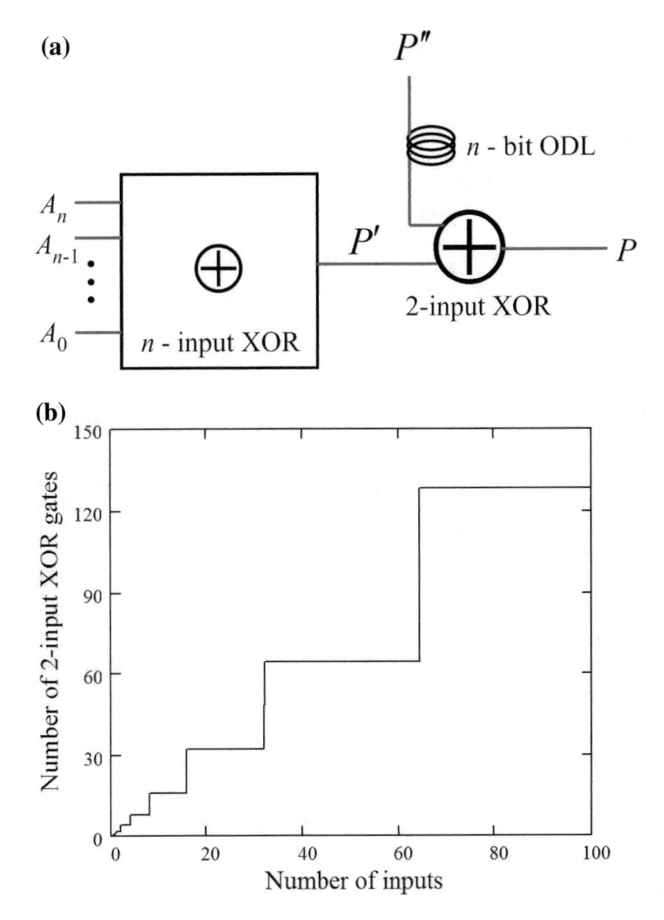

Fig. 9 **a** n—bit parity generator/checker circuit (block diagram). **b** Number of 2-input XOR gates used in the n-bit parity generator/checker circuit

Acknowledgements The corresponding author is grateful to Dr. Kousik Mukherjee about the suggestion for submission of this book chapter.

References

1. H.J. Caulfield, S. Dolev, Why future supercomputing requires optics. Nat. Photon. **4**, 261–263 (2010)
2. A.K. Ghosh, A. Basuray, Trinary flip-flops using Savart plate and spatial light modulator for optical computation in multivalued logic. Optoelectron. Lett. **4**, 443 (2008)
3. S. Mandal, D. Mandal, M.K. Mandal, S.K. Garai, Design of optical quaternary multiplier circuit using polarization switch. Opt. Wireless Tech. **1546**, 111–119 (2020)
4. A.J. Poustie, K.J. Blow, A.E. Kelly, R.J. Manning, All-optical parity checker with bit differential delay. Opt. Commun. **162**(1–3), 37–43 (1999)

5. K.R. Chowdhury, D. De, S. Mukhopadhyay, Parity checking and generating circuit with nonlinear material in all-optical domain. Chin. Phys. Lett. **22**(6), 1433–1435 (2005)
6. S. Samanta, S. Mukhopadhyay, All-optical method of developing parity generator and checker with polarization encoded light signal. J. Opt **41**(3), 167–172 (2012)
7. N. Pahari, All-optical even and odd parity bit generator and checker with optical nonlinear meterial. J. Opt. **46**(3), 336–341 (2016)
8. V.K. Srivastava, V. Priye, All-optical 4-bit parity checker design. Opt. Appl. **41**(1), 157–164 (2011)
9. J.K. Rakshit, J.N. Roy, T. Chattopadhyay, Design of micro-ring resonator based all-optical parity generator and checker circuit. Opt. Commun. **303**, 30–37 (2013)
10. F.K. Law, M.R. Uddin, N.M. Masir, Y.H. Won, Digital photonic even parity bit generator, in *2018 © IEEE photonics conference (IPC)*, 30 Sept.–4 Oct. 2018, Reston. https://doi.org/10.1109/IPCon.2018.8527192
11. S. Kaur, M.K. Shukla, All-optical parity generator and checker circuit employing semiconductor optical amplifier-based Mach Zehnder interferometer. Opt. Appl. **XLVII**(2), 263–271 (2017)
12. L. Singh, A. Bedi, S. Kumar, Modeling of all-optical even and odd parity generator circuit using metal insulator-metal plasmonic waveguides. Photon. Sens. **7**(2), 182–192 (2017)
13. S. Kumar, Chanderkanta, A. Amphawan, Design of parity generator and checker circuit using electro-optic effect of Mach-Zehender interferometers. Opt. Commun. **364**, 195–224 (2016)
14. E. Dimitriadou, K.E. Zoiros, T. Chattopadhyay, J.N. Roy, Design of ultrafast all-optical 4-bit parity generator and checker using quantum-dot semiconductor optical amplifier-based Mach-Zehnder interferometer. J. Comput. Electron. **12**, 481–489 (2013)
15. F. Wang, Z. Gong, X. Hu, X. Yang, H. Yang, Q. Gong, Nanoscale on-chip all-optical logic parity checker in integrated plasmonic circuits in optical communication range. Sci. Rep. **6**(24433), 1–8 (2016)
16. N. Nair, S. Kaur, R. Goyal, All-optical integrated parity generator and checker circuit using optical tree architecture. Curr. Opt. Photon. **2**(5), 400–406 (2018)
17. A. Bhattacharyya, D.K. Gayen, T. Chattopadhyay, All-optical parallel parity generator circuit with the help of semiconductor optical (SOA)-assisted Sagnac switches. Opt. Commun. **313**, 99–105 (2014)
18. D. Mandal, S. Mandal, M.K. Mandal, S.K. Garai, Theoretical approach of developing a frequency-encoded reversible optical arithmetic and logic unit using semiconductor optical amplifier-based polarization switches. Opt. Eng. **58**(1), 015104 (2019)
19. A. Raja, K. Mukherjee, J.N. Roy, K. Maji, Analysis of polarization encoded optical switch implementing cross polarization modulation effect in semiconductor optical amplifier. Int. J. Photon. Opt. Tech. **5**(1), 1–5 (2019)
20. D. Samanta, Implementation of polarization-encoded quantum Fredkin gate using Kerr effect. J. Opt. Commun. Retrieved 4 Jul 2019. https://doi.org/10.1515/joc-2019-0077
21. D. Samanta, Implementation of polarization-encoded quantum toffoli gate. J. Opt **48**(1), 70–75 (2019)
22. L. Brzozowski, R.H. Sargent, Optical signal processing using nonlinear distributed feedback structures. J. Quantum Electron. **36**(5), 550–555 (2000)
23. J.H. Vella, J.H. Goldsmith, A.T. Browning, N. Limberopoulous, H. Vitebskiy, E. Makri, T. Kottos, Experimental realization of a reflective optical limiter. Phys. Rev. Appl. **5**(6), 064010-1-7 (2016)
24. S. Valligatla, A. Chiasera, S. Varas, P. Das, B.N. Shivakiran Bhaktha, A. Łukowiak, F. Scotognella, D. Narayana Rao, R. Ramponi, G.C. Righini, M. Ferrari, Optical field enhanced nonlinear absorption and optical limiting properties of 1-D dielectric photonic crystal with ZnO defect. Opt. Mater. **50**, 229 (2015)
25. B.E.A. Saleh, M.C. Teich, *Fundamental of Photonics* (Wiley, New York, 1991)
26. R.W. Boyd, *Nonlinear Optics*. 3rd ed. (Academic Press © & Elsevier Inc., 2008)
27. L. Brzozowski, R.H. Sargent, All-optical analog-to-digital converters, hardlimiters, and logic gates. J. Lightwave Technol. **19**(1), 114–119 (2001)

28. T.A. Moniem, M.H. Saleh, Fuzzy logic membership implementation using optical hardware components. Opt. Commun. **285**, 4474–4482 (2012)
29. T.E. Murphy, J.T. Hastings, H.I. Smith, Fabrication and characterization of narrow-band Bragg-reflection filters in Silicon-on-insulator ridge waveguides. J. Lightwave Technol. **19**(12), 1938–1942 (2001)
30. T.A. Moniem, M.H. Saleh, Fuzzy logic membership implementation using optical hardware components. Opt. Commun. **285**, 447–4482 (2012)
31. T.A. Moniem, Adaptive integrated optical bragg grating in semiconductor waveguide suitable for optical signal processing. Fiber Integr. Opt. **35**(3), 110–127 (2016)
32. S. Ghosh, S. Keyvaninia, W.V. Roy, T. Mzumoto, G. Roelkens, R. Baets, Adhesively bonded Ce:YIG/SOI integrated optical circulator. Opt. Lett. **38**(6), 965–967 (2013)
33. T. Chattopadhyay, All-optical simultaneous XOR-AND operation using 1-D periodic nonlinear material. J. Opt. Commun. 000010151520210145 (2021). https://doi.org/10.1515/joc-2021-0145
34. D. Samanta, S. Mukhopadhyay, All-optical method for maintaining a fixed intensity level of a light signal in optical computation. Opt. Commun. **281**, 4851–4853 (2008)

Section I: Wide Bandgap (WBG) Semiconductors as Terahertz Radiation Generator

Suranjana Banerjee

Abstract IMPATT device with double drift layer created on WBG (wide bandgap) materials in the terahertz region (0.1–10 THz) of the RF spectrum will be inspected in this chapter to study their reliability as the generator of terahertz (THz) radiation. Above W-band (75–110 GHz), avalanche response time (ART) poses a limitation on the RF performance of various semiconductors which can be overcome by the use of WBG semiconductors like SiC and GaN which will be discussed in details here. ART analysis provides a novel method to determine the suitability of the WBG semiconductors to operate in THz regime. Analysis of RF signal of large amplitude using NSVE model will also be presented in this section. Based on this model, various RF parameters like conductance, susceptance, negative resistance, quality factor etc. of WBG semiconductors like SiC and GaN will be extracted to get a comprehensive idea of the acceptability of these semiconductors as reliable sources of THz radiation. Results indicate that below 1.0 THz, both SiC and GaN act as promising tera source but above 1.0 THz, GaN excels SiC as a source of THz radiation. Simulation results presented here will be of substantial use to fabricators and experimenters for practically realizing double drift layer Avalanche Transit Time (ATT) diodes with Silicon Carbide or Gallium Nitride as the base material for operation in the THz domain.

Keywords Double drift layer · IMPATT · Avalanche response time

1 Introduction

The terahertz realm (0.1–10 THz) is the propitious yet vexing part of the RF spectrum that lies between the microwave and the optical regions and covering a frequency range from 0.1 to 10 THz. Terahertz radiation or Tera rays are safer than X-rays or gamma rays as they are non-ionizing. This means tera ray photons are not energetic enough to break-off electrons from the atoms or molecules of human tissue which could give rise to harmful chemical reactions. Still, this regime remained

S. Banerjee (✉)
Department of Electronics, Dum Dum Motijheel College, Kolkata, India

S. Das et al. (eds.), *Advances in Terahertz Technology and Its Applications*,
https://doi.org/10.1007/978-981-16-5731-3_9

underdeveloped due to the dearth of efficient, coherent and compact tera sources and detectors and the goal of turning laboratory outcomes in tera domain into real-world applications has proved evasive. Legions of researchers have struggled with that challenge for decades. Hence, this region is also referred to as *Terahertz gap*. Another challenge in the THz band is high losses. THz waves have high absorption in the atmospheric situation and the moist environment. So, signal degradation in THz region is appreciably mòre than the neighbouring microwave and optical or infrared band.

Intensified effort has started since 1990s to tame and harness the utility of terahertz domain. Tera spectrum offers high data rate which is effectively used in another rising field of wireless communication. The legendary works of Prof. J. C. Bose declared in May 1895 at Calcutta (Asiatic Society of Bengal) is a landmark in the history of millimeter waves. He demonstrated the generation of electromagnetic waves with a wavelength of 6 mm using Galena detector and coined the word 'millimeter wave' for the first time. The solid state millimeter wave and terahertz sources have almost replaced the vacuum tube sources due to their low cost, weight and reliability. The objective is to realize simple, reliable, efficient, powerful and cost effective solid state sources for terahertz communication system.

DDR (Double Drift Region) IMPATT (Impact Avalanche Transit Time) Diode based on Wide band gap (WBG) semiconductors like SiC and GaNwill be explained in this chapter as a dependable tera radiation generator. This chapter will also provide an in-depth knowledge of the design and establishment of a high power-low noise reliable IMPATT device as a source of THz radiation. IMPATT is basically a reverse biased negative resistance device and has established itself as a source of elevated power and efficiency and can plunge into the submillimeter or THz domain efficiently by using proper base material. Above W-band (75–110 GHz), the performance of IMPATT diode is limited by avalanche response time which depends on the carrier ionization rates in the base semiconductor which will be explained later in this chapter. This limitation can be overcome by using WBG semiconductors like SiC or GaN whose material parameters like breakdown voltage, carrier saturation velocity and thermal conductivity are highly favourable to realize high power-high frequency semiconductor device like IMPATT diode at Terahertz frequency region.

2 Operation Principle of IMPATT and Its Development as a Terahertz Radiation Generator

IMPATT is an acronym for Impact ionization and Avalanche Transit Time. IMPATT diode is a negative resistance device which operates at microwave, millimeter wave and terahertz frequencies where the negative resistance mainly arises due to avalanche multiplication and the transit time effect. These two phenomena produces total 180° phase lag between the input excitation and the output response: avalanche delay causing a 90° phase lag and transit time delay giving rise to another 90° lag. IMPATT diode produced appreciable power at the output when properly implanted in RF cavity resonator. The research and development of IMPATT diodes have been prodigious over a wide dimension of frequency from microwave to due to variations introduced in structure, base semiconductor, doping conditions and other important properties of the diode. Different remarkable experimental outcomes propelled many researchers to dive into the THz field and inspect it from various aspects. Shockley in 1954 [1] conceived the idea of obtaining negative resistance from a reverse-biased *p–n* junction diode. He showed that if an electron bunch is injected into the drift region of the reverse-biased diode then the phase delay involved in the transit of electrons through the high field drift layer gives rise to a negative resistance. He proposed that a forward-biased junction can be used to inject electron bunch into the drift region. However, the negative resistance of the diode was low and consequently the microwave output power was also low. In 1958 Read [2] put forward the idea of introducing an intrinsic layer in between n–n^+ with p^+ at the other end forming the p^+n junction avalanche growth region near the high field p^+-n junction. Read diode gradually withered to obscurity for a few years as feasibility of the diode became difficult. In 1965, first pragmatic Read diode was accomplished by Lee et al. [3] at lower microwave frequency with low output power. In 1965 Johnston and his co-researchers [4] demonstrated oscillation at microwave frequencies from a single-drift region (SDR) diode structure. In the later part of the same year Misawa [5] developed small signal theory for IMPATT diode.

In 1967, Gummel and Blue [6] carried out noise analysis of the IMPATT diode with a small amplitude RF signal. Scharfetter et al. in 1970 [7] proposed a structure with two drift regions called the DDR structure (p^+-p-n-n^+)comprising an avalanche layer sandwiched between the electron and the hole drift layer. Their observation reveals better efficiency and more high frequency power generation of IMPATT diode with two drift layers over a single drift layer. RF power is mainly contributed by the drift layer of the diode, DDR structure having two drift layers obviously contributes more power than SDR structure. Works of Seidel et al. in 1971 divulge practical realization of both double drift layer and single drift layer structures of the diode with successful delivery of 1 W and 0.53 WRF power respectively at a frequency of 50 GHz. Post these works, multitude of researchers established the fact through their works that diodes incorporated with double drift layer provides better RF performance than with single drift layer. At THz frequencies, wide bandgap semiconductors like GaN having a bandgap of 3.4 eV or 3C-SiC with bandgap of 2.24 eV or 4H-SiC with bandgap of 3.26 eV are highly suitable as base material for IMPATT diode. Reports on microwave frequency synthesizers [8] show that they can generate few microwatts of power and hence finds application as local oscillators in radio astronomy receivers. Additionally, IR lasers using multiple quantum wells in the structure, HEMTs [9] and HBTs [10] have emanated as probable THz sources.

Low cost terahertz sources operational at room temperature are in great demand but not yet available commercially. The novel idea of using high energy electronic bandgap semiconductors as the base material for IMPATT is a demanding research domain which if scrutinized methodically can provide large amount of power at higher millimeter and terahertz frequency regime. These terarays can be used for various applications in medical imaging, bioscience and surveillance. Terahertz domain is still in the dawning stage due to paucity of electronic components, detectors and generators in this particular frequency range while those which are available do not satisfy the state-of-the-art need of researchers as they are voluminous and operate below the ambient temperature. These sources find important applications both in civilian and defence sectors like satellite communication, remote sensing, collision avoidance systems, wireless television and worldwide personal communication systems due to their several advantages like greater bandwidth, larger communication capacity for data, higher spatial resolution, smaller component and antenna size, compact and light weight equipment, lower probability of intercept or interference compared to microwave systems. Further terahertz waves can penetrate smoke, dust, rain and clouds much more effectively than infrared or optical waves. THz sources also find important applications in tracking radars and missile guidance. Hence, the terahertz realm has fascinated researchers to investigate the prospect of the domain to be used for various applications.

3 Wide Bandgap (WBG) Semiconductors as Reliable THz Source

The prospect of high energy electronic bandgap semiconductors as base materials for IMPATT diode has been reported [11–20] recently as a prospective source of tera radiation. Reports in [17] and [18] manifest the performance of 4H-SiC as a suitable base material for IMPATT. To the best of author's knowledge, dearth of superior wafer for GaN has hindered its prospect of fabricating IMPATT diode, though physical properties of the material is highly approving. Diamond is another propitious material with high energy bandgap for use in terahertz domain due to its highly commending characteristics and modernized growth process [19, 20].

4 THz Operation Based on Avalanche Response Time (ART) Analysis

Terahertz operation of IMPATT device based on WBG semiconductors depends greatly on ART. Dependence of THz operation of the device on ART will be discussed here in this section and tunneling will be discussed in Section II. The threshold frequency of the device can be predicted faithfully from the ART analysis as will be presented in the following section where a comparison is made between ART and time needed for the mobile charge carriers to transit through the drift zone, abbreviated as TT in IMPATT diode based on large gap semiconductors as Silicon Carbide and Gallium Nitride. This analysis effectively proves Avalanche Transit Time device as a generator of THz radiation. Rate of ionization (α_n, α_p) and average velocities (v_{sn},

v_{sp}) of mobile carriers determined by electric field considerably regulate the dc and RF features of the device. The significance of these properties lie in the fact that they control the rise of current generated due to avalanche multiplication process and the time needed to move through the drift region. Thus, proper choice of the base semiconductor is very important for determining both ART and TT. The following analytical method helps to find out the limiting frequency from ART and TT.

The avalanche response times initiated by electrons, τ_{An} and holes, τ_{Ap} are expressed as

$$\tau_{An} = \frac{1}{(v_{sn} + v_{sp})} \int_{-x_{A_1}}^{x_{A_2}} \exp\left[- \int_{-x_{A_1}}^{x} (\alpha_n - \alpha_p) \mathrm{d}x' \right] \mathrm{d}x. \tag{1}$$

$$\tau_{Ap} = \tau_{An} \exp\left[\int_{-x_{A_1}}^{x_{A_2}} (\alpha_n - \alpha_p) \mathrm{d}x \right]. \tag{2}$$

τ_A is the avalanche response time due to both the carriers, electrons and holes and is given by

$$\tau_A = \tau_{An} \left\{ (1 - k) + k.\exp\left[- \int_{-x_{A_1}}^{x_{A_2}} (\alpha_n - \alpha_p) \mathrm{d}x \right] \right\}^{-1}. \tag{3}$$

where the parameter $k = J_{ps}/J_s$ and $(1 - k) = J_{ns}/J_s$.

In Eq. (3), J_{ns} and J_{ps} are the currents generated due to minority carrier injection and $J_s = J_{ps} + J_{ns}$ being the total current The avalanche response times are obtained from Eqs. (1–3) with the realistic material parameters from [21–24].

4.1 ART Method to Find the Suitability of Silicon Carbide and Gallium Nitride

Method: Firstly, ART (τ_A) of the base materials are obtained from Eq. (3). For that, ($\alpha_n(x)$) for electrons and ($\alpha_p(x)$) for holes which gives the space domain variations of the carrier ionization rates are obtained from the dc simulation outputs and then (τ_A) is obtained by solving Eqs. (1) and (2) numerically. The time taken by the carriers to drift across the active region that is transit times (τ_T) depends on the frequency of operation and is obtained from [25].

Results: Changes of τ_A and τ_T with optimum frequency for WBG semiconductors are clear from Fig. 1. The graph shows that ART is less than TT below the operating frequency which should be the case for obtaining power output from the DDR

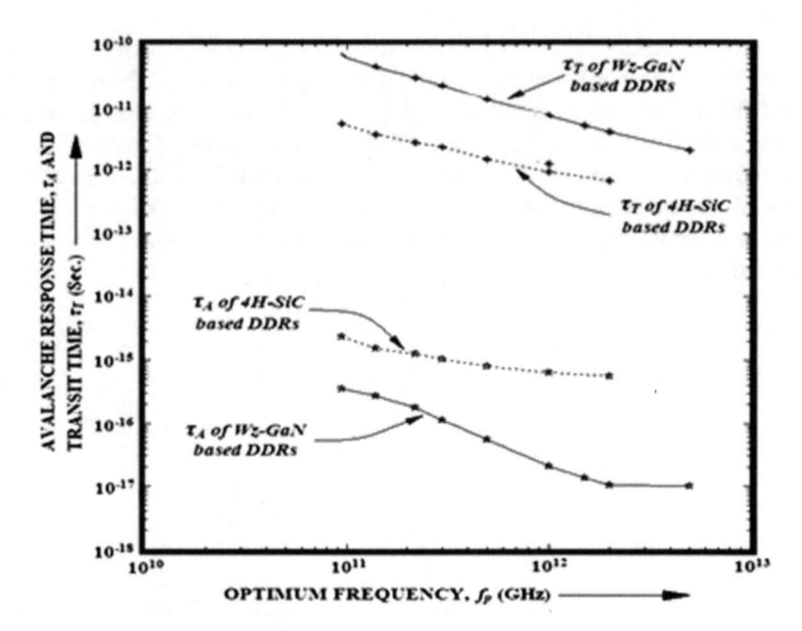

Fig. 1 Comparison plots of ART and TT against optimum frequency forGaN and 4H-SiC as the base materials

IMPATT diode. It also shows that the choice of the base semiconductor governs ART which in turn depends upon v_{sn}, v_{sp} and α_n, α_p.

Figure 1 shows that GaN based diode gives less ART with respect to 4H-SiC. It is also visible from the graph that operating frequency upto 5.0 THz is possible for GaN as the base material as its τ_A is of the order of 10^{-17} s while its TT is almost seven order more. Outcomes also indicate sufficient power at higher frequency of 1 THz.

5 Physical Parameters of 4H-Silicon Carbide and Wurtzite-Gallium Nitride

The significant physical parameters that control the steady state and RF characteristics of the IMPATT diode are rate of ionization of mobile carriers, drift velocities and mobilities which are strongly dependent on E-field and temperature. The reported experimental values of ionization rates of the mobile carriers, velocity of drift through the drift layer and mobilitiesof different base semiconductorsof IMPATT diodes i.e., Wz-GaN and 4H-SiC are used here. In this section these material parameters, field dependency and temperature dependencies are discussed. Johnson's and Baliga's figure of merit (FOM) [26, 27] on power and frequency parameters for terahertz application helps us to judge the RF properties of the device.

5.1 Ionization Rates

Rate of ionization for electrons and holes given as α_n and α_p corresponding to electrons and holes respectively are very significant in determining various static and higher frequency properties of IMPATT. These are also very much dependent on temperature and increase exponentially with electric field as obvious from Eq. (4). Rate of ionization is the probability of ionization of a mobile carrier per unit time and per unit distance travelled under the influence of electric field [28]. The rate of ionization is given by Shockley's low field model as defined below except for 4H-Silicon Carbide.

$$\alpha_{n,p}(\xi) = A_{n,p} \exp\left[\left(\frac{-B_{n,p}}{\xi}\right)^m\right]. \tag{4}$$

where, ξ is the electric field and the exponent, m is 1 for GaN or SiC. The ionization coefficients A_n, B_n and A_p, $\underline{B_p}$ are obtained from the experimental reports of Kunihiro et al. [29] and Konstantinov et al. [30] respectively at different electric field values as given in Table 1.

Ionization rates for 4H-Silicon Carbide are given below shows that they are rapidly increasing functions of temperature.

$$\alpha_n(\xi) = \left(\frac{\xi}{A_n}\right) \exp\left(\frac{-B_n}{\xi^2}\right). \tag{5}$$

$$\alpha_p(\xi) = \left(\frac{\xi}{7}\right) \exp\left[\frac{-1}{\left(A_p\xi^2 + B_p\xi\right)}\right]. \tag{6}$$

The following empirical relations for the field and temperature variations of these ratesfit very well with $W.\ N$ Grant's experimental results [21].

In the lower field range, 2.4×10^7 V m^{-1} < ξ < 5.3×10^7 V m^{-1},

Table 1 Coefficient of ionization for the charge carriersat 300 K

Semiconductor material	Field range, ξ ($\times 10^7$ V m^{-1})	A_n ($\times 10^9$ m^{-1})	B_n ($\times 10^9$ V m^{-1})	A_p ($\times 10^9$ m^{-1})	B_p ($\times 10^9$ V m^{-1})
Wz-GaN	4.00–10.00	13.8000	1.4280	0.6867	0.8720
	>10.00	12.2700	1.3630	0.3840	0.7950
4H-SiC	1.00–10.00	10.0000	4.0268	4.1915	4.6428

$$\alpha_n(\xi, T) = 6.2 \times 10^7 \left[\exp\left\{ -\frac{(1.08 \times 10^8 + 1.3 \times 10^5(T - 22))}{\xi} \right\} \right]$$
$$\alpha_p(\xi, T) = 2.0 \times 10^8 \left[\exp\left\{ -\frac{(1.97 \times 10^8 + 1.1 \times 10^5(T - 22))}{\xi} \right\} \right]. \quad (7)$$

In the higher field range, 5.3×10^7 V m^{-1}<ξ< 7.7×10^7 V m^{-1},

$$\alpha_n(\xi, T) = 5.0 \times 10^7 \left[\exp\left\{ -\frac{(9.90 \times 10^7 + 1.3 \times 10^5(T - 22))}{\xi} \right\} \right]$$
$$\alpha_p(\xi, T) = 5.6 \times 10^7 \left[\exp\left\{ -\frac{(1.32 \times 10^8 + 1.1 \times 10^5(T - 22))}{\xi} \right\} \right]. \quad (8)$$

where, T is the junction temperature in Kelvin. The junction temperature of IMPATT diode operating in CW mode rises above the ambient temperature. The rise of junction temperature can be thermally modeled by proper heat sink arrangement to keep it below or close to 500 K.

5.2 *Drift Velocity of Charge Carriers*

Drift velocity vs. field variation shows negative mobility in Wz-GaN [31] which is considered for simulation of the diode uses the following relation.

$$v_n(\xi) = \frac{\left[\mu_n \xi + v_{sn}(\xi/\xi_c)^4\right]}{\left[1 + (\xi/\xi_c)^4\right]}. \quad (9)$$

where, ξ is any value of electric field and ξ_c is the critical field at which electron reaches the peak velocity before saturation. In 4H-SiC, carrier drift velocity is given below [32].

$$v_{n,p}(\xi) = \left[\frac{\mu_{n,p}\xi}{\left(1 + \left(\mu_{n,p}\xi / v_{sn,sp}\right)^{\kappa}\right)}\right]^{1/\kappa}. \quad (10)$$

where, the constant $\kappa = 1.20$. Saturated drift velocities of the mobile carriers are strongly temperature-dependent.

Table 2 shows the experimental values of v_{sn} and v_{sp} for 4H-SiC andWz-GaN at 300 K.

Table 2 Saturated drift velocities in Wz-Gallium Nitride and 4H-Silicon Carbide

Saturated drift velocity	$T = 300$ K	
	Wz-GaN	4H-SiC
v_{sn} ($\times 10^5$ m s^{-1})	3.0000	2.1200
v_{sp} ($\times 10^5$ m s^{-1})	0.7500	1.0800

5.3 Various Important Parameters

Table 3 gives the values of other important material parameters for Wz-GaN and 4H-SiC which are available in the published reports [22, 23]. Those important parameters are: (i) the bandgap energy (E_g), (ii) relative permittivity (ε_r), (iii) intrinsic carrier concentration (n_i), (iv) effective density of states of conduction and valance bands (N_c, N_v), (v) density of state effective mass ($m_d{}^*$), (vi) mobilities (μ_n, μ_p), (vii) diffusion coefficients (D_n, D_p), (viii) diffusion lengths (L_n, L_p) and (ix) critical field corresponding to the peak drift velocity (ξ_c) of electrons in 4H-SiC and Wz-GaN.

Table 3 Other Important parameters of Wz-GaNand 4H-SiC

Material parameter	$T = 300$ K	
	Wz-GaN	4H-SiC
E_g (eV)	3.4691	3.1934
$m_d{}^*$ ($\times m_0$)	1.5000	0.7700
ε_r	10.4000	8.5884
n_i ($\times 10^{13}$ m^{-3})	3.6114×10^{-17}	0.0161×10^{-13}
N_c ($\times 10^{25}$ m^{-3})	0.2234	1.6887
N_v ($\times 10^{25}$ m^{-3})	4.6246	2.4942
μ_n (m^2 V^{-1} s^{-1})	0.1000	0.1000
μ_p (m^2 V^{-1} s^{-1})	0.0034	0.0100
D_n ($\times 10^{-4}$ m^2 s^{-1})	2.6000	2.7100
D_p ($\times 10^{-4}$ m^2 s^{-1})	0.8798	2.5875
L_n ($\times 10^{-6}$ m)	6.5000	12.0000
L_p ($\times 10^{-6}$ m)	2.1000	1.5000
ξ_c ($\times 10^5$ V m^{-1})	0.5000	–

– $m_0 = 9.1 \times 10^{-31}$ Kg is the rest mass of an electron
$\varepsilon_s = \varepsilon_r \varepsilon_0$ is the permittivity of the semiconductor material; where $\varepsilon_0 = 8.85 \times 10^{-12}$ F m^{-1} is the permittivity of vacuum
$m_d{}^*$ represents "effective mass" of the material which is different from the rest mass of the material

6 Design and Methodology

In the first step, the parameters controlling the structural design and doping ofGaN IMPATTs are preconceived at the desired THz frequency. Sze and Ryder's empirical formula given below will be used to estimate the input doping parameters corresponding to the design frequency and base semiconducting material of IMPATT didoes. The formula is:

$$W_{n,p} = 0.37\, v_{sn\,,\,sp}/f. \tag{11}$$

where, $W_{n,p}$ = depletion layer thickness for the n and p sides, $v_{sn,sp}$ = carrier saturation velocity and f = operating frequency.

6.1 Device Model

Classical drift–diffusion model is used preferably over energy relaxation model for simulation of DC and RF properties GaN based DDR IMPATT as both the models provide comparable result so far as execution of the device performance at high frequency is concerned as reported by Dalle et al. [33].

A device model will be used in the simulation software to obtain the aforementioned dc properties. The computation is initiated from the location where electric field gives the peak value in the active layer depleted of mobile carriers near p–n junction. Concentration of impurities per unit volume in the n-side and p-side of the junction are considered as the input parameters assuming symmetrical doping profile, doping functions in the n and p regions are uniform and those at the epitaxy-substrate interfaces are considered to be exponential functions given as:

$$\begin{aligned} N(x) &= N_{n^+}\exp\left(-1.08\lambda_n(x) - 0.78\lambda_n(x)^2\right) && x \leq 0 \\ &= N_D && 0 > x \geq x_1 \\ &= N_D\left[1 - \exp\left(\tfrac{x}{s}\right)\right] && x_1 > x \geq x_j \\ &= -N_A\left[\exp\left(-\tfrac{x}{s}\right) - 1\right] && x_j > x \geq x_2 \\ &= -N_A && x_2 > x \geq W \\ &= -N_{p^+}\exp\left(-1.08\lambda_p(x) - 0.78\lambda_p(x)^2\right) && x > W. \end{aligned} \tag{12}$$

where, N_A is the acceptor concentration and W_p is the width of p-epitaxial layer, x_j is the junction location measured from the surface.

Near the junction the doping profiles are considered to be steeply rising exponential functions because the junctions are produced by MBE (Molecular beam epitaxy) technique. The important physical parameters governing the material behavior are carrier saturation velocity, rate of ionization of the carrier, permittivity of the semiconductors, diffusion constant, carrier mobility etc.

6.2 DC Analysis

The important one-dimensional device equations under dc conditions will be solved numerically. These equations are Poisson's equation, Carrier transport equation, current density equations and space charge equations. The effects of cloud of carrier charge accumulated in the active layer and carrier diffusion due to concentration gradient will be incorporated in the simulation program. All these equations are simultaneously solved numerically by using Finite Difference Technique. A double iterative field maximum method reported in [34] will be used here. The method used for simulation of the steady state characteristics of the device is based on double iteration commencing from the location of the maximum electric field [34].

The computation starts from the point having peak electric field value, then it continues to the rightmost extremity of the active layer, returning to the centre peak field position and then pursuing the same to the leftmost extremity of the active layer (Fig. 2).

The next step is to test whether the boundary conditions of the electric field and the difference of carrier current density normalized with respect to the total current density match with those obtained from the computation at the boundaries of the depletion layer. Iteration is done over two parameters: location of the peak field and the peak field value subject to proper limiting conditions at the edges of the active layer and this continues till the conditions are satisfied. The simulation program enables the convergence to reach within 20 iterations. Once the convergence is obtained, the solution for distributions of field, $\xi(x)$ and current normalized with respect to the maximum value, $P(x)$ can be obtained. The avalanche zone width can also be obtained from the current density profile.

E-field integrated over the entire width of the active region gives the breakdown voltage V_B and that integrated over the thickness of the avalanche zone gives the avalanche layer voltage V_A. The drift zone voltage V_d is the difference of V_B and V_A. Efficiency of conversion from static to high frequency can be obtained using Scharfetter-Gummelformula given below:

$$\eta = 2m/\pi \times V_D/V_B \tag{13}$$

where, m = modulation factor, V_D = drift voltage and V_B = breakdown voltage.

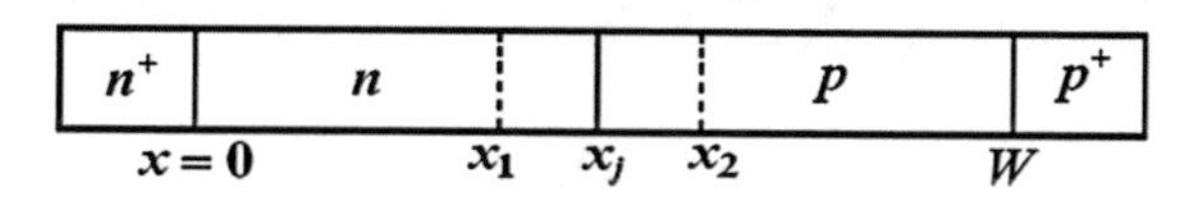

Fig. 2 1-D structure of DDR IMPATT diode

6.3 Analysis for Small Amplitude RF Signal

The analysis for small amplitude RF signal will be carried out using the dc parameters obtained from the dc analysis by using Gummel-Blue model described in [6]. The admittance (G-B; G-conductance, B-susceptance) plots and the negative impedance (R and X; R-resistance and X-reactance) plots will be produced for this small amplitude signal by varying the external current excitation. The plots will provide a deep physical insight into the RF power generation in the depletion layer. The frequency at which the negative G reaches the maximum value is the threshold optimum frequency while the frequency at which B makes a transition from positive (inductive) to negative (capacitive) is the avalanche resonance frequency. For oscillation to occur, negative G and positive B condition must be fulfilled. Integration of the impedance profiles over the active layer using numerical method gives the total resistance and reactance from which the G and B can be obtained using the equations given below. Thus,

$$Z_R(\omega) = \int_0^W R(x,\omega)\mathrm{d}x; \quad Z_X(\omega) = \int_0^W X(x,\omega)\mathrm{d}x. \tag{14}$$

$Z_D(\omega)$ and $Y_D(\omega)$ is obtained from Eqs. (15) and (16) respectively as a function of a specified frequency $\omega = 2\pi f$ and for an injected current density J_0.

$$Z_D(\omega) = \int_0^W Z(x,\omega)\mathrm{d}x = Z_R(\omega) + i\,Z_X(\omega). \tag{15}$$

$$Y_D(\omega) = \frac{1}{Z_D(\omega)} = G(\omega) + i\,B(\omega) = \frac{1}{(Z_R(\omega) + i\,Z_X(\omega))}. \tag{16}$$

The small signal simulation provides the admittance characteristics of the device which enables determination of the peak optimum frequency, f_p corresponding to peak negative conductance, G_p. Then the negative resistance plots and Q-factor are simulated at the peak frequency. Immediate necessary conditions for obtaining proper design parameters are procured from the outcomes of the device simulation at millimeter and terahertz frequency regime by modulating the injected current density. The expressions given below helps procurement of the conductance (−ve) and susceptance (+ve).

$$\left.\begin{aligned} |-G(\omega)| &= Z_R(\omega)\big/\left(Z_R(\omega)^2 + Z_X(\omega)^2\right) \\ |B(\omega)| &= -Z_X(\omega)\big/\left(Z_R(\omega)^2 + Z_X(\omega)^2\right) \end{aligned}\right\}. \tag{17}$$

Power obtained at THz frequency operation is given as

$$P_{RF} = \frac{1}{2}(V_{RF})^2 |G_p| A_j. \tag{18}$$

where, V_{RF} (=$m_x V_B$) is the voltage obtained at the particular THz frequency, m_x is the percentage of modulation (<30%), V_B is the static voltage at breakdown and A_j is the circular cross-sectional area of the device.

6.4 *Large Signal Analysis*

The above analysis under low signal amplitude gives an immediate idea of the design of the device and the various static and high frequency properties but under actual operating conditions of the device embedded into the resonant cavity, the large signal analysis is necessary to obtain power and efficiency at different voltage modulation.

NSVE model which uses a non-sinusoidal voltage to excite the device is used in this analysis. This voltage is given by

$$v_{rf}(t) = V_B \sum_{p=1}^{n} (m_x)^p \sin(2p\pi f_d t). \tag{19}$$

where, m_x is the percentage of modulation which is the amount of voltage swing over the breakdown component of voltage. The primary component of the ac voltage corresponding to $p = 1$ is given by $v_{rf1}(t) = m_x V_B \sin(2\pi f_d t)$. In Eq. (19), f_d is the primary component of frequency which is exactly the same as the frequency at which the device has been designed.

Figure 3 shows the IMPATT device controlled by the ac voltage $v_{rf}(t)$, current excitation I_0 serving as the biasing input which is A_j times the current density J_0 and C is the capacitor which couples the response of the device to the output nodes. The time varying voltage, $v_{rf}(t)$ is applied due to which the current response of the device is $J_t(t) = J_0 + j_t$. The current response is obtained from large signal simulation.

The structural components i.e., $W_n, W_p, N_D, N_A, N_{n+}, N_{p+}$, the bias current density (J_0), the design frequency f_d and the physical parameters of the base material serve as the feed in parameters for the simulation.

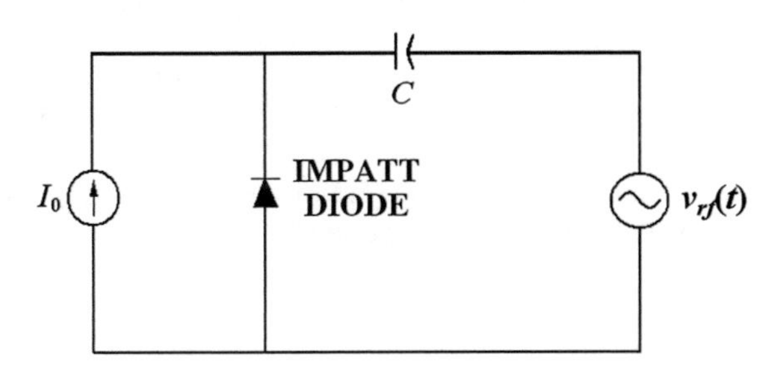

Fig. 3 NSVE model of IMPATT diode oscillator

Steady state parameters are obtained at the very outset by considering zero modulation factor which leads to zero ac voltage which finally makes the basic device equations to be independent of time but dependent on space which can be solve by numerical method obeying the conditions to be satisfied at the edges of the active layer. Different static parameters obtained from the dc simulation are ξ_P, V_B, V_A, x_A at the operational frequency f_d.

Taking the static parameters into account, simulation with large amplitude signal commences with an initial value of voltage modulation factor, m_x. The large-signal program continues till a full phase cycle is attained. Then this continues for some consecutive phase cycles to prove the steadiness of oscillation. A current excitation is provided to bias the device at a fixed value. The amplitude of the primary component of ac voltage is V_{RF}. The snap-shots of electric field, terminal voltage, carrier current are obtained at various junctures of time and phase for the complete phase cycle. Fourier analysis of the end-stage voltage and current over a complete cycle gives the necessary frequency domain information. The admittance profile gives many important parameters under large amplitude ac signal as is provided in the result section. Modulation percentage is risen in paces of Δm_x where $\Delta m_x = 0.05$, i.e. 5% and this continues till the terminating value of m_x i.e. 0.7 or 70% is attained. Eventually the important parameters will be obtained as described below.

Impedance of the device is obtained as

$$Z_D(\omega) = \frac{1}{Y_D(\omega)} = \frac{1}{[G(\omega) + jB(\omega)]A_j} = Z_R(\omega) + i\,Z_X(\omega). \tag{20}$$

where, $Z_R(\omega)$ and $Z_X(\omega)$ are

$$Z_R(\omega) = \frac{G(\omega)}{\left[G(\omega)^2 + B(\omega)^2\right]A_j}; \quad Z_X(\omega) = \frac{-B(\omega)}{\left[G(\omega)^2 + B(\omega)^2\right]A_j} \tag{21}$$

The AC output power is

$$P_{RF} = \frac{1}{2}(V_{RF})^2 |G_p| A_j. \tag{22}$$

where, $|G_p|$ is the amplitude of peak conductance for large amplitude signal. Efficiency of conversion for the same large amplitude signal is

$$\eta_L = \frac{P_{RF}}{P_{DC}}. \tag{23}$$

The DC input power is given by $P_{DC} = J_0 V_B A_j$, where, J_0 is the static current value used for biasing the device.

7 Millimeter Wave and THz Simulation Results

The different parameters obtained as large signal simulation output are shown in Table 4.

The admittance profiles for 4H-SiC and Wz-GaN at different millimeter wave and THz frequencies are shown in Figs. 4, 5, 6 and 7. Table 4 shows that the *Q*-factor ($Q_p = -B_p/G_p$) for 4H-SiC is lower than Wz-GaNand is quite near to unity (<1) for 4H-SiC which indicates better rise of oscillation and better efficiency of conversion from dc to ac. The results shown in Table 4 plays a vital role in selection of the proper base material for DDR IMPATT diode and design the device effectively for THz operation.

Table 4 also shows that 4H-SiC delivers more RF power as compared to Wz-GaN as obvious from the negative resistance values which are higher for 4H-SiC at all the frequencies considered which indicates acceptability of the materials chosen to form the base layer of the device in the THz domain. Simulation results in Table 4 also proves the worthiness of Wz-GaN as the base material for IMPATT diode above 1 THz as a high power device. Well within the THz range, not even SiC can keep pace with GaN as the desired base material for IMPATT.

8 Conclusion

Finally, the prospective of WBG materials like 4H-Silicon Carbide and Wz-Gallium Nitride as the base semiconductor for DDR IMPATT device is quite apparent from the various outcomes of MATLAB simulation and the Avalanche Response Time (ART) analysis. ART analysis is very significant to find the limiting frequency for a particular base material to be used for the IMPATT diode. ART analysis shows that avalanche response time (ART) should be less than transit time (TT) for proper operation of the device and for obtaining RF power output. ART of GaN is about 10^{-16} s and the TT is nearly five order higher which warrants operation of the device as far as 5 terahertz. Hence, it can be inferred that modest amount of power is procured from GaN based IMPATT diode even above 1.0 THz but at 1.0 THz or below that (300 or 500 GHz), both 4H-SiC and Wz-GaN based IMPATT diode act as reliable source of RF power. Simulation results also show that below 1.0 THz, 4H-SiC has lower Q-factor (near to 1.0) which is required for better growth rate of oscillation and also more Z_R (negative resistance) which is needed for more power output. Operation principle of IMPATT diode, design methodology in dc, small signal and large signal domain have been presented in this chapter too. A rigorous review has been given on the various works previously done on IMPATT diode. So, it is assured form the various simulation results that at lower THz domain, 4H-SiC gives better performance but above 1.0 THz, Wz-GaN is almost without competition with any other base material, even it can go to much higher THz which is limited for 4H-SiC. Simulation outputs presented in this chapter infer that 4H-Silicon Carbide

Table 4 Significant RF parameters

Base material	Serial NUMBER	f_p (GHz)	G_p ($\times 10^7$ Sm^{-2})	B_p ($\times 10^7$ Sm^{-2})	$Q_p = -(B_p/G_p)$	Z_R ($\times 10^{-9}$ Ωm^{-2})	P_{RF} (mW)	D_j (μm)
4H-SiC	1	94	−1.9711	0.5574	0.28	−46.9756	30,290.98	35.0
	2	140	−4.2152	1.3511	0.32	−21.5132	16,180.81	25.0
	3	220	−9.1919	3.7236	0.41	−9.3454	13,315.09	20.0
	4	300	−20.4917	8.3277	0.41	−4.1883	13,130.85	15.0
	5	500	−51.1584	14.350	0.28	−1.8121	9355.78	10.0
	6	1000	−121.6192	46.149	0.38	−0.7187	1982.31	5.0
Wz-GaN	1	94	−0.1638	1.7788	10.86	−5.1361	3117.78	35.0
	2	140	−0.3953	3.7869	9.58	−2.7262	2818.41	25.0
	3	220	−1.1101	9.1187	8.21	−1.3155	2622.73	20.0
	4	300	−2.3439	16.9007	7.21	−0.8051	1806.27	15.0
	5	500	−8.1282	46.659	5.74	−0.3632	1162.62	10.0
	6	1000	−47.1109	180.2237	3.82	−0.1358	504.16	5.0
	7	1500	−126.7654	408.2341	3.22	−0.0693	192.09	2.5
	8	2000	−211.6412	730.3715	3.45	−0.0366	47.54	1.2
	9	5000	−445.7865	1045.0674	2.34	−0.0345	7.62	0.7

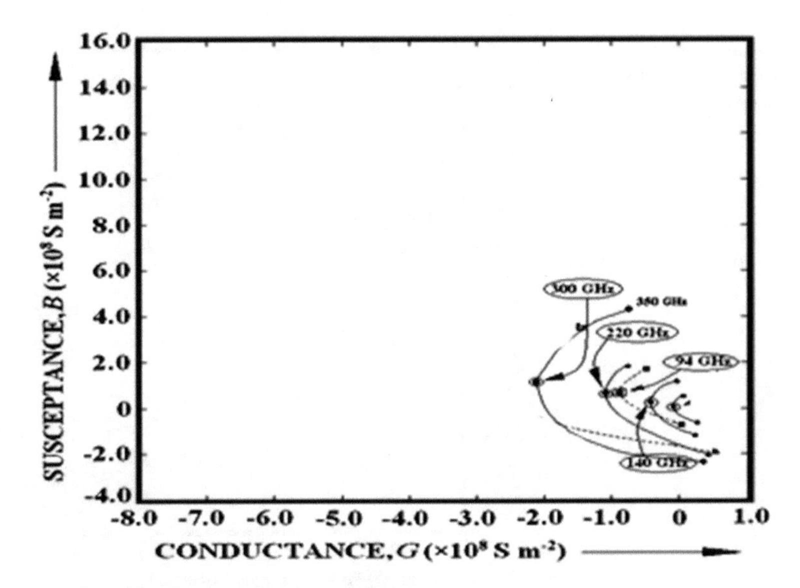

Fig. 4 Admittance profiles for 4H-SiC at millimeter wave frequencies

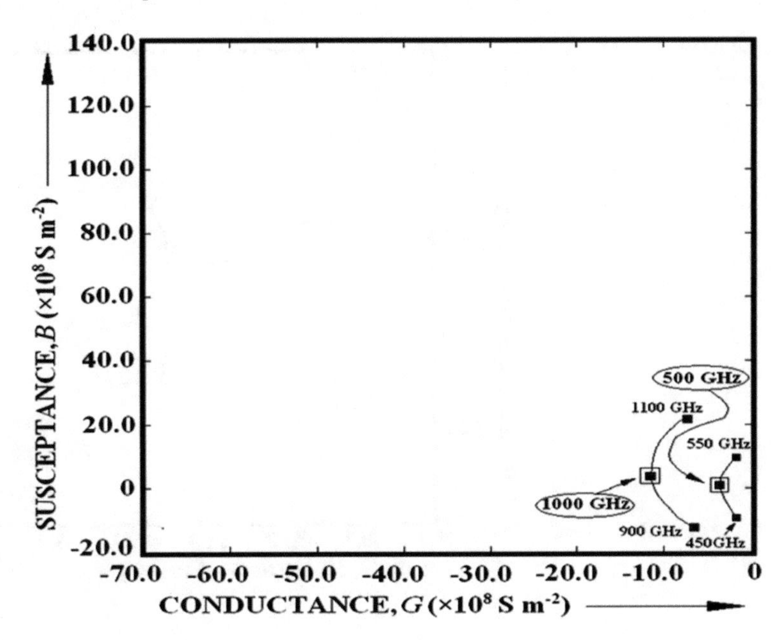

Fig. 5 Admittance profiles for 4H-SiC for THz frequencies

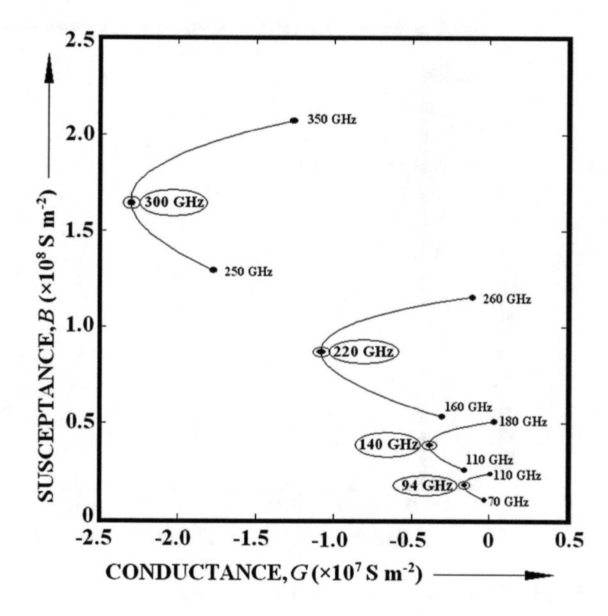

Fig. 6 Admittance profiles for Wz-GaN for millimeter wave frequencies

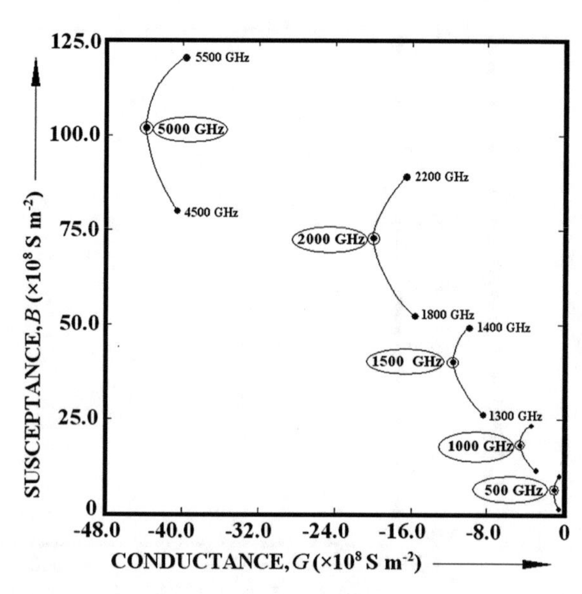

Fig. 7 Admittance profiles for Wz-GaN for THz frequencies

and Wz-Gallium Nitride will be of immense use for fabrication, experimentation and practical realization of double drift layered IMPATT diode for operation in the terahertz regime.

References

1. W. Shockley, Negative resistance arising from transit time in semiconductor diode. Bell. Syst. Tech. J. **33**, 799 (1954)
2. W.T. Read, A proposed high-frequency negative-resistance diode. Bell. Syst. Tech. J. **37**, 401 (1958)
3. C.A. Lee, R.L. Batdrof, W. Wiegmann, G. Kaminsky, The Read diode-an avalanching transit time negative resistance oscillator. Appl. Phys. Lett. **6**, 89 (1965)
4. R.L. Johnston, B.C. Deloach, B.G. Cohen, A silicon diode microwave oscillator. Bell Syst. Tech. J. **44**, 369 (1965)
5. T. Misawa, The negative resistance in p-n junctions under avalanche breakdown conditions, Part I and II. IEEE Trans. Electron Devices **ED-13**, 137 (1966)
6. H.K. Gummel, J.L. Blue, A small-signal theory of avalanche noise in IMPATT diodes. IEEE Trans. on Electron Devices **14**, 569–580 (1967)
7. D.L. Scharfetter, W.J. Evans, H.L. Johnson, Double drift region p^+pnn^+ avalanche diode oscillators. Proc. IEEE (Lett.) **50**, 1131 (1970)
8. J. Ward, E. Schlecht, G. Chattopadhyay, A. Maestrini, J. Gill, F. Maiwald, H. Javadi, I. Mehdi, Capability of THz Sources based on Schotiky diode frequency multiplier chains, in *IEEE MTT-S Digest*., pp. 1587–1590 (2004)
9. R. Lai, X. Mei, W. Deal, W. Yoshida, Y. Kim, P. Liu, J. Lee, J. Uyeda, V. Radisic, M. Lange, T. Gaier, L. Samoska, A. Fung, Sub 50 nm InP HEMT device with f_{max} greater than 1 THz, in *Proceedings of the IEEE International Electron Devices Meeting*, 2007, pp. 609–611
10. M. Urteaga, M. Seo, J. Hacker, Z. Griffith, A. Young, R. Pierson, P. Rowell, A. Skalare, M. Rodwell, InP HBT integrated circuit technology for terahertz frequencies, in *Proceedings of the IEEE Compound Semiconductors Integrated Circuit Symposium*, 2010, pp. 1–4
11. M. Mukherjee, N. Mazumder, S.K. Roy, Prospects of 4H-SiC double drift region IMPATT device as a photo-sensitive high-power source at 0.7 Terahertz frequency regime. Act. Passive Electron. Compon. **2009**, 1–9 (2009)
12. A. K. Panda, R. K. Parida, N. C. Agarwala, G. N. Dash, A comparative study on the high band gap materials(GaN and SiC)-based IMPATTs, in *Proceedings of the Asia-Pacific Microwave Conference*, pp. 1–4 (2007)
13. A.K. Panda, D. Pavlidis, E. Alekseev, DC and high-frequency characteristics of GaN-based IMPATTs. IEEE Trans. Electron Devices **48**, 820–823 (2001)
14. S. Banerjee, M. Mukherjee, J.P. Banerjee, Bias current optimization of Wurtzite-GaN DDR IMPATT diode for high power operation at THz frequencies. Int. J. Adv. Sci. Technol. **16**, 12–20 (2010)
15. A. Acharyya, J.P. Banerjee, Prospects of IMPATT devices based on wide bandgap semiconductors as potential terahertz sources. Appl. Nanosci. **4**, 1–14 (2014)
16. A. Acharyya, J. P. Banerjee, Potentiality of IMPATT devices as terahertz source: an avalanche response time based approach to determine the upper cut-off frequency limits. IETE J. Res. (in press, publication schedule: March-April 2013)
17. L. Yuan, A. James, J.A. Cooper, M.R. Melloch, K.J. Webb, Experimental demonstration of a silicon carbide IMPATT oscillator. IEEE Electron Device Lett. **22**, 266–268 (2001)
18. K. V. Vassilevski, A. V. Zorenko, K. Zekentes, K. Tsagaraki, E. Bano, C. Banc, A. Lebedev, 4H-SiC IMPATT diode fabrication and testing, in *Technical Digest of International Conference on SiC and Related Materials*, Tsukuba, Japan, pp. 713–714 (2001)
19. R.J. Trew, J.B. Yan, P.M. Mock, The potentiality of diamond and SiC electronic devices for microwave and millimeter-wave power applications. Proc. IEEE **79**(5), 598–620 (1991)
20. P.M. Mock, R. J. Trew, RF performance characteristics of double-drift MM-wave diamond IMPATT diodes, in *Proceedings of the IEEE/Cornell Conf. Advanced Concepts in High-Speed Semiconductor Devices and Circuits*, pp. 383–389 (1989)
21. W.N. Grant, Electron and hole ionization rates in epitaxial Silicon. Solid State Electron. **16**, 1189–1203 (1973)

22. Electronic Archive: New Semiconductor Materials, Characteristics and Properties. Available from: http://www.ioffe.ru/SVA/NSM/Semicond/index.html, August 2014
23. B.V. Zeghbroeck, *Principles of Semiconductor Devices* (Colorado Press, USA, 2011)
24. C. Canali, G. Ottaviani, A.A. Quaranta, Drift velocity of electrons and holes and associated anisotropic effects in silicon. J. Phys. Chem. Solids **32**, 1707–1720 (1971)
25. S.M. Sze, R.M. Ryder, Microwave avalanche diodes, in *Proceedings of the IEEE, Special Issue on Microwave Semiconductor Devices*, vol. 59, pp. 1140–1154 (1971)
26. E.O. Johnson, Physical limitations on frequency and power parameters of transistors. RCA Rev. **26**, 163–177 (1965)
27. B.J. Baliga, Power semiconductor device *figure of merit* for *high-frequency* applications. Electron Device Lett. **10**(10), 455–457 (1989)
28. G. Gibbons, *Avalanche-diode Microwave Oscillators* (Oxford University Press, Oxford, 1973), pp. 13, 53
29. K. Kunihiro, K. Kasahara, Y. Takahashi, Y. Ohno, Experimental evaluation of impact ionization coefficients in GaN. IEEE Electron Dev. Lett. **20**(12), 608–610 (1999)
30. A.O. Konstantinov, Q. Wahab, N. Nordell, U. Lindefelt, Ionization rates and critical fields in 4H-Silicon Carbide. Appl. Phys. Lett. **71**, 90–92 (1997)
31. S.C. Shiyu, G. Wang, High-field properties of carrier transport in bulk wurtziteGaN: Monte Carlo perspective. J. Appl. Phys. **103**, 703–708 (2008)
32. K.V. Vassilevski, K. Zekentes, A.V. Zorenko, L.P. Romanov, Experimental determination of electron drift velocity in 4H-SiC p^+-n-n^+ avalanche diodes. IEEE Electron Dev. Lett. **21**, 485–487 (2000)
33. C. Dalle, P.A. Rolland, Drift-diffusion versus energy model for millimetre-wave IMPATT diodes modelling. Int. J. Numer. Model. Electron. Networks Dev. Fields **2**, 61–73 (1989)
34. S.K. Roy, M. Sridharan, R. Ghosh, B.B. Pal, Computer method for the dc field and carrier current profiles in the IMPATT device starting from the field extremum in the depletion layer, in *Proceedings of the 1st Conference on Numerical Analysis of Semiconductor Devices (NASECODE I)*, ed. By J.H. Miller, Dublin, Ireland, pp. 266–274 (1979)

Section II: Prospect of Heterojunction (HT) IMPATT Devices as a Source of Terahertz Radiation

Suranjana Banerjee

Abstract This chapter will present the outcomes of inspection of the IMPATT device under the influence of large amplitude RF signal in Terahertz (THz)regime with heterojunction (HT) semiconductor. The significant RF paarmeters derived from this inspection will be useful to authenticate the worthiness of the device for operation in the THz domain. As already mentioned in Section I, tunneling greatly degrades the high frequency performance of homojunction (HM) devices which can be abated by the introduction of heterojunctions. Avalanche noise reduction is another important factor leading to the choice of heterojunctions. The first HT semiconductor taken is 3C-SiC ~ Si whose RF characteristics will be analysed at 300 and 500 GHz. The other HT combination taken for study at 1 THz is $Al_mGa_{1-m}N$ ~ GaN with composition of Al or mole fraction of Al in the ternary alloy is 40% i.e., $m = 0.4$.RF performances of these materials are compared with their homojunction equivalents. Results show that the HT devices outstandsHM devices concerning efficiency of conversion and the power obtained at the output due to the impact of large amplitude ac signal in Terahertz domain. These studies will further enable researchers and analyzers to practically realize these heterojunction devices for producing reliable terasource.

Keywords Heterojunction · High frequency performance · Dc to RF conversion efficiency

1 Introduction

Heterojunctions are formed when two semiconductors of different bandgaps and lattice constants are brought together, generally by epitaxy to provide carrier confinement. These heterojunctions, if incorporated in IMPATT diode can produce better RF and noise performance of the device. The important characteristic of heterojunction (HT) is the formation of depletion-accumulation region at the interface of the two semiconductors which leads to conduction band and valence band discontinuity and intensity of the maximum value of electric field at the junction of the diode.

S. Banerjee (✉)
Department of Electronics, Dum Dum Motijheel College, South Dumdum, India

S. Das et al. (eds.), *Advances in Terahertz Technology and Its Applications*,
https://doi.org/10.1007/978-981-16-5731-3_10

Tunneling causes deterioration of RF performance of homojunction (HM) IMPATT devices at higher frequencies [as mentioned in Section I]. But if HM IMPATTs are replaced by HT devices then the RF performance improves considerably due to significant reduction of tunneling current. Further, HT IMPATTs have lower noise figure than their HM counterparts. Various reports [1–5] have established significant performance improvements, superior dc characteristics and less phase noise of heterojunction IMPATT devices. But a high voltage is required for HTIMPATTs to actuate avalanche breakdown [6] caused by saturation current in the reverse biased condition. But the transformation from normal operating condition to breakdown is not well-defined in a standard homojunction device since different regions of the diode and the avalanche zone are involved in the carrier generation process under reverse biased condition while the transformation in heterojunction device is quite abrupt due to the narrowing of the avalanche layer. Report [7] provides theoretical study on heterojunction IMPATT which shows its proficiency over its homojunction equivalent concerning efficiency of conversion and noise generated due to avalanche multiplication phenomenon. The above reports are encouraging to enhance the RF performance of HT diode. With this objective, the author has included the end results of MATLAB simulation of the HT diode under the influence of large amplitude ac signal taking different base semiconductors like 3C-SiC ~ Si and $Al_mGa_{1-m}N$ ~ GaN. These outcomes are obtained considering the concentration of Al in AlGaN i.e., m to be 40% or 0.4 at terahertz frequencies.

2 Potentiality of 3C-SiC ~ Si and $Al_mGa_{1-m}N$ ~ GaN Based HT IMPATTs as Promising Sources of THz Radiation

At higher mm-wave windows above W-band and terahertz frequency regions, the performance of IMPATTs is limited by avalanche response as discussed previously in Section I. This limitation can be overcome by adopting suitable base semiconductor for the double drift layered IMPATT diode. The wider band gap of SiC and its material parameters facilitate terahertz frequency operation.Silicon Carbide (SiC) is a wide band-gap semiconductor (bandgap $E_g = 2.24$ eV at 300 K) with appreciable material characteristics like high voltage at breakdown, velocity attained by the charge carrier is very high at saturation point in the presence of very high electric field and thermal conductivity is also quite high. Thus, the material can be used to realize practically different high power-high frequency semiconductor devices like MESFETs, MOSFETs and IMPATTs. Various polymorphic crystal forms of SiC are available like 4H ~ SiC, 6H ~ SiC and 3C ~ SiC. Among these, fabrication of IMPATT became possible only with 4H ~ SiC. This particular polytype produced oscillation at microwave frequency as exhibited by Vassilevski et. al. [8] and Yuan et al. [9] but with nominal amount of power at the output. This observation, however deviated a lot from the theoretical and simulation studies [10, 11]. The possible reason behind this disparity is the series resistance and various parasitic resistances ensuing from the undepletedepilayer which is quite high. But if 3C ~ Silicon Carbide is used instead as the base semiconductor then the disparity emerging from the parasitic resistances will be substantially reduced due to higher mobility of carriers in 3C ~ Silicon Carbide. Further, 3C-SiC ~ Si HT has lower conduction band discontinuity than 4H-SiC ~ Si HTand therefore the formation of 3C-SiC ~ Si HT will be simpler. These revelations have motivated the author to carry on investigations of the double drift layered heterojunction IMPATT with 3C polytype of SiC as the base semiconductor in the terahertz regime.

Another large bandgap material GaN with bandgap of 3.4 eV at room temperature is also a suitable material to act as the base semiconductor for IMPATT at terahertz frequencies owing to its various appreciative material properties like high voltage at breakdown, velocity attained by the charge carrier is very high at saturation point in the presence of very high electric field and mobility of the carriers is also quite high. Various reports [12, 13] confirm the capability of this material based diode to operate reliably in the THz realm. Its ternary alloy, $Al_mGa_{1-m}N$ also have wide band gap energy exceeding 3 eV. Thus, IMPATT diode with double drift layer and heterojunction subjected to $Al_mGa_{1-m}N$ ~ GaN base material can generate appreciable amount of power at terahertz frequency. Al in the ternary alloy $Al_mGa_{1-m}N$ has 40% composition i.e., m = 0.4. The HT structures considered here are *N*-$Al_mGa_{1-m}N$ ~ *p*-GaN and *n*-GaN ~ *P*-$Al_mGa_{1-m}N$. Reports [14, 15] confirm the use of this particular heterojunction in various semiconductor devices operational at high frequency. This influenced the usage of heterojunction in IMPATT diode and study its RF and avalanche noise properties in the terahertz domain. In this chapter, studies and discussions on the DC, RF and noise properties will be carried on DDR IMPATTs based on two complementary HTs: N-3C-SiC ~ p-Si and n-Si ~ P-3C-SiC are 3C-SiC ~ Si at two frequencies in THz band (0.3 and 0.5 THz) and those results will be compared with Si and 3C-SiC HM. The other HTsconsidered for discussion are N-$Al_{0.4}Ga_{0.6}$ N ~ p-GaN and n-GaN ~ P-$Al_{0.4}Ga_{0.6}$ N, then the outcomes of the whole inspection are compared with their HM equivalents. It is established from the end results that N-$Al_{0.4}Ga_{0.6}$ N ~ p-GaN HT outshines all other HTs and HMs considered for analysis of IMPATT at 1 THz. Therefore, these discussions divulge the competency of these heterostructures for realization of low noise-high power devices. Static and RF properties of IMPATT with large amplitude signal are obtained using the same methodology already discussed in Section I.

THz frequency band has numerous applications particularly in broadband wireless communication, THz imaging [16], spectroscopy [17] and astronomy [18], biosensing [19], quality inspection in various industries [20–22], medical and pharmaceutical [23, 24] applications. Still this frequency band remains uninspected due to dearth of dependable sources that can produce adequate RF power at terahertz frequency.. Not only that, detection and processing of RF signals are also indispensable for a complete THz system which is also not well-founded as per previous reports.

3 Device Design

The device structure is considered to be one-dimensional since the physical phenomenon takes place in the semiconductor bulk along the symmetry axis (i.e. x-axis) of the device. Complementary error function is considered at the juncture of the epitaxial layer and the substrate layer while exponential distribution at the metallurgical juncture which help to realize the doping distribution of simple double drift layered IMPATT. 1-D IMPATT structure is considered for the present discussion as shown in Fig. 1. Two complementary DDR HT structures of IMPATTs based on N-3C-SiC ~ p-Si and n-Si ~ P-3C-SiC are considered in this study along with their HM counterparts (Si and SiC). Similarly, the other heterostructures considered are: N-$Al_mGa_{1-m}N$ ~ p-GaN (HTDD1) and n-GaN ~ P-$Al_mGa_{1-m}N$ (where m = 0.4) (HTDD2) and homojunction devices based on GaN (HMDD1) and AlGaN (HMDD2).

W is the depletion layer width where $W = W_n + W_p$. The junction between the p and n side is located at $x = x_o$, the thickness of the avalanche region is x_A and that of the drift regions on both sides of the junction for electron and hole are d_n and d_p respectively as visible from Fig. 1.

The device representation considers the effect of the cloud of charge accumulating in the space charge layer. The design parameters (W_n, W_p, N_D, N_A) are primarily selected from the Sze-Ryder equation $W_{n,p} = 0.37$ vsn, sp/fd then they are optimized subject to largest efficiency of conversion at the design frequency and input current excitation J_0 using NSVE model for simulation [25] (previously discussed in Section I).

All material parameters required for simulation for the HTs considered here are taken from the reports [26–30]. The parameters describing the structure and the doping concentrations of HM and HT DDR IMPATTs as mentioned before are distinctly considered for design at 0.3 and 0.5 THz for 3C-SiC/Si and at 1.0 THz for $Al_mGa_{1-m}N$ ~ GaN which are tabulated in Tables 1a and b. A meticulous scrutiny is carried out to study the thermal performance of heat sink [31] required to limit the temperature increase at the junction to avoid thermal runaway. An iterative method [32] is used for the dc simulation of the HT and HM devices considered here. The method is laid on two factors: location of the maximum electric field and its value at that point. The effective junction areas (A_j) of the devices at different design frequencies are obtained from the above mentioned thermal analysis and given in Table 1a and b.

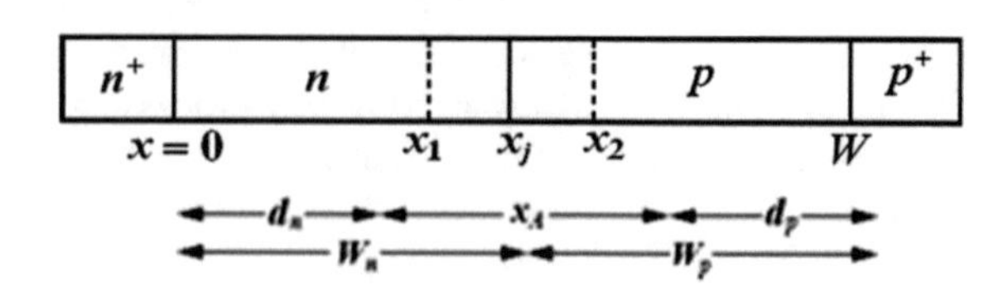

Fig. 1 D representation of double drift layer IMPATT

Table 1 **a** Design parameters, **b** Design parameters

(a) Design parameters

Device	f_d (GHz)	W_n (μm)	W_p (μm)	N_D (× 10^{23} m^{-3})	N_A (× 10^{23} m^{-3})	N_{n+} (× 10^{25} m^{-3})	N_{p+} (× 10^{25} m^{-3})	D_j (μm)
Si HMDDR	300	0.118	0.110	6.50	7.300	5.000	2.700	13.5
	500	0.068	0.066	15.500	16.700	5.000	2.700	10.0
3C-SiC HMDDR	300	0.160	0.160	13.500	13.500	5.000	2.700	13.5
	500	0.103	0.103	16.500	16.500	5.000	2.700	10.0
N-3C-SiC ~ *p*-Si HTDDR	300	0.220	0.210	3.100	4.600	5.000	2.700	13.5
	500	0.140	0.130	5.200	7.400	5.000	2.700	10.0
n-Si ~ *P*-3C-SiC HTDDR	300	0.114	0.125	5.000	4.920	5.000	2.700	13.5
	500	0.063	0.065	5.400	5.300	5.000	2.700	10.0

(b) Design parameters

Structure	Symbol	W_n (nm)	W_p (nm)	N_D (× 10^{23} m^{-3})	N_A(× 10^{23} m^{-3})	N_{n+} (× 10^{25} m^{-3})	N_{p+} (× 10^{25} m^{-3})	D_j (μm)
n-GaN ~ p-GaN HMDD	HMDD1	185.00	185.00	6.84	7.20	5.00	2.70	5.00
n-Al_mGa_{1-m} N~ p-GaN HTDD	HTDD1	174.00	185.00	6.97	7.15	5.00	2.70	5.00
n-GaN ~ p-$Al_mGa_{1-m}N$ HTDD	HTDD2	185.00	172.00	6.98	7.25	5.00	2.70	5.00
n-$Al_mGa_{1-m}N$ ~ p-$Al_mGa_{1-m}N$ HMDD	HMDD2	176.00	173.00	6.96	7.30	5.00	2.70	5.00

4 Simulated Outputs

End results of all the HM and HT structures considered here are designed using a drift–diffusion model and double-iterative field maximum simulation method [32]. The methodology is described in details in Section I. Primary device equations are solved considering the constraints at the boundary of the depletion layer [25] and substituting modulation factor to be zero to obtain the static properties of the device.

4.1 Results for HT 3C-SiC/Si and Its HM Counterparts

4.1.1 DC Properties

The static outcomes of the device obtained after simulation are listed below in Table 2.

Integration of the field profile over the entire space charge layer width ($-x_1$ to x_2) can be obtained by numerical method to give the voltage at breakdown i.e., V_B. Similarly, integration of the field profiles of the avalanche layer (x_{A1} to x_{A2}) gives the avalanche layer voltage drop i.e., V_A and the difference of these voltage drops i.e., $V_B - V_A$ is the drift layer voltage drop (V_D). At any design frequency, the simulated value of V_B is highest in 3C-SiC HM device and lowest in Si HM device. Between the two complementary HT devices, the magnitude of V_B is higher in N-3C-SiC ~ p-Si than in n-Si ~ P-3C-SiC. This holds good for all five design frequencies. Efficiency of conversion from static to RF can be estimated from (i) thickness of the avalanche layer (x_A) (ii) x_A/W and (iii) V_D/V_B. The calculated values of all these parameters are shown in Table 2. Similar to breakdown voltage, the magnitude of V_A is highest, one order higher in 3C-SiC HM DDR device followed by that in Si HM device, HT n-Si/P-3C-SiC and N-3C-SiC/p-Si heterojunctions. V_D/V_B ratiois however larger in HT devices than in their HM counterparts. Between the two HT devices, this ratio is slightly higher in n-Si ~ P-3C-SiC DDR IMPATTs. Again, the ratio x_A/W is less for HT devices than HM devices. The avalanche zone width of HT n-Si ~ P-3C-SiC is narrower than that of its HT counterpart. HT n-Si ~ P-3C-SiC has larger efficiency thanHT N-3C-SiC ~ p-Si due to constriction of the avalanche layer width and larger V_D/V_B ratio. An important observation from Table 2 is that the maximum junction field grows with rise of the design frequency while both V_B and V_A decrease for all the four HT considered here detailed study.

4.1.2 Large Signal Properties

Several notable parameters drawn out from the MATLAB simulation of IMPATT with Si ~ 3C-SiC HT and its equivalent HM under the impact of large amplitude ac

Table 2 DC parameters

Device	f_d (GHz)	J_0 ($\times 10^8$Am^{-2})	ξ_p ($\times 10^7$Vm^{-1})	V_B (V)	V_A (V)	V_D/V_B (%)	x_A (μm)	x_A/W (%)
Si HMDDR	300	26.500	9.3341	11.52	9.29	19.34	0.138	60.52
	500	59.000	11.2603	8.78	7.75	11.68	0.090	66.18
3C-SiC HMDDR	300	44.500	54.7200	122.77	74.16	39.59	0.174	54.38
	500	76.000	56.2550	93.49	61.74	33.96	0.136	66.02
N-3C-SiC/p-Si HTDDR	300	25.590	9.7311	20.99	7.78	62.92	0.116	26.98
	500	50.180	12.1445	16.76	6.94	58.61	0.092	34.07
n-Si/P-3C-SiC HTDDR	300	35.300	9.8871	15.84	5.78	63.51	0.061	25.31
	500	64.500	12.7040	11.31	4.64	58.97	0.039	30.08

signal with 60% modulation over the dc level at 0.3 and 0.5 THz are tabulated below in Table 3. These parameters are.

(i) f_p–optimum value of frequency
(ii) f_a—avalanche resonance frequency
(iii) $-G_p$—maximum negative conductance
(iv) B_p—maximum susceptance at the optimum frequency and
(v) Q_p—quality factor.

The large signal negative resistance per unit junction area(Z_R/A_j), powerobtained for the RF signal (P_{RF}) andefficiency of conversion ($\eta 3_L$) of the four HT structures are then obtained at the design frequencies. Table shows the above mentioned parameters at a voltage modulation of 60%.

The optimum frequency is close to the design frequency for all the structures as obvious from Table 3. This indicates that the design formulation is accurate. Table 3 also establishes the fact that the difference between optimum and avalanche frequency is lowest in HT n-Si ~ P-3C-SiC and highest in3C-SiC HM structure. Thus, the band width of IMPATT oscillation is narrowest in HTn-Si ~ P-3C-SiC and widest in3C-SiC HM structure.

Table 3 also shows that the $|G_p|$ is largest in HT n-Si ~ P-3C-SiC DDR structure followed by HM Si, HT N-3C-SiC ~ p-Si and HM 3C-SiC DDR structure. The Quality factor, Q is lowest in the HT N-3C-SiC ~ p-Si which is a good attestation of stable oscillation from the diode, henceforth the device can provide better RF performance.

Table 3 shows that highest peak CWoutput power (18.284 W) is obtained from HM 3C-SiC DDR device followed by HT devices based on n-Si ~ P-3C-SiC (1.360 W) and N-3C-SiC ~ p-Si (1.16 W). Among the four DDR HT and HM devices, HM Si DDR IMPATTs provide the lowest peak output power of 0.653 W.

The results given in Table 3 show that HT N-3C-SiC ~ Si double drift layer IMPATT delivers slightly lower power than its complementary counterpart due to the lower value of G_p in it. The results shown in Table 3 also reveal that the output power decreases with increasing optimum frequency. Figure 2 shows the plots of efficiency under the impact of large amplitude signal versus voltage modulation and optimum frequency respectively forHM Si and SiC, HT *N*-3C-SiC ~ *p*-Si and *n*-Si ~ *P*-3C-SiC with 60% voltage modulation. The efficiency of all four DDR structures of IMPATT device initially increases with increase of voltage modulation, attains peak value at about 60% voltage modulation and then decreases. These results ensure HT diodes to be more efficient than HM diodes.

Table 3 Parameters for large amplitude AC Signal for $m_x = 60\%$

Device	f_d (GHz)	f_a (GHz)	f_p (GHz)	G_p ($\times 10^7$ S m^{-2})	B_p ($\times 10^7$ S m^{-2})	$Q_p = -(B_p/G_p)$	Z_R/A_j ($\times 10^{-9}$ Ω m^{-2})	V_{RF} (V)	P_{RF} (mW)	η_L (%)
Si HMDDR	300	137.9	302.7	−5.0867	43.5257	8.56	−0.2649	6.91	173.9	3.97
	500	193.2	501.1	−9.3138	126.374	13.57	−0.0580	5.27	101.7	2.49
3C-SiC HMDDR	300	75.9	300.3	−0.6317	28.6743	45.39	−0.0768	73.66	2453.3	3.14
	500	96.7	502.6	−0.7585	76.3646	100.6	−0.0130	56.09	957.3	1.68
N-3C-SiC ~ p-Si HTDDR	300	206.4	302.1	−5.0963	15.4966	3.04	−1.9151	12.59	578.7	7.83
	500	284.2	502.3	−9.6537	46.5228	4.82	−0.4276	10.06	383.4	5.80
n-Si ~ P-3C-SiC HTDDR	300	179.7	300.3	−10.518	32.9046	3.13	−0.8814	9.50	680.0	8.49
	500	246.2	500.1	−22.875	125.278	5.48	−0.1413	6.79	413.7	7.22

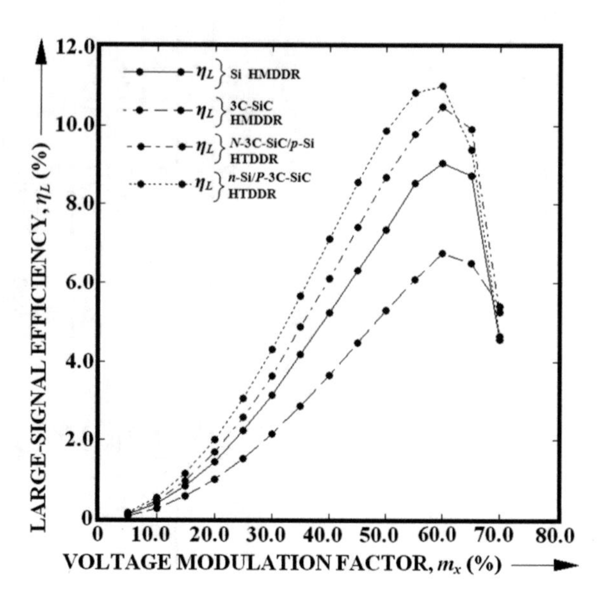

Fig. 2 Variations of efficiency of conversion for large amplitude RF signal for HM Si and SiC,HT *N*-3C-SiC ~ *p*-Si and *n*-Si ~ *P*-3C-SiC

4.2 *Results for HT $Al_{0.4}Ga_{0.6}$ N ~ GaN and Its HM Counterparts*

4.2.1 DC Properties

Table 4 shows the simulated DC parameters for heterojunction DDR IMPATTs based on *N*-$Al_mGa_{1-m}N$ ~ p-GaN (HTDD1) and n-GaN ~ P-$Al_mGa_{1-m}N$ (HTDD2) and homojunction devices based on GaN (HMDD1) and AlGaN (HMDD2). Table 4 shows that both V_B and V_A increase with injected input bias J_0. A fascination revelation is that for all injected bias currents, $Al_{0.4}Ga_{0.6}$ N HM gives maximum breakdown voltage while GaN HM provides the minimum. On the other hand, V_B for both HTs i.e., HTDD1 and HTDD2 lie in between the the HMs considered for all the bias currents. But the HT IMPATTs show lower voltage across the avalanche layer with respect to the HM IMPATTs as avalanche layer gets constricted in heterojunctions. V_D/V_B ratio is higher in heterojunctions with respect to homojunctions as perceived from Table 4 which leads to larger efficiency of HTs over HMs since efficiency is directly related to V_D/V_B as stated by the formula $\eta = (1/\pi) \times (V_D/V_B)$ [7]. Furthermore, x_A/W is lower in heterojunctions with respect to homojunctions as it is already revealed that avalanche layer thickness x_A gets reduced in HTs. This constriction of avalanche layer restrains random carrier generation by avalanche breakdown, thus less noise is generated in heterojunctions.

Table 4 Simulated DC parameters

Structure	Symbol	J_0 ($\times 10^8$ Am^{-2})	ξ_P ($\times 10^7$ Vm^{-1})	V_B (V)	V_A (V)	V_D/V_B (%)	x_A (nm)	x_A/W (%)
n-GaN ~ p-GaN	HMDD1	25.00	21.321	28.65	14.74	48.57	121.0	32.58
		30.00	21.245	29.24	14.90	49.05	122.2	33.04
		35.00	21.145	29.75	15.58	47.63	129.9	35.08
		40.00	21.045	30.18	16.29	46.03	137.5	37.15
		45.00	20.945	30.64	16.98	44.54	145.1	39.18
N-$Al_mGa_{1-m}N$ ~ p-GaN	HTDD1	25.00	21.728	31.49	13.67	56.58	99.11	27.77
		30.00	21.603	32.32	14.08	56.39	103.4	28.96
		35.00	21.478	33.23	14.43	56.57	107.1	29.99
		40.00	21.345	33.88	14.89	56.05	111.8	31.32
		45.00	21.225	34.74	15.34	55.84	115.0	32.21
n-GaN ~ P-$Al_mGa_{1-m}N$	HTDD2	25.00	21.3580	30.52	14.47	52.61	101.1	28.17
		30.00	21.2760	31.17	14.65	53.02	104.6	29.13
		35.00	21.2330	31.83	15.07	52.65	108.0	30.09
		40.00	21.1330	32.22	15.89	50.64	115.8	32.26
		45.00	21.0600	32.74	16.46	49.73	121.0	33.70
n-$Al_mGa_{1-m}N$ ~ p-$Al_mGa_{1-m}N$	HMDD2	25.00	21.8070	35.05	17.55	49.94	157.90	45.25
		30.00	21.7230	35.46	17.79	49.82	159.80	45.77
		35.00	21.6130	36.29	18.24	50.47	163.40	46.81
		40.00	21.5040	36.88	18.23	50.54	167.00	47.84
		45.00	21.3540	37.25	18.89	49.29	176.00	50.43

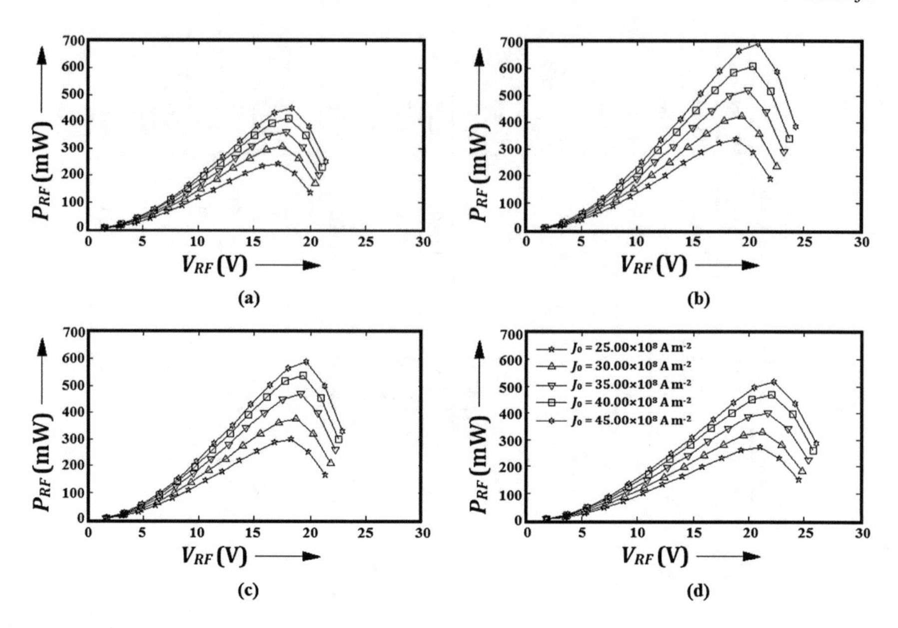

Fig. 3 Variation of P_{RF} with V_{RF} at 1.00 THz for various injected current bias: **a** n-GaN ~ p-GaN, **b** n-$Al_mGa_{1-m}N$ ~ p-GaN, **c** n-GaN ~ p-$Al_mGa_{1-m}N$ and **d** n-$Al_mGa_{1-m}N$ ~ p-$Al_mGa_{1-m}N$

4.2.2 Results for Large Amplitude RF Signal

Injected current bias is varied from 25×10^8 to $45 \times 10^8 \text{Am}^{-2}$ and the results are investigated minutely for various HM and HT structures. Alterations of power output P_{RF} and efficiency of conversion η_L with different injected current bias for the HMs and HTs like *n*-GaN ~ *p*-GaN, *n*-$Al_mGa_{1-m}N$ ~ *p*-GaN, *n*-GaN ~ *p*-$Al_mGa_{1-m}N$ and *n*-$Al_mGa_{1-m}N$ ~ *p*-$Al_xGa_{1-x}N$ at 1 THz are respectively exhibited in Figs. 3 and 4. Figure 3 indicates that at first the power at the output from IMPATT rises with rise of m_x, reaches the maximum value at 60%m_x, then declines. So, it is apparent that 60% modulation gives optimum results. This perception remains true for each and every injected current bias and all HMs and HTs. Moreover, graphical results in Figs. 3 and 4 confirms the acceptance of *n*-$Al_mGa_{1-m}N$ ~ *p*-GaN heterojunction as the most efficient HT for operation at 1 THz with respect to output power and the efficiency of conversion from steady state to RF.

Figure 5 indicates admittance profiles of various HT and HM IMPATTs for a particular current bias $J_0 = 45 \times 10^8 \text{Am}^{-2}$ where the peak of the profiles gives the optimum frequency f_p. Negative conductance at this optimum value for HTDD 1is larger than all other structures. The power obtained at the output for large amplitude ac signal and the efficiency of conversion are calculated from the expressions provided: $P_{RF} = (1/2) \times (V_{RF})^2 \times |G_p| \times A_j$, and $\eta_L = P_{RF}/P_{DC}$ respectively where the input power in the steady state condition is $P_{DC} = V_B \times J_0 \times A_j$ and $A_j = \pi(D_j/2)^2$. It is also inferred from Figs. 3 and 4 that these power and efficiency for HTDD 1 is also larger than all the other HT and HM structures.

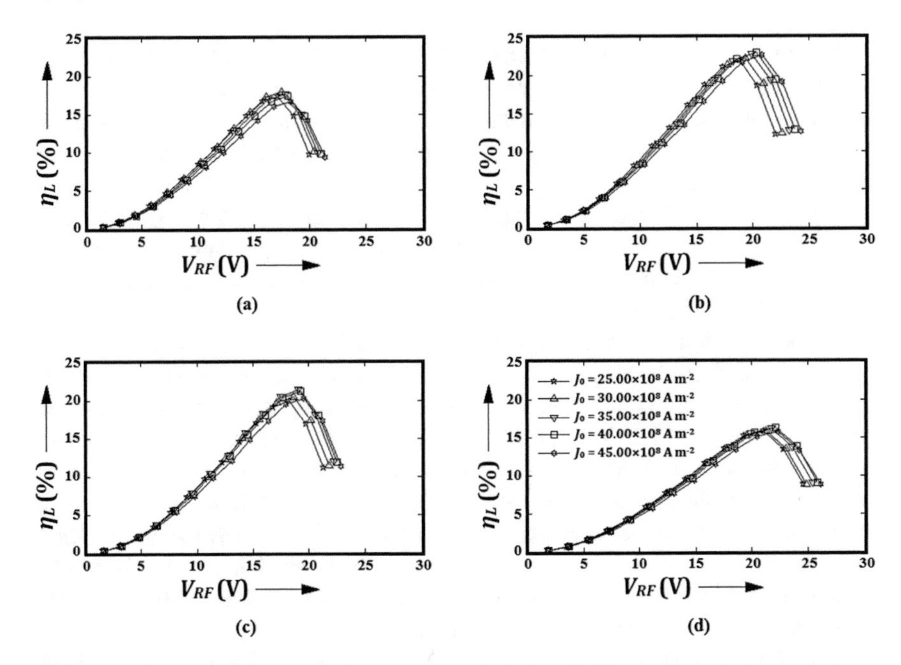

Fig. 4 Variations of η_L with V_{RF} of: **a** n-GaN ~ p-GaN, **b** n-$Al_mGa_{1-m}N$ ~ p-GaN, **c** n-GaN ~ p-$Al_mGa_{1-m}N$ and **d** n-$Al_mGa_{1-m}N$ ~ p-$Al_mGa_{1-m}N$ DDR IMPATT diodes at 1.00 THz for various injected current bias

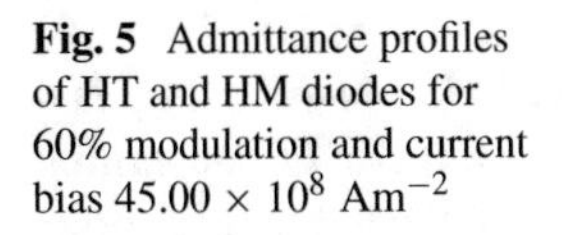

Fig. 5 Admittance profiles of HT and HM diodes for 60% modulation and current bias 45.00×10^8 Am^{-2}

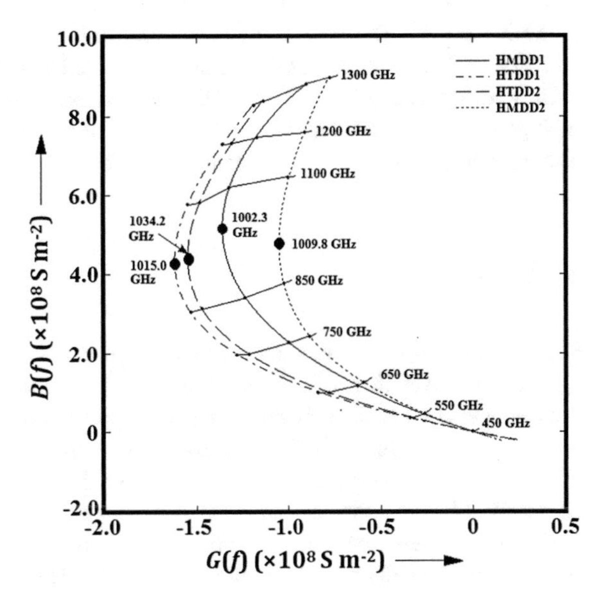

Quality factor ($Q_p = -B_p/G_p$) of HTDD1, HMDD1, HTDD2 and HMDD2 are 5.10, 6.47, 5.42 and 8.58 respectively at an optimum frequency close to 1 THz and at a bias current density of $J_0 = 45 \times 10^8$ Am^{-2}. Furthermore, HTDD 1 outshines all HT and HM diodes concerning Q_p as the comparison parameter which is the lowest for this particular HT.

The specifications obtained using the NSVE model under the impact of large amplitude signal for various HT and HM diodes at 1 THz are tabulated above in Table 5. The resonant frequency under avalanche breakdown condition i.e.,f_a indicates the transition of conductance from positive to negative value and this is larger for HTs with respect to HMs and this increase is even more at larger values of injected current bias. Hence, at these larger current bias, oscillation commences at rising frequency for the HT structures. Quality factor regulates the stability of this oscillation whose desirable value should be lower, more precisely should be nearer to 1. This Quality factor is also seen to fulfill the above oscillatory condition for the heterojunctions but not for homojunctions at any particular frequency. In addition to that, negative resistance which is also a noteworthy specification as is essential for generation of output power also gives larger values for HTs over HMs which is obvious from Table 5. These revelations ensure that HT structurespresent themselves with much better prospective at high frequency than HM structures.

Table 5 also clearly indicates that HTDD 1 outclasses all the other HT and HM structures concerning power at the output and the efficiency of conversion at every injected current bias as was also seen from Figs. 3 and 4. Thus, the current bias values which provides optimum execution of high frequency properties for HTDD 1, HTDD 2, HMDD 1 and HMDD 2 are respectively 40×10^8, 35×10^8, 30×10^8 and 40×10^8 Am^{-2}. Eventually, it can be remarked that HTDD 1 or N-$Al_{0.4}Ga_{0.6}$ N ~ p-GaN surpasses all other HTs and HMs as the best material for IMPATT concerning high power at the output and higher efficiency of conversion in the terahertz regime, although here outcomes are presented for 1 THz.

5 Conclusion

Steady state and RF characteristics of heterojunction (HT) and homojunction (HM) double drift layer IMPATT diodes under the influence of large amplitude ac signal are inspected in this Chapter using MATLAB simulation to establish their worthiness as a generator of power in the terahertz regime with low avalanche noise and large efficiency of conversion from steady state to RF. Firstly, HTs N-3C-SiC ~ p-Si and n-Si ~ P-3C-SiC are considered as base semiconductors for IMPATT for performance analysis and their outcomes are compared with their HM equivalents at 0.3 and 0.5 THz. End results reveal that n-Si ~ P-3C-SiC HT surpasses the ac performance of the other structures concerning power outcome and efficiency of conversion.

Secondly, HTs N-$Al_{0.4}Ga_{0.6}$ N ~ p-GaN and n-GaN ~ P-$Al_{0.4}Ga_{0.6}$ N are taken as base semiconductors for the diode for realization of their desirability as a propitious generator of ac power at 1 THz and then their outcomes are minutely analyzed

Table 5 Parameters derived for large amplitude AC signal with 60% modulation

Structures	J_0 (×10^8 Am^{-2})	f_a (GHz)	f_p (GHz)	G_p (×10^8 S m^{-2})	B_p (× 10^8 S m^{-2})	$Q_p = -(B_p/G_p)$	Z_R (×10^{-9} Ω m^{-2})	P_{RF} (mW)	η_L (%)
n-GaN ~ p-GaN(HMDD1)	25.00	393.7	981.0	−0.8390	8.7969	10.49	−0.1074	243.56	17.31
	30.00	403.9	988.6	−1.0243	8.8465	8.63	−0.1292	309.62	17.97
	35.00	414.2	997.3	−1.1535	8.8518	7.67	−0.1448	360.58	17.64
	40.00	422.4	1000.0	−1.2786	8.8323	6.91	−0.1605	411.22	17.34
	45.00	431.4	1002.3	−1.3618	8.8134	6.47	−0.1712	451.54	16.68
n-$Al_mGa_{1-m}N$ ~ p-GaN (HTDD1)	25.00	389.9	973.0	−0.9654	8.2751	8.57	−0.1391	338.13	21.88
	30.00	408.7	981.5	−1.1465	8.3218	7.26	−0.1625	423.11	22.23
	35.00	420.0	992.3	−1.3285	8.3267	6.27	−0.1869	518.51	22.70
	40.00	438.9	1006.9	−1.5049	8.3089	5.52	−0.2111	610.55	22.94
	45.00	456.6	1015.0	−1.6238	8.2906	5.10	−0.2275	692.62	22.56
n-GaN ~ p-$Al_mGa_{1-m}N$ (HTDD2)	25.00	401.3	996.90	−0.9064	8.3967	9.26	−0.1271	298.43	19.92
	30.00	418.6	1004.2	−1.0967	8.4440	7.69	−0.1513	376.63	20.51
	35.00	432.6	1011.1	−1.3065	8.4491	6.46	−0.1787	467.80	21.38
	40.00	442.9	1022.4	−1.4642	8.4309	5.75	−0.2000	537.12	21.23
	45.00	451.2	1034.2	−1.5513	8.4124	5.42	−0.2120	587.68	20.32
n-$Al_mGa_{1-m}N$ ~ p−$Al_mGa_{1-m}N$ (HMDD2)	25.00	379.3	971.20	−0.6260	9.0178	14.41	−0.0766	271.84	15.79
	30.00	387.7	983.40	−0.7434	9.0686	12.19	−0.0898	330.28	15.81
	35.00	399.0	992.60	−0.8614	9.0740	10.53	−0.1037	400.86	16.08
	40.00	414.5	1001.7	−0.9758	9.0540	9.27	−0.1177	469.11	16.19
	45.00	426.1	1009.8	−1.0529	9.0347	8.58	−0.1273	515.34	15.69

with respect to their HM equivalents. End results clearly establish N-$Al_{0.4}Ga_{0.6}$ N ~ p-GaN HT as the premier base semiconductor for IMPATT or other solid state source for producing considerable power with large efficiency at the output in the terahertz regime. All these outcomes combined together will support researchers and fabricators to pioneer the development of IMPATT diodes using the perfect base semiconductor to make it compatible with the real world scenario.

References

1. J.C. De Jaeger, R. Kozlowski, G. Salmer, High efficiency GaInAs/InP heterojunction Impatt diodes. *IEEE Trans. Electron. Devices*, **ED 30**, 790 (1983)
2. D. Lippens, J.L. Nieruchalski, E. Constant, Multilayered heterojunction structure for millemeter wave Impatt devices. Physics **134 B**, 72 (1985)
3. N.S Dogan, J,R. East, M. Elta, G.I. Haddad, Millimeter wave heterojunction MITATT diodes. IEEE Trans. Microw. Theory Tech. MTT **35**, 1304 (1987)
4. M.J. Kearney, N.R. Couch, J. Stephens, R.S. Smith, Heterostructure impact avalanche transit time diodes grown by molecular beam epitaxy. Semicond. Sci. Tech. **8**, 560 (1993)
5. G.N. Dash, S.P. Pati, Computer aided studies on the microwave characteristics of InP/GaInAs and GaAs/GaInAsheterostructure single drift region impact avalanche transit didoes. J. Phys. D. Appl. Phys. **27**, 1719 (1994)
6. M.J. Bailey, Hetrojunction IMPATT diodes. IEEE Trans. Electron. Devices **39**, 1829 (1992)
7. J.K. Mishra, A.K. Panda, G.N. Dash, An extremely low-noise heterojunction IMPATT. IEEE Trans. Electron. Devices, **ED-44**(12), 2143–2148 (1997)
8. K.V. Vassilevski, A.V. Zorenko, K. Zekentes, Experimental observation of microwave oscillations produced by pulsed Silicon Carbide IMPATT diode. Electron. Lett. **37**, 466 (2001)
9. L. Yuan, A. James, J.A. Cooper, M.R. Melloch, K.J. Webb, Experimental demonstration of a silicon carbide IMPATT oscillator. IEEE Electron. Device Lett. **22**, 266–268 (2001)
10. S. Mukhopadhyay, S. Banerjee, J. Mukhopadhyayand, J.P. Banerjee, Mobile space charge effect in 4H-SiC IMPATT diodes, in *Proceeding of the Fourteenth International workshop on the Physics of Semiconductor Devices (IWPSD 2007)* (IIT Bombay, 2007)
11. J. Mukhopadhyay, J.P. Banerjee, Mobile space charge effect in Silicon Carbide high-low SDR IMPATTs, in *Proceeding of the International Conference on Computers and Devices for Communication (CODEC-04)* (Hyatt Regency, Kolkata, 2004)
12. A. Acharyya, J.P. Banerjee, Prospects of IMPATT devices based on wide bandgap semiconductors as potential terahertz sources. Appl. Nanosci. **4**, 1–14 (2014)
13. A. Acharyya, A. Mallik, D. Banerjee, S. Ganguli, A. Das, S. Dasgupta, J.P. Banerjee, IMPATT devices based on group III-V compound semiconductors: Prospects as potential terahertz radiators. HKIE Trans. (2014)
14. F. Sacconi, A.D. Carlo, P. Lugli, H. Morkoç, Spontaneous and piezoelectric polarization effects on the output characteristics of AlGaN/GaN heterojunction modulation doped FETs. IEEE Trans. Electron. Devices **48**(3), 450–457 (2001)
15. H.G. Xing, U.K. Mishra, Temperature dependent I-V characteristics of AlGaN/GaN HBTs and GaN BJTs, in *Proceedings of IEEE Lester Eastman Conference on High Performance Devices*, 4–6 August, vol. 48(3) (2004), pp. 195–200
16. W.L. Chan, J. Deibel, D.M. Mittleman, Imaging with terahertz radiation. Rep. Prog. Phys. **70**, 1325–1379 (2007)
17. D. Grischkowsky, S. Keiding, M. Exter, C. Fattinger, Far-infrared time-domain spectroscopy with terahertz beams of dielectrics and semiconductors. J. Opt. Soc. Am. B **7**, 2006–2015 (1990)
18. P.H. Siegel, THz instruments for space. IEEE Trans. Antenn. Propag. **55**, 2957–2965 (2007)

19. C. Debus, P.H. Bolivar, Frequency selective surfaces for high sensitivity terahertz sensing. Appl. Phys. Lett. **91**, 184102–184103 (2007)
20. T. Yasui, T. Yasuda, K. Sawanaka, T. Araki, Terahertz paintmeter for noncontact monitoring of thickness and drying progress in paint film. Appl. Opt. **44**, 6849–6856 (2005)
21. C.D. Stoik, M.J. Bohn, J.L. Blackshire, Nondestructive evaluation of aircraft composites using transmissive terahertz time domain spectroscopy. Opt. Express **16**, 17039–17051 (2008)
22. C. Jördens, M. Koch, Detection of foreign bodies in chocolate with pulsed terahertz spectroscopy. Opt. Eng. **47**, 037003 (2008)
23. A.J. Fitzgerald, B.E. Cole, P.F. Taday, Nondestructive analysis of tablet coating thicknesses using terahertz pulsed imaging. J. Pharm. Sci. **94**, 177–183 (2005)
24. P.H. Siegel, Terahertz technology in biology and medicine. IEEE Trans. Microwave Theor. Tech. **52**, 2438–2447 (2004)
25. A. Acharyya, S. Banerjee, J.P. Banerjee, Influence of skin effect on the series resistance of millimeter-wave IMPATT devices. J. Comput. Electron. **12**, 511–525 (2013)
26. T. Tut, M. Gokkavas, B. Butun, S. Butun, E. Ulker, E. Ozbay, Experimental evaluation of impact ionization coefficients in $Al_xGa_{1-x}N$ based avalanche photodiodes. Appl. Phys. Lett. **89**, 183524–1–3, (2006)
27. K. Kunihiro, K. Kasahara, Y. Takahashi, Y. Ohno, Experimental evaluation of impact ionization coefficients in GaN. IEEE Electron. Device Lett. **20**(12), 608–610 (1999)
28. S.C. Shiyu, G. Wang, High-field properties of carrier transport in bulk wurtziteGaN: monte carlo perspective. J. Appl. Phys. **103**, 703–708 (2008)
29. Electronic Archive: New Semiconductor Materials, Characteristics and Properties. Available from: http://www.ioffe.ru/SVA/NSM/Semicond/index.html, August 2014
30. T.W. Kim, D.C. Choo, K.H. Yoo, M.H. Jung, Y.H. Cho, J.H. Lee, Ju. H. Lee, Carrier density and mobility modifications of the two-dimensional electron gas due to an embedded AlN potential barrier layer in $Al_xGa_{1-x}N$/GaNheterostructures. J. Appl. Phys. **97**, 103721–1–5 (2005)
31. A. Acharyya, J. Mukherjee, M. Mukherjee, J.P. Banerjee, Heat sink design for IMPATT diode sources with different base materials operating at 94 GHz. Arch. Phys. Res. **2**, 107–126 (2011)
32. A. Acharyya, S. Banerjee, J.P. Banerjee, Large-signal simulation of 94 GHz pulsed DDR silicon IMPATTs including the temperature transient effect. Radioengineering **21**(4), 1218–1225 (2012)

Advanced Materials-Based Nano-absorbers for Thermo-Photovoltaic Cells

Sajal Agarwal, Yogendra Kumar Prajapati, and Ankur Kumar

Abstract Thin-film solar cells are very popular because of their sustainable feature of energy generation/conversion. Since the first solar cell was proposed, a number of advancements took place for the improvement of conversion efficiency of the cell. Although silicon solar cells have a very acute disadvantage of limited electromagnetic range, lower terahertz region. These cells are limited to the part of the visible region, and thus, the improvement of the efficiency is also limited in these cases. Thin film/nano-absorbers are appeared to overcome this disadvantage of the solar cell and started a whole new field for research. The major advantage of nano-absorbers is their wide wavelength bandwidth (wide frequency coverage in terahertz) and high absorption. There are numerous research groups working on this aspect of energy devices to improve the efficiency by using different materials and techniques. Advanced materials, like graphene, metamaterial, and transition metal dichalcogenides, etc., are proved to be very efficient to expand the horizon of the energy devices, i.e., thermo-photovoltaic solar. In this chapter, a detailed study is given for advanced material-based nano-absorbers in the field of thermo-photovoltaic cells.

Keywords Nano-absorber · Thermo-photovoltaic cells · Transition metal dichalcogenides (TMD) · Metamaterial · Graphene · Terahertz (THz)

S. Agarwal (✉)
Department of Electronics Engineering, Rajiv Gandhi Institute of Petroleum Technology, Jais, Amethi, India
e-mail: sagarwal@rgipt.ac.in

Y. K. Prajapati
Department of Electronics and Communication Engineering, Motilal Nehru National Institute of Technology Allahabad, Prayagraj, India
e-mail: yogendrapra@mnnit.ac.in

A. Kumar
Department of Electronics and Communication Engineering, Institute of Engineering and Technology, Lucknow, India
e-mail: ankur.ecd.cf@ietlucknow.ac.in

S. Das et al. (eds.), *Advances in Terahertz Technology and Its Applications*,
https://doi.org/10.1007/978-981-16-5731-3_11

1 Introduction

Sun is the most important and prime power source of our earth and solar system. It radiates whole earth in 1 h for year's energy consumption of society. However, the direct use of solar energy is not possible without the conversion of solar energy into electrical energy. Device which is used to convert the solar energy into electrical energy is known as solar cell or photovoltaic cell. Use of photovoltaic (PV) power was first discovered by Edmond Becquerel in 1839. However, Charles Fritts first demonstrated the solar cell in 1882, made of selenium, coated with gold. First silicon (Si) solar cell was produced by Bell laboratory in 1954. Conversion efficiency of the proposed cell was 4%. After that D. M. Chapin et al. proposed [1] Si PV cell with increased conversion efficiency of solar cell, i.e., 6%. Continued work of the eminent researchers of this field enhanced the conversion efficiency of the solar cells up to 20%. However, in the late 90s awareness of global communities like United Nations (UN), South Asian Association for Regional Cooperation (SAARC), etc. toward the global environment changes push the limits further to produce more and more efficient solar cells. With continuous refinement and development of photovoltaic cells, a conversion efficiency of 24% is achieved through Si PV cells in early 2000. By 2007, researchers are able to produce the Si PV cell with 28% efficiency. Although all the reported works above are intended to use Si as the main cell material, and this is the key downside of these cells. Since absorption range of the cell depends on the base material, this absorption range actually refers to the covered wavelength range of solar spectrum.

Solar spectrum is very broad in nature and varies from 200 to 3000 nm, whereas the absorption limit of Si PV cell is 300–1100 nm [2]. Thus, most of the solar spectrum range gets wasted because of the material limitation, as incident photons of the lower energy (high wavelength) cannot be absorbed, and extra energy is lost as the heat when high energy photons (higher than bandgap energy) incident on the materials, and electrons of high energy fall to the edge of band [3]. These two effects can limit the upper range to 31% for single junction Si PV cell without concentration and maximum 41% with concentration [4]. Shockley-Queisser limit is a well-known concept which sets the theoretical upper limit of solar cells [5]. Si PV cell are limited by the temperature factor of the cell which causes drop of efficiency when module temperature rises. To further improve PV cell efficiency of single junction/conventional Si cells, various models are developed and these models are divided into three main categories;

(a) First generation
(b) Second generation
(c) Third generation.

Details of the generations of cells will be discussed later in this chapter. Basically, all of the solar cell generation pointed out about the various structures/materials of the Si PV cells. However, all the efforts made by the research of this field are dedicated toward the reduction of size, thickness and weight of the PV cells [6–10]. No doubt;

reduced thickness of the PV cells has improved the cell efficiency and controls the cost, but there is a major challenge with these thin-film solar cells is the weak absorption of solar light. Thinness of these solar cells causes weak light absorption because of the Shockley–Queisser limit. Thus, it is required to propose a light-trapping solar cell structure which may enhance the light absorption as well as the wavelength range of the cell. Thermo-photovoltaic (TPV) is an efficient technology which can exceed Shockley-Queisser limit, by enhancing the absorption of solar spectrum [11]. In the next section, a short review about the Si PV cell generations is given with the features which are lacking in these structures. Third section comprises the technical details about the TPV cell and importance of absorber in TPV cell. Along with that, importance of constituent material of absorber is also discussed. In fourth section, various proposed absorbers are discussed based on the advanced constituent materials. In last section, chapter is concluded based on the arguments discussed in whole chapter.

1.1 Generations of PV Cells

Si-based solar cells have a long history to evolution. First generation of PV cells mainly concentrated toward the crystalline solar cells, whereas second generation is mainly concentrated for thin-film solar cell and third-generation solar cell has various complex materials as the cell layer.

1.1.1 First-Generation PV Cell

First-generation solar cells are mainly classified into three types;

(a) Mono-crystalline
(b) Poly-crystalline
(c) Ribbon silicon.

Primarily, crystalline Si is the material used in first-generation PV cells. Till now, most of the commercial market is ruled by the crystalline Si PV cells. Efficiency of first-generation cells ranges from 14 to 19% [12]. Domination of first-generation solar cell in the market is because of its low cost and well-established manufacturing units. Since the cost of PV cell is the foremost component which can further be reduced with the advanced manufacturing and material processing technologies [13, 14].

1.1.2 Second-Generation PV Cell

Thin-film solar cells are the advanced stage of crystalline solar cells with reduced cost of manufacturing [15]. In thin-film-based solar cells, multiple layers of same/different material are deposited one over another, approximately 1–4 μm thickness is achieved

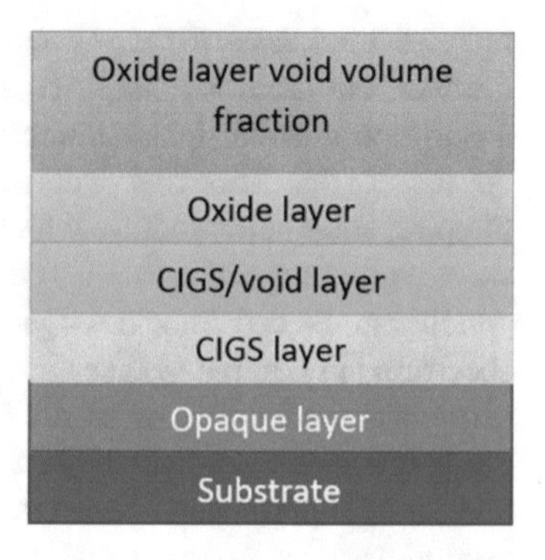

Fig. 1 Example of a multi-layer thin-film solar cell based using CIGS and glass substrate

over an inexpensive substrate like glass, cheap metal, cadmium, etc. The foremost advantage of thin-film-based solar cells is their ability of packaging which make these flexible and light-weighted to be installed as an integral part of various components. Most used thin-film solar cells are, amorphous silicon cell [16, 17], cadmium-telluride [18], copper-indium-selenide [19], and copper-indium-gallium-diselenide (CIGS) [20] solar cell (Fig. 1).

On reviewing the second-generation solar cell in detail, it is found that these are attractive due to their low cost and other flexible manufacturing advantages. However, lower efficiency of these cells balanced out all the good and thin-film technology is not mature enough to compete with the first-generation solar cell. Although second-generation solar cells are commercialized but still facing issues related to the durability and toxicity of the used materials.

1.1.3 Third-Generation PV Cell

Third-generation solar cells are at research stage and very different from the previous two generations, in terms of structure as well as technology used. Some of the third-generation solar cells use the concept of quantum mechanics. However, on material front, third-generation solar cells can be organic [21] or dye-sensitized [22]. The most attractive cell of this generation is concentrating PV cell in which a lens is used to concentrate the solar radiation on the active surface of cell. These cells require heat transport layer, as heating of the structure is a big problem. Figure 2 depicts the concept of the concentration solar cell structure.

Third-generation solar cells are yet to be commercialized, some of these cells can challenge the efficiency of previous generation whereas organic and other PV cells of third-generation may have low-efficiency but with low cost. However, these cells can fill the gap market where these structures are required.

Solar cells studied and discussed till now in this chapter are thin, cost-effective, and flexible but still the efficiency is not good enough. Lacking of efficiency in discussed cells is because of their limited wavelength range. Mostly all the solar cells are concentrated from visible to near infrared (NIR) region, i.e., 300–1100 nm

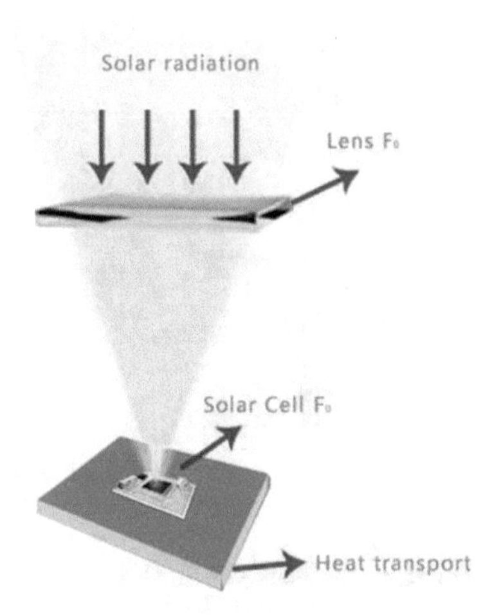

Fig. 2 Schematic of concentration PV cell [23]

at maximum. Whereas the solar spectrum varies from 200 to 3000 nm thus, most of the spectrum get wasted which possess most of the heat from sun. Here comes the TVP cells, these cells covert the thermal energy from sun to useful electrical energy [11] for much higher temperature and wavelength range than a typical solar cell. It is suggested by United Nations in its World Population Prospects that the human population will increase by 2.2 billion by 2050 [24]. This is an alarming figure, as with the increased population, requirement of energy will also increase, and this will lead us far away from the sustainability. However, use of renewable energy to address this issue is a very good way to keep the earth environment sustainable. And to achieve this, it is important to move toward the alternative PV cells such as TPV which can improve the overall efficiency of the conversion with almost same or lower manufacturing cost and toxicity.

2 Basics of Thermo-Photovoltaic Cell

TPV cell was first proposed by B. D. Wedlock in 1963 [25] as the high temperature and efficiency replacement of thermionics. This field undergone a numerous evolution, yet it is not commercialized due to some to its inherent shortcoming of substantial loss at all conversion points. To understand this, first we should understand the working of the TPV cell. Figure 3 shows a basic schematic diagram of TPV cell indicating the most important layers of the cell and the energy conversion sites. It is to be noted that the basic principle of the TPV cell is same as the conventional PV cell where optical energy is absorbed by the *p–n* junction to generate electricity. The elementary difference between the two is that direct sun radiation is not used to fall upon the *p–n* junction instead two extra components are used namely absorber and emitter to do the job. Work of the absorber in the TPV cell structure to absorb as much as radiation

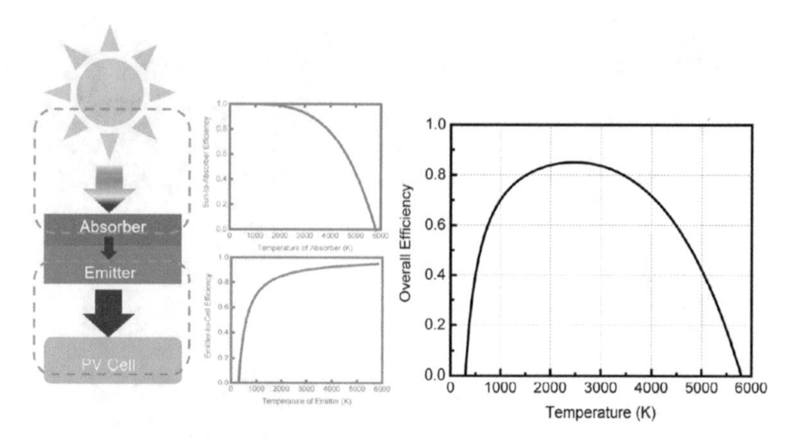

Fig. 3 Basic schematic of TPV cell with the conversion efficiency of TPV cell at different stages with respect to the temperature [3]

from sun is possible and for as much as wide wavelength range like a black body radiation.

By absorbing such amount to radiation absorber gets heated up to a very high temperature (300–6000 K). This temperature is used to heat up the emitter of the TPV cell to emit the light [26] which is basically is concentrator, the upper limit is governed by the Carnot efficiency [27]. Temperature of the emitter is normally higher than 1000 K, cause the thermal radiation to be absorbed by the PV cell and converted in electricity. Thus, efficiency of the TPV cell depends on the temperature of various layers of the cell as,

$$\eta = \left(1 - T^4/T_s^4\right) \cdot (1 - T_e/T) \quad (1)$$

in Eq. (1); T, T_s, and T_e are the temperature of absorber, emitter, and sun, respectively.

2.1 Important Parameters, Issues and Constituent Materials

Since TPV cell recycles unabsorbed photons to get heated up for energy conversion, this has a number of advantages including static conversion process, quiet, separation of heat conduction and electricity generation processes make it more durable, and lack of fundamental temperature gradients across materials [3]. However, the structure of TPV cell is much more complex and losses are there at energy conversions sites which may cause lower efficiency than mere PV cell (Fig. 4).

Since the efficiency of TPV cell solely depends on different layers, i.e., absorber and emitter, it is important to design these layers properly by choosing appropriate material, thickness, and structure. It is evident from Eq. (1) that controlling of the

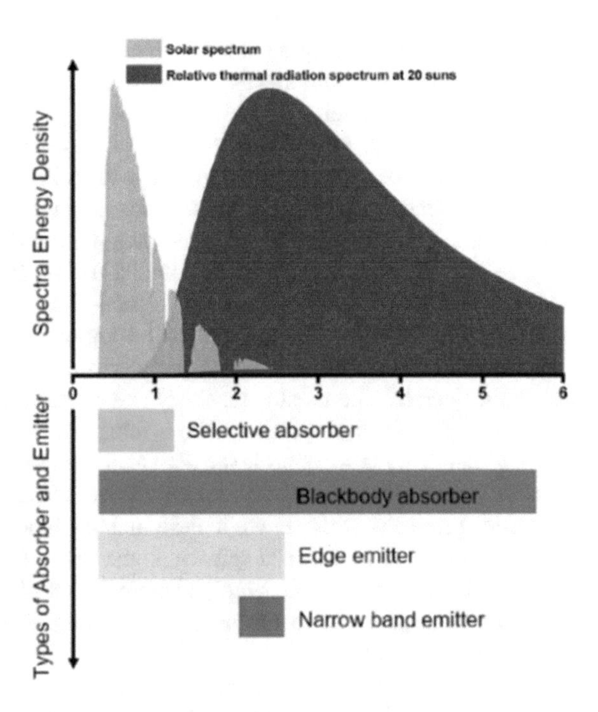

Fig. 4 Spectra and different types of absorbers and emitters [3]

temperature of absorber, emitter, and PV cell is most important parameter to operate TPV cell effectively. Along with this, solar radiation collection and conversion is also an imperative factor. Essentially, designers work is to advance the efficiency of cell for substantial cost, to do so, absorber should capture photons from sun radiation only [28]. This can be done by concentrating sunlight on absorber surface and by selecting appropriate material for absorber construction. Normally multiple materials and/or nanostructured surfaces are used to keep the fragility at high temperature. Characteristics of conventional materials depend on their temperature, as expansion coefficient depends on the temperature and thus the use of conventional material is not wise in high temperature applications such as TPV cells. Lower melting point of nanoscale structure compared to bulk material further limits the variety of useable material in TPV cells.

In conclusion, there are two vital parameters to ensure the satisfying performance of TPV cell; first is the temperature control and second is enough solar radiation capturing/transportation. And both of these parameters depend on the material and structure of the layer. Since high temperature is required at absorber/emitter for high efficiency and low temperature at PV cell to ensure optimum performance, it is highly anticipated to discuss all the parts separately. Although this chapter is dedicated to the absorber of TPV cell. Authors are likely to discuss the realization of absorber for TPV cell on the basis of the constituent material. Since it is already established that the natural and conventional materials may not live up to the aspirations which are required in current energy and sustainability crises scenario, thus advanced materials such as metamaterials and two-dimensional materials, graphene, and TMDs are discussed in this chapter.

Artificial material, i.e., metamaterial has displayed incredible features such as high absorption, wide operating wavelength range tunability, light-weighted, and exquisite property of electric and optical tuning. As these materials, i.e., metamaterials possess tremendous physical, elastic, and electromagnetic properties thus, it is possible to achieve altered refractive index, negative stiffness, etc. which is not available in conventional materials. These innovative properties make advanced nano-materials a perfect candidate for the energy harvesting devices. Apart from metamaterials, two-dimensional (2D) materials have unique electronic and structural properties which are again tunable. The most extensively researched 2D material is Graphene, and found that its unique properties make it most interesting material for many nano-applications. This is a single layer 2D sheet that has high electrical and thermal conductivity, flexibility, charge mobility, large surface area, etc. [29]. And these properties proved very useful while designing a nano-absorber for TPV cell.

As a rising star in material science, some other TMD 2D material sheets, like molybdenum disulfide (MoS_2), molybdenum diselenide ($MoSe_2$), etc. are considered to have a high potential to further improve the absorber performance for energy conversion. From all the above argument, it is believed that the use of advanced materials to manufacture absorber, expected critical parameters of TPV cell can be achieved with optimum cost while mass manufactured.

3 Advanced Material-Based TPV Cell Absorber

Concept of the TPV cell is solely based on the conversion of solar radiation into heat/thermal energy by the absorber (first layer in the TPV cell). Then after, thermal energy is used to energize the electrons of PV cell to convert this energy into electrical form. The main challenge of solar absorbers is their manufacturing for wide wavelength region and re-radiation suppression for higher temperatures.

For TPV cell, various selective solar absorbers are available, including composite material [30–32], semiconductor–metal tandems [33], plasmonic material [34], and one-dimensional (1D)/2D/three-dimensional (3D) materials.

Figure 5 depicts the structure and the proposed absorption for the plasmonic and tandem absorber. From above spectrum, it is depicted that composite materials can improve the wavelength range but absorbance is not up to the mark and further enhancement is possible. To achieve better performance of TPV cell absorber, advanced materials; metamaterial, 1D/2D/3D materials can be used. Thus, in this section authors have discussed several advanced material-based absorbers and then a comparative analysis is done to conclude the study.

To begin with, metamaterial-based solar absorbers are discussed. Although metamaterial-based absorber was first proposed in 2008 by Landy et al. [35], this study used a split ring resonator to realize the metamaterial surface. Peak absorbance

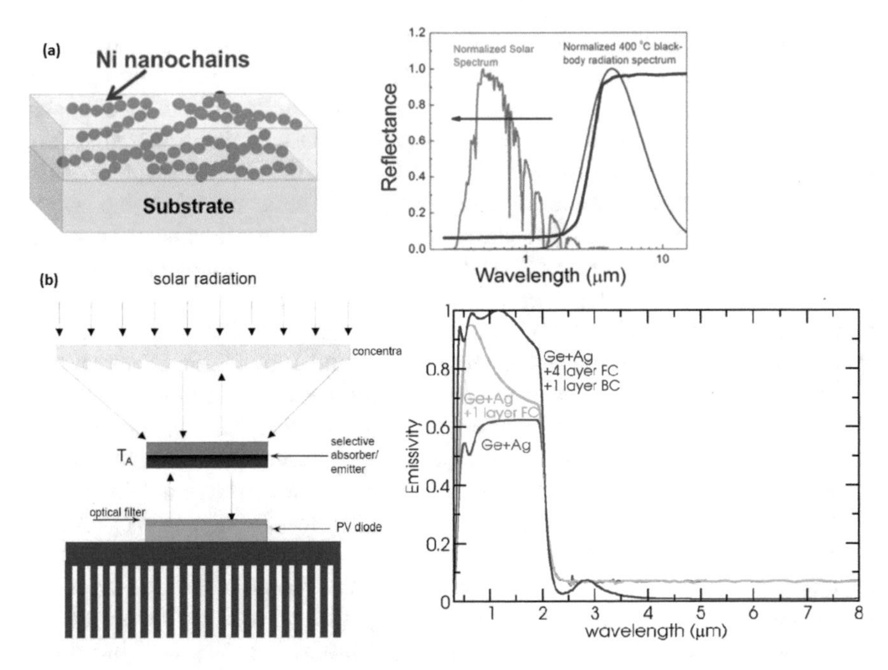

Fig. 5 Schematic structure and absorption spectrum for **a** semiconductor–metal tandem absorber [33] **b** plasmonic absorber [34]

of more than 88% is achieved at 11.5 gigahertz (GHz). However, it is a single band absorber having very small width of absorbance but it gave a direction to various researchers to work in this field. Thereafter, numerous researchers started working for different application of metamaterial absorbers, such as antenna and communication. Since absorbers have various applications, thus authors selected limited studies dedicated to TPV cell absorber only. However, most of the metamaterial absorbers proposed in the starting phase operates in GHz region. Thus, utilization of these absorbers for solar cell application was not a wise idea. Although with the advancement and continuous research in this field made metamaterial size smaller even further to make it compatible with the higher frequency range, such as from few terahertz (THz) to hundreds of THz and thus metamaterial proved useful for the solar cell applications. In 2012, Chihhui Wu et al. [36] proposed a metamaterial-based plasmonic absorber for TPV system. In the proposed study roll-to-roll imprint lithography is suggested for system construction. For high thermal stability, proposed model is based on the tungsten (W) metal layer with aluminum nitride (AlN) as the spacer layer. A cylindrical cone is used to wrap the geometry as the heat exchange element in the proposed system.

Solar radiation is illuminated through a small slit in coaxial shield, and absorbed energy is converted into heat which dissipates as radiative emission toward the emitter. It depicts the absorption spectrum for two different polarizations, i.e., *p* and *s* polarization and different incident angles. It is detected that the above structure is not polarization independent, for *p* polarization absorption is better than the other one. However, absorption is high for wide wavelength range but spectrum has fluctuation which in turn affects the overall performance of the proposed system. Proposed efficiency of the system is 31% for T > 1200 K. After some time in 2014, Wei Li et al. [37] proposed a three-layer metamaterial structure, where top layer is made of Titanium Nitride (TiN) square structure having 30 nm thickness. These unit cells are repeated with a pitch of 300 nm in *x* and *y* direction. Second layer is made of silicon dioxide (SiO_2) layer with 60 nm thickness, this layer is optically transparent and melting point is 1600 °C, after that 35 nm thick silicon nitride (Si_3N_4) spacer layer is used with higher melting point around 1900 °C, and back layer of TiN of 150 nm thickness is used over sapphire substrate. This structure possesses the polarization independent nature for normal incidence of the solar radiation on the top layer of the cell. It is depicted that TiN upper layer provides high and fluctuation free absorption, whereas gold (Au) and silver (Ag) based structure has large fluctuations which degrades the absorber performance. However, the wavelength range for the proposed absorber is from 400 to 800 nm, which is very small compared to another metamaterial-based absorber. Thus, it can be said that absorption is improved but the wavelength range is very narrow and the overall performance is not up to mark.

In 2015, Wang et al. [38] proposed an angle-insensitive integrated broadband absorber for TPV cell. TiN nanostructured (plus shaped) are used as the unit cell for the metamaterial below this, dielectric and lower TiN layers are stacked for approximately 93% light absorption. Operating wavelength range varies from 300 to 900 nm which is better than the above study but not wide enough. Absorption of the proposed absorber is excellent in visible band and reduced abruptly after that. In 2016, Hao Wang et al. [39] published a research study for simple metamaterial absorber made of SiO_2 and W as shown in Fig. 6. It is depicted that the upper layer structure and the lower most layer is made of W, whereas SiO_2 layer is used as the spacer dielectric layer. This simple metamaterial absorber displayed very promising wavelength range up to 2 μm from 0.2 μm. However, absorption spectrum has large fluctuations which degrades the overall performance.

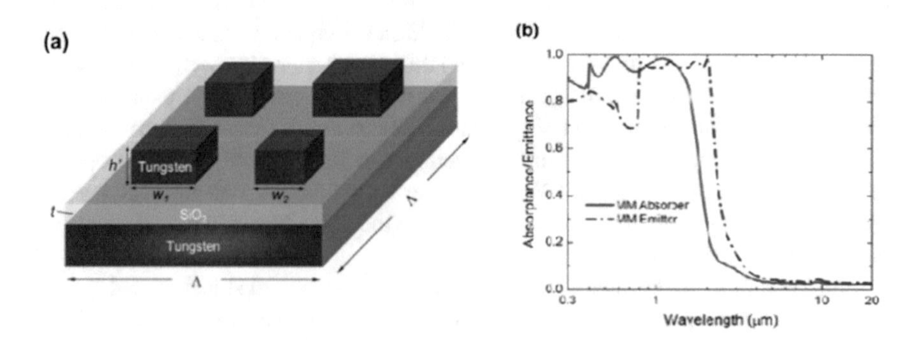

Fig. 6 **a** Graphical representation of proposed absorber **b** absorption spectrum [39]

In proposed study, authors discussed various aspects of the absorber performance based on temperature, cell size, concentration factor, etc. 8 to 10% conversion efficiency is achieved for the proposed absorber, which is good but not as desired. Quing et al. [40] proposed a three-layer Ag grating-based metamaterial absorber with, ultrathin gallium antimonide (GaSb) layer, and Ag substrate. In this numerical study, absorption of visible and NIR photons is achieved by the ultrathin GaSb layer. This study discussed different parameters and respective absorption spectrum for transverse magnetic (TM) and transverse electric (TE) light radiation. On observing the absorption spectrum, it is seen that there are a number of dips present in absorption spectrum which degrade overall performance of the absorber. In 2016, Agarwal and Prajapati [41], published a study for a 3-D metamaterial structure having three/four intertwined helices made of aluminum (Al). A unit cell consists of two sets of left-handed and right-handed helices each as shown in Fig. 7. Placement of opposite handed helices in one unit cell confirms the polarization (*s* and *p* polarization) independent nature of proposed absorber. This structure also ensures the better absorbance due to better interaction of incident light with the structure. The helix unit is placed over a thin glass substrate, glass substrate is transparent to the light thus, would not affect the absorption of the absorber. It is stated in the study that high absorption through the structure is a combined effect of absorption (ohmic loss) and the end-fire emissions (transmittance and reflectance).

Geometrical parameters are also varied such as length of the pitch of helix, number of pitches in the helix, diameter of helix, etc. It is observed that the geometrical parameter affects the absorption widely and longer length of the helix promises better absorber. However, increasing the length of helix beyond some point is not advised due to the stability of the helix in free space. Thus, optimizing achieved overall performance of the absorber is very good with 93% absorbance for a wavelength range from 340 to 1680 nm. From Fig. 7, it is also seen that the fluctuations are very less in the proposed absorber; however, fabrication of this structure is bit tricky, but with advanced fabrication techniques, it is possible. In 2020, Qin et al. [42] proposed a triple band metamaterial absorber using numerical approach, i.e., finite-difference

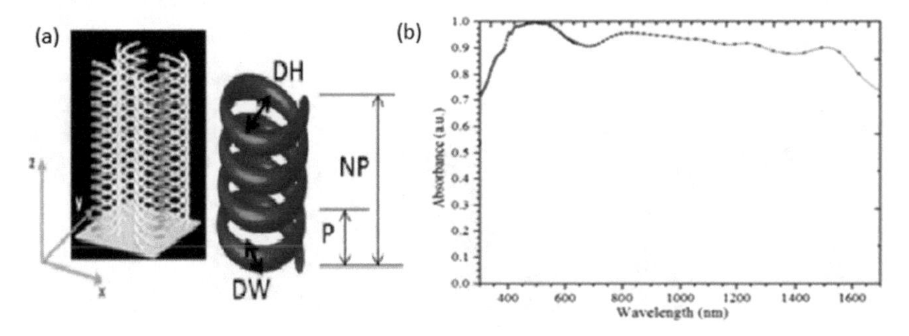

Fig. 7 **a** Schematic diagram of the unit cell of the absorber **b** overall absorption through the absorber for four intertwined helices [41]

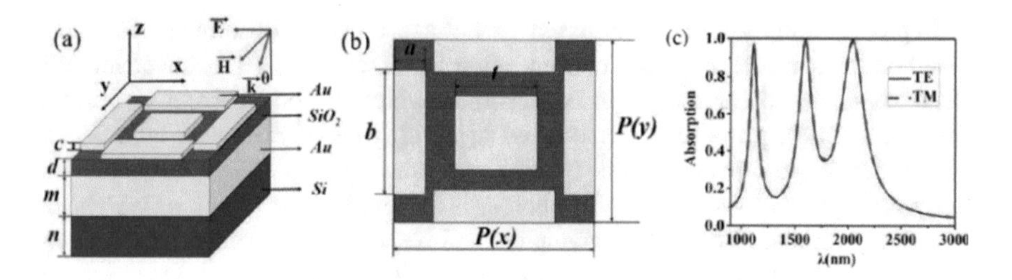

Fig. 8 **a** 3-D view of proposed metamaterial unit cell **b** top view of the absorber **c** absorption for TE and TM polarized incident light [42]

time domain (FDTD) method. Figure 8 has the absorber structure with the absorption spectrum for TM and TE polarization.

It is depicted from Fig. 8, that the wavelength range varies from 1000 to 3000 nm but the absorption after 2500 nm is almost null. It is seen that peaks are varied from 1100 to 2100 nm thus, can be used for various applications based on the wavelength requirement. This study has a full detailed study for angle, geometrical parameter and polarization variation, and observed that the proposed absorber has high tolerance for angle and polarization variation. High angle and polarization tolerance are resulted due to Au nano-cuboids array. And thus, it can be said that based on used material and structure of the metamaterial, absorption spectrum and wavelength range can be tuned for optimum and different performance. Now, a detailed study is done for 2D material-based absorbers.

In 2013, Xu et al. [43] proposed a tunable absorber by using a very simple geometry of thin film with graphene layer on top as shown in Fig. 9.

The absorber has a dielectric graphene stack to observe the tunable absorber. Tunability of the graphene layer conductivity made the proposed absorber tunable. From detailed study, it is observed that the absorption depends on the graphene sheet properties such as thickness and chemical potential. The proposed absorber provides tunability for both polarizations over a wide frequency range. It is also observed that for various stacking approaches and lattice parameters, absorption

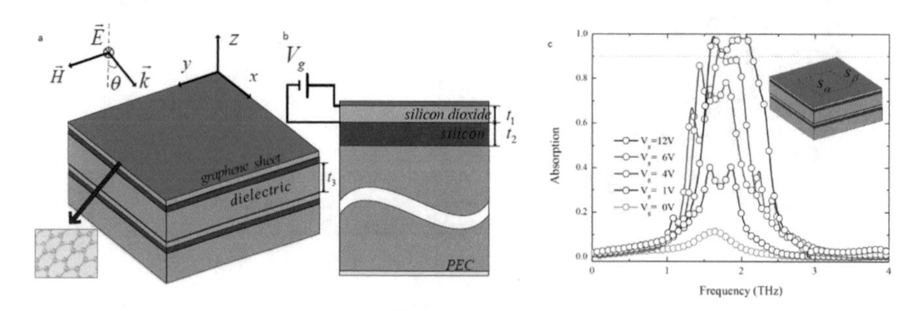

Fig. 9 **a** 3D view **b** cross-sectional view **c** Absorption of the structure by the bias voltage from 0 to 12 V [43]

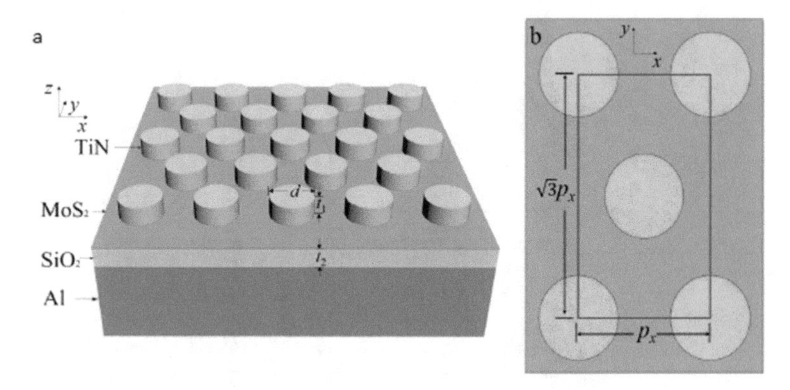

Fig. 10 **a** 3D view **b** cross-sectional view of the proposed absorber [45]

varied and for high chemical potential absorption is highest. Then in 2016, Fardoost et al. [44] proposed a multi-layer graphene ultra-wideband THz absorber. This study quantified the scattering parameters of plasmonic waves for the calculation of the absorbance, which propagates on graphene layers. Here again, the detailed study is done for various combinations of geometrical parameter and chemical potential to optimize the final absorber structure. It is observed that frequency region is 2.7 THz wide whereas the central frequency is 3 THz, from the results it is evident that the proposed absorber can be implemented in TPV cells for wide absorption range. In 2017, MoS_2 based [45] absorber is proposed by Dewang Huo et al., this absorber is composed of hexagonally arranged single sized TiN nano-disk. For the dielectric medium SiO_2 layer is used, which is sandwiched between the nano-disk structure and the Al substrate. MoS_2 layer is placed just below the nano-disks. Placement of the TiN disks is given in Fig. 10, and it is observed that the fabrication of the structure is easy with the advanced fabrication techniques. Thickness of MoS_2 monolayer is assumed to be 0.625 nm which is a standard thickness.

Simulation of the proposed structure is done using FDTD method. On observing the absorption closely, it is found that for the wavelength range from 400 to 850 nm, an average absorption of 96.1% is achieved. It is also observed that introduction of MoS_2 layer provides high absorption at lower wavelengths and thus, enhancement in the absorption is seen of 98.1%, and wavelength range now varies from 400 to 850 nm, which can be seen from Fig. 11. The most important finding of this study is its polarization insensitivity, which makes this absorber extremely useful for many applications where polarization insensitivity is required.

In 2018, Agarwal and Prajapati [46] proposed a combinational absorber having a stack of graphene and other TMD material sheets over Au substrate for wide absorption range with high absorption.

Transition metal atoms has high absorption of optical spectrum due to their dipolar transitions with high joint density of states and oscillator strength in localized states with strong spatial overlap. Oscillation strengthens are constructively super positioned in visible region which in turn gives better radiation absorption. The proposed

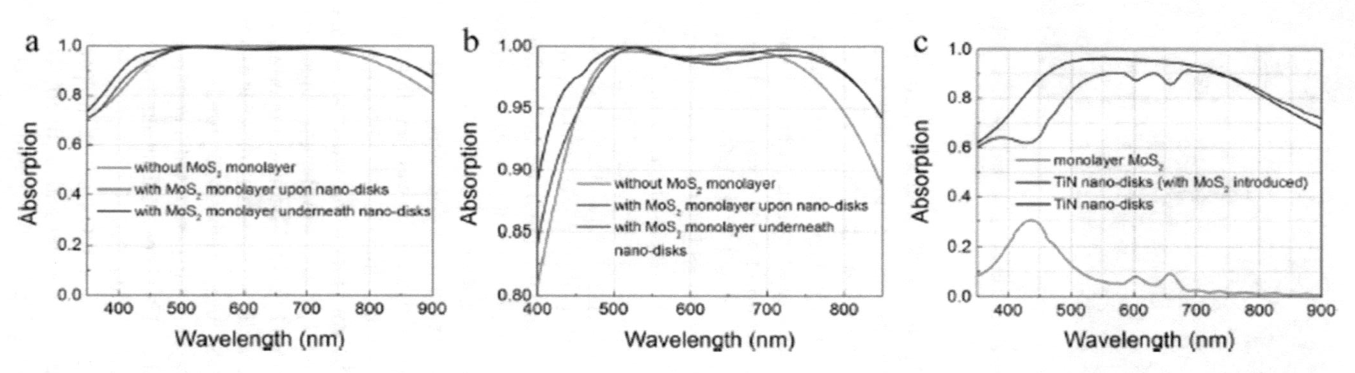

Fig. 11 Absorption spectrum **a** enlarged **b** MoS_2 is introduced underneath at different positions **c** nano-disks and monolayer MoS_2 separately [45]

absorber having wide operating range from 200 to 1000 nm with perfect absorbance (Fig. 12) for all over the desired range and only 26 nm overall thickness. This makes the proposed study very interesting as in any previous studies such high absorbance is not found for this much thin absorber.

Recently, Mostaan and Saghaei [47] proposed a metamaterial structure which is made of graphene and it is a very important approach which combining the metamaterial and graphene, this approach provides more tunability of parameters and wider wavelength range which is required in TPV cell applications. Figure 13 has the schematics of the proposed structure.

It is observed that the geometry is simple as only nanometer range 'U' rings are made to get the metamaterial effect. Gold is used as the substrate material and between Au and graphene layers dielectric glass is used as the spacer. In the proposed absorber, three graphene layers are used to make the top layer and it provides increased interaction of light with graphene and thus, improved bandwidth is achieved. This study is done in detail by observing parametric effect on the absorption and operating region

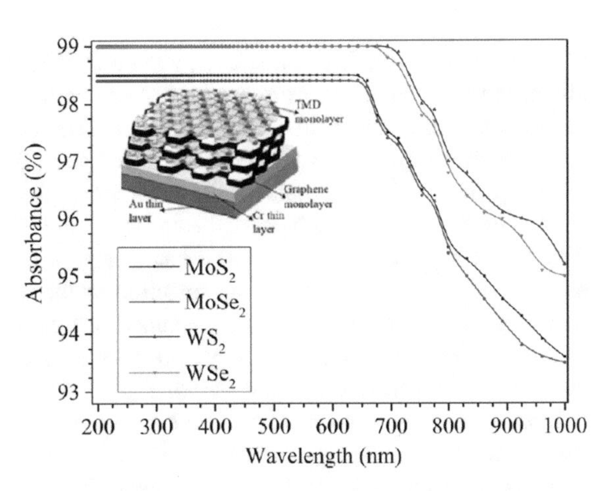

Fig. 12 Schematic of the proposed absorber with the optimized absorbance spectrum [46]

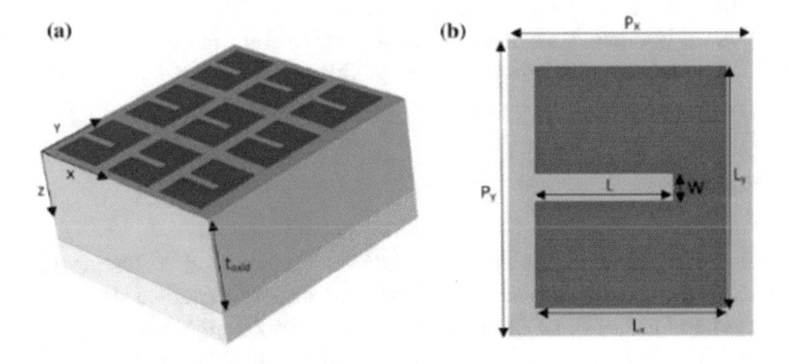

Fig. 13 **a** 3D view **b** top view of proposed absorber [47]

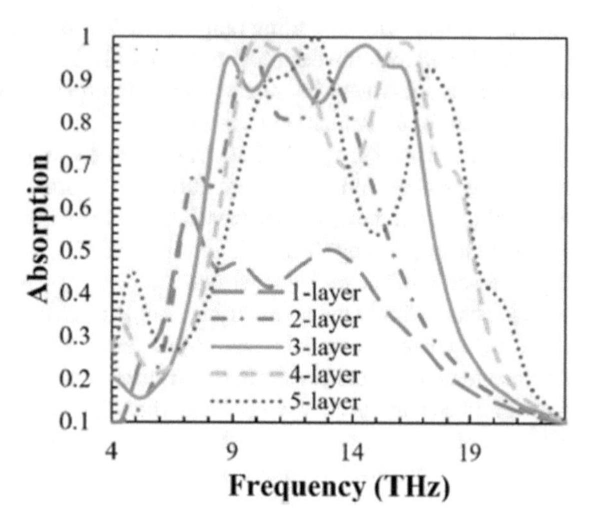

Fig. 14 Effect of number of graphene layers on proposed absorption spectrum [47]

of the absorber. From Fig. 14, it is depicted that stack of number of graphene layers affects the absorption spectrum severely. Along with this, it is also observed that the absorption bandwidth of this absorber is 8.48 THz which is fairly large.

There are a number of other studies reviewed based on the constituent material, i.e., metamaterial and 2D materials for the absorber application; however, in this chapter, authors pick the work which are dedicated to solar/TPV cell application specifically. And it can be said that the material has a huge effect on the performance of the absorber for any application, thus, selection of the material is the most important step of absorber design. However, comment on the fabrication and realization of these absorbers for TPV cells is very important part, since fabrication of some of the complex structure discussed above is still not possible due to the practical limitations of fabrication technologies. Some of the discussed 2D nano-sheets like MoS_2, WS_2, etc. are still under development and large-scale fabrication is not achieved yet. Cost of realization of complex and advanced materials/nano-sheets is very large; however, this can be optimized by bulk realization of these materials. And it is also believed that with the advancements in material processing field and fabrication technologies, realization and cost optimization will be possible in near future.

4 Conclusion

The review study done in this chapter is dedicated to the absorbers having advanced materials as the constituent one. This chapter consists of a detailed study about the various generations of the solar absorbers; their shortcomings, advantages, and the used materials. It is deduced that the conventional materials and simple few layer structure could not do the job for wide bandwidth and high absorption, thus, advanced materials came into picture to improve the performance of the optical absorber. Although optical absorbers can be used for many applications, but in this chapter, authors focused on the thermo-photovoltaic application. It is observed from different studies that 2D materials proved to be an efficient way to improve the absorber performance however, it is not true every time as the geometry of the absorber also affects the performance. Thus, to achieve good performance out of an absorber for TPV cell application, it is important to optimize all the geometrical parameters as well as the constituent material of the absorber. Although synthesis of the advanced material is bit tricky, with advanced synthesis techniques, it is possible to do the large-scale production.

References

1. DM Chapin CS Fuller GL Pearson 1954 A new silicon p-n junction photocell for converting solar radiation into electrical power J. Appl. Phys. 25 5 676 677
2. P.M. Ushasree, B. Bora, *CHAPTER 1: Silicon Solar Cells, in Solar Energy Capture Materials* (2019), pp. 1–55
3. Y. Wang, H. Liu, J. Zhu, Solar thermophotovoltaics: Progress, challenges, and opportunities. APL Mater. **7**(8), 080906 (2019)
4. E Rephaeli S Fan 2009 Absorber and emitter for solar thermo-photovoltaic systems to achieve efficiency exceeding the Shockley-Queisser limit Opt. Express 17 17 15145 15159
5. LC Hirst NJ Ekins-Daukes 2011 Fundamental losses in solar cells Prog. Photovoltaics Res. Appl. 19 3 286 293
6. A. Emrani, P. Vasekar, C.R. Westgate, Effects of sulfurization temperature on CZTS thin film solar cell performances. Solar Energy **98**(PC), 335–340 (2013)
7. W. Wang, M.T. Winkler, O. Gunawan, T. Gokmen, T.K. Todorov, Y. Zhu, D.B. Mitzi, Device characteristics of CZTSSe thin-film solar cells with 12.6% efficiency. Adv. Energy Mater. **4**(7), 1301465 (2014)
8. J Ge J Jiang P Yang C Peng Z Huang S Zuo L Yang JA Chu 2014 5.5% efficient co-electrodeposited ZnO/CdS/Cu2ZnSnS4/Mo thin film solar cell Sol. Energy Mater. Sol. Cells 125 20 26
9. Y.S. Lee, T. Gershon, O. Gunawan, T.K. Todorov, T. Gokmen, Y. Virgus, S. Guha, Cu2ZnSnSe4 thin-film solar cells by thermal coevaporation with 11.6% efficiency and improved minority carrier diffusion length. Adv. Energy Mater. **5**(7), 1401372 2015
10. TP Dhakal CY Peng TR Reid R Dasharathy CR Westgate 2014 Characterization of a CZTS thin film solar cell grown by sputtering method Sol. Energy 100 23 30
11. Z Zhou E Sakr Y Sun P Bermel 2016 Solar thermophotovoltaics: reshaping the solar spectrum Nanophotonics 5 1 1 21
12. S Sharma KK Jain A Sharma 2015 Solar cells: in research and applications—a review Mater. Sci. Appl. 6 12 1145
13. Energy Market Authority, Handbook for solar photovoltaic (PV) systems, pp. 8
14. E. Sachs, J. Cryst, String ribbon growth technique, (Evergreen Solar, Sovello) Growth **82**, 117 (1987)
15. KL Chopra PD Paulson V Dutta 2004 Thin-film solar cells: an overview Prog. Photovoltaics Res. Appl. 12 2–3 69 92
16. S Mehta 2010 PV Technology, Production and Cost, Outlook: 2010–2015 Greentech Media Research Boston, MA
17. A.H. El-din, C.F. Gabra, A.H. Ali, March. A comparative analysis between the performances of monocrystalline, polycrystalline and amorphous thin film in different temperatures at different

locations in Egypt. in *1st Africa Photovoltaic Solar Energy Conference and Exhibition*, Durban (2014)
18. Dena German Energy Agency, Information about German renewable energy, industries, companies and product (Federal Ministry of Economics and Technology). ISIN: B002MNZE4U (2014), pp. 41
19. I Konovalov 2004 Material requirements for CIS solar cells Thin Solid Films 451 413 419
20. M Kaelin D Rudmann AN Tiwari 2004 Low cost processing of CIGS thin film solar cells Sol. Energy 77 6 749 756
21. S Abermann 2013 Non-vacuum processed next generation thin film photovoltaics: towards marketable efficiency and production of CZTS based solar cells Sol. Energy 94 37 70
22. MK Nazeeruddin E Baranoff M Grätzel 2011 Dye-sensitized solar cells: a brief overview Sol. Energy 85 6 1172 1178
23. A.H. Eldin, M. Refaey, A. Farghly, A review on photovoltaic solar energy technology and its efficiency. in *Conference paper* (2015)
24. United Nations, Department of Economic and Social Affairs P D 2017 World Population Prospects: The 2017 Revision, Key Findings and Advance Tables
25. BD Wedlock 1963 Thermo-photo-voltaic energy conversion Proc. IEEE 51 5 694 698
26. C.J. Crowley, N.A. Elkouh, S. Murray, D.L. Chubb, Thermophotovoltaic converter performance for radioisotope power systems. in *AIP Conference Proceedings*, vol. 746(1). (American Institute of Physics, 2005), pp. 601–614
27. S Buddhiraju P Santhanam S Fan 2018 Thermodynamic limits of energy harvesting from outgoing thermal radiation Proc. Natl. Acad. Sci. 115 16 E3609 E3615
28. A Datas DL Chubb A Veeraragavan 2013 Steady state analysis of a storage integrated solar thermophotovoltaic (SISTPV) system Sol. Energy 96 33 45
29. JA Paradiso T Starner 2005 Energy scavenging for mobile and wireless electronics IEEE Pervasive Comput. 4 1 18 27
30. QC Zhang DR Mills 1992 Very low-emittance solar selective surfaces using new film structures J. Appl. Phys. 72 7 3013 3021
31. P Gao LJ Meng MP Santos Dos V Teixeira M Andritschky 2000 Study of ZrO2–Y2O3 films prepared by rf magnetron reactive sputtering Thin Solid Films 377 32 36
32. QC Zhang 1999 High efficiency Al-N cermet solar coatings with double cermet layer film structures J. Phys. D Appl. Phys. 32 15 1938
33. P Bermel M Ghebrebrhan W Chan YX Yeng M Araghchini R Hamam CH Marton KF Jensen M Soljačić JD Joannopoulos SG Johnson 2010 Design and global optimization of high-efficiency thermophotovoltaic systems Opt. Express 18 103 A314 A334
34. X. Wang, H. Li, X. Yu, X. Shi, J. Liu, High-performance solution-processed plasmonic Ni nanochain-Al2O3 selective solar thermal absorbers. Appl. Phys. Lett. **101**(20), 203109 (2012)
35. N.I. Landy, S. Sajuyigbe, J.J. Mock, D.R. Smith, W.J. Padilla, Perfect metamaterial absorber. Phys. Rev. Lett. **100**(20), 207402 (2008)
36. C. Wu, B. Neuner III, J. John, A. Milder, B. Zollars, S. Savoy, G. Shvets, Metamaterial-based integrated plasmonic absorber/emitter for solar thermo-photovoltaic systems. J. Opt. **14**(2), 024005 (2012)
37. W Li U Guler N Kinsey GV Naik A Boltasseva J Guan VM Shalaev AV Kildishev 2014 Refractory plasmonics with titanium nitride: broadband metamaterial absorber Adv. Mater. 26 47 7959 7965
38. H Wang Q Chen L Wen S Song X Hu G Xu 2015 Titanium-nitride-based integrated plasmonic absorber/emitter for solar thermophotovoltaic application Photonics Res. 3 6 329 334
39. H Wang JY Chang Y Yang L Wang 2016 Performance analysis of solar thermophotovoltaic conversion enhanced by selective metamaterial absorbers and emitters Int. J. Heat Mass Transf. 98 788 798
40. Q Ni H Alshehri Y Yang H Ye L Wang 2018 Plasmonic light trapping for enhanced light absorption in film-coupled ultrathin metamaterial thermophotovoltaic cells Front. Energy 12 1 185 194

41. S Agarwal YK Prajapati 2017 Analysis of metamaterial-based absorber for thermo-photovoltaic cell applications IET Optoelectron. 11 5 208 212
42. F Qin Z Chen X Chen Z Yi W Yao T Duan P Wu H Yang G Li Y Yi 2020 A tunable triple-band near-infrared metamaterial absorber based on Au nano-cuboids array Nanomaterials 10 2 207
43. BZ Xu CQ Gu Z Li ZY Niu 2013 A novel structure for tunable terahertz absorber based on graphene Opt. Express 21 20 23803 23811
44. A Fardoost FG Vanani AA Amirhosseini R Safian 2016 Design of a multilayer graphene-based ultrawideband terahertz absorber IEEE Trans. Nanotechnol. 16 1 68 74
45. D Huo J Zhang H Wang X Ren C Wang H Su H Zhao 2017 Broadband perfect absorber with monolayer MoS2 and hexagonal titanium nitride nano-disk array Nanoscale Res. Lett. 12 1 1 8
46. S Agarwal YK Prajapati 2018 Design of broadband absorber using 2-D materials for thermo-photovoltaic cell application Opt. Commun. 413 39 43
47. SMA Mostaan H Saghaei 2021 A tunable broadband graphene-based metamaterial absorber in the far-infrared region Opt. Quant. Electron. 53 2 1 14

Broadband SIW Traveling Wave Antenna Array for Terahertz Applications

Nabil Cherif, Mehadji Abri, Fellah Benzerga, Hadjira Badaoui, Junwu Tao, Tan-Hoa Vuong, and Sarosh Ahmad

Abstract A substrate integrated waveguide (SIW) traveling wave antenna with single element was proposed in this paper using novel configuration. This novel configuration is then used to construct a two-way 1 × 2 linear antenna array feed by Y-junction SIW power divider for Terahertz applications. The radiation pattern, gain and return loss were analyzed. We used Rogers RT5880 substrate which has a relative permittivity of 2.2 for construct the antenna array. Microstrip to SIW transition is used to provide smooth transition between the structures. The simulation results obtained by CST simulator show that the 1 × 2 antenna array operated in the range 0.135–0.205 THz with an average gain is about of 8 dBi and the lower return loss is about of −45 dB. The proposed antenna array is featured by low size, wideband and high gain which make it a good component for Terahertz applications.

Keywords Substrate integrated waveguide (SIW) · Two-way 1 × 2 linear antenna array · Terahertz application · Y-junction SIW power divider

N. Cherif (✉) · F. Benzerga
Electrotechnical Department, Mustapha Stambouli University, Mascara, Algeria
e-mail: nabil.cherif@univ-mascara.dz

F. Benzerga
e-mail: fellahbenzerga@gmail.com

M. Abri · H. Badaoui
Telecommunication Department, Abou Bekr Belkaid University, Tlemcen, Algeria
e-mail: abrim2002@yahoo.fr

J. Tao · T.-H. Vuong
Laplace Laboratory, Ensheeiht ,Toulouse, France
e-mail: tao@laplace.univ-tlse.fr

T.-H. Vuong
e-mail: tan-hoa.vuong@enseeiht.fr

S. Ahmad
Signal Theory and Communications Department, Universidad Carlos III de Madrid (UC3M), Madrid, Spain
e-mail: saroshahmad@ieee.org

S. Das et al. (eds.), *Advances in Terahertz Technology and Its Applications*,
https://doi.org/10.1007/978-981-16-5731-3_12

1 Introduction

Terahertz technologies have been one of the topical subjects that have great importance in industrial and academic research [1]. High gain antennas are essential for radar imaging and detection devices and wireless communication due to the high propagation loss of the mm-wave and THz wave transmission [2]. The Substrate Integrated Waveguide (SIW) is used for the design of various microwave components especially millimeter wave antennas due to its low loss characteristics, low manufacturing costs, and its ease of integration into the planar circuit [3]. Our research effort in this article focuses on the design of a broadband two-way 1×2 linear antenna array based on SIW technology operating from 0.135 to 0.205 THz using novel configuration. Initially, a single element antipodal antenna was proposed and designed ranging from 0.133 to 0.243 THz. Next a 1×2 SIW power divider having *Y*-junction was proposed, designed and optimized. Finally, the proposed antenna is integrated with the optimized 1×2 SIW power divider for realizing the antenna array [4].

2 Single SIW Antenna Design

In this part, we are going to design a new SIW traveling wave antenna (TWA) with single element used to construct the antenna arrays. Using the CST we designed a SIW TWA structure with modified comb-shaped profile configuration. The geometry of the proposed SIW antenna element is shown in Fig. 1. The used substrate is Rogers 5880 characterized by a loss tangent (tg∂) about of 0.0009, a dielectric constant of 2.2, and a thickness h of 0.508 mm [5].

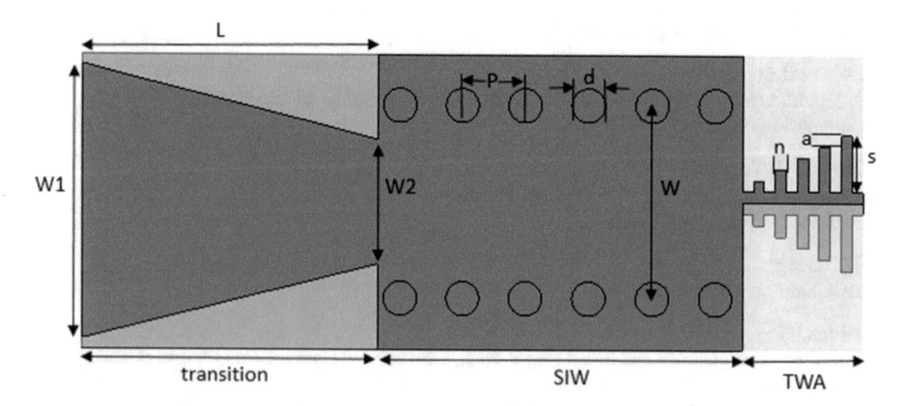

Fig. 1 The proposed antenna element configuration

Table 1 Optimized parameters of the antenna

Parameters	Value (mm)	Parameters	Value (mm)
W1	2.4	d	0.3
W2	1.1	n	0.1
L	2.7	a	0.1
W	1.7	s	0.5
p	0.6	h	0.508

As shown in Fig. 1, the modified antipodal comb-shaped antenna is comprised of three parts: the taper transition, the SIW feed and the antipodal radiating arms. The optimized value of the taper and the required vias conditions are determined by using the equations in [6].

The input impedance is characterized by 50 Ω. The optimized parameters value of the antenna is presented in Table 1.

The main design Equation of a SIW is given by Eq. (1).

$$w = w_{\mathrm{d}} + \frac{d^2}{0.95p} \tag{1}$$

w_{d} represent the dielectric filled waveguide width.

$$w_{\mathrm{d}} = \frac{a}{\sqrt{\varepsilon_r}} \tag{2}$$

For $TE_{1.0}$ mode the cutoff frequency is given by Eq. (3).

$$f_c = \frac{c}{2a} \tag{3}$$

From Fig. 2, it can be noticed that the proposed SIW antenna is well adapted and it operates in the band ranging from 0.133 THz to 0.243 THz and its lower return loss value is exceeded −60 dB at the frequency of 0.198 THz, with the appearance of the several resonance frequencies.

3 Single SIW Antenna Design

In this section, we designed a 1 × 2 SIW power divider having *Y*-junction used to feeding the array antenna and for dividing the power by the same amount. The vias P1 and P2 are positioned at the right angle to enhance the input reflection coefficient [7]. We used Rogers RT5880 substrate for construct the power divider and we used the parametric study with CST simulator for optimized the simulation results (Fig. 3).

Figure 4 shows the simulation results of the *S* parameters of the SIW power divider. The simulation results of S11, S21 and S31 point out that the power divider

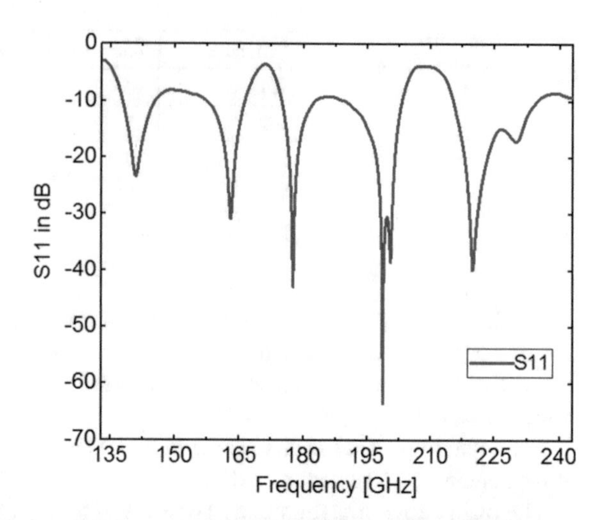

Fig. 2 Reflection coefficient of the proposed antenna

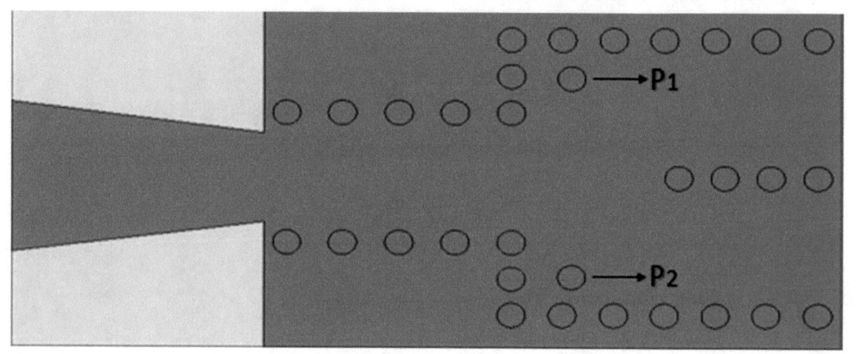

Fig. 3. 1 × 2 SIW power divider

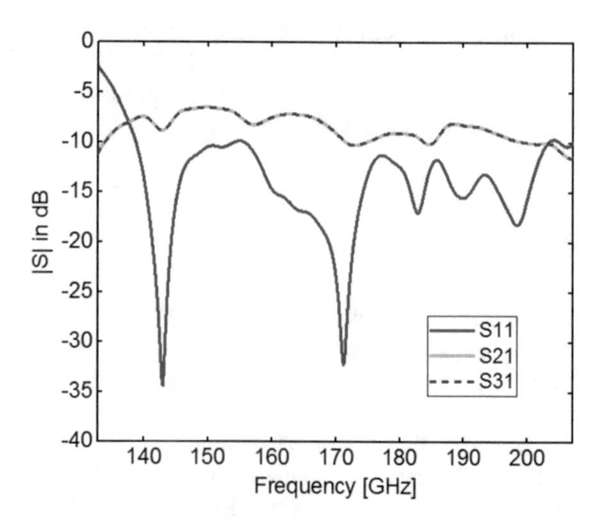

Fig. 4 The Simulated S-Parameters of the 1 × 2 SIW power divider

is operational over 0.135 to 0.205 THz with S11 < −10 dB with lower return loss about of −33 dB at the frequency of 143 GHz. it is also noted that the transmission coefficient is above than −10 dB in the whole functional band.

4 The Proposed Two-Way Linear SIW Antenna Array

The proposed antenna element is integrated with the optimized SIW power divider to construct a 1 × 2 linear antenna array. The overall size of the proposed SIW array antenna is 10 mm × 4.3 mm × 0.6 mm which make it a compact component (Fig. 5).

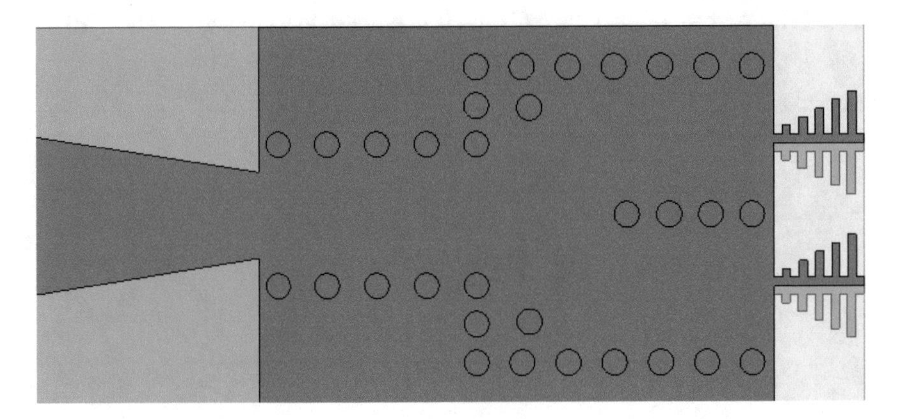

Fig. 5. 1 × 2 SIW antenna array

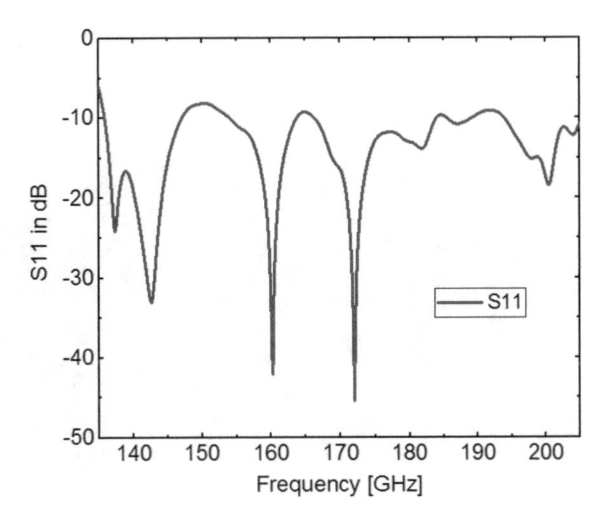

Fig. 6 The 1 × 2 SIW array antenna return loss

From Fig. 6, we notice that the value of S11 lower than −10 dB ranging over 0.135–0.205 THz and it reached −45 dB at 172 GHz. Some deteriorations appeared around the 0.150 THz caused by the resonance between the two elements of the antenna array.

Regarding to the simulated gain shown in Fig. 7, we notice that the gain values is acceptable with fair flatness, it exceeds 7.5 dBi from 0.135 to 0.155 THz and from 0.177 to 0.205 THz and its highest value is 10.3 dBi at both frequency of 0.137 THz and 0.152 THz.

Fig. 8 shows the electric distribution in the radiating arms for the frequency of 0.172 THz. We notice that the field is well distributed in the radiating arms which mean that the antenna array radiates perfectly with equiamplitude excitation.

Fig. 9 represents the 1 × 2 SIW array antenna pattern for the frequency of 0.182 THz. We can see that the radiation is maximum in the plane (Theta = 90°) in

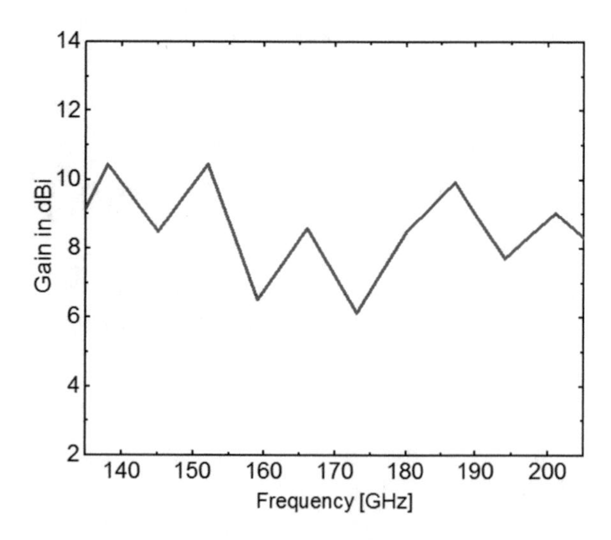

Fig. 7 The 1 × 2 SIW array antenna simulated gain

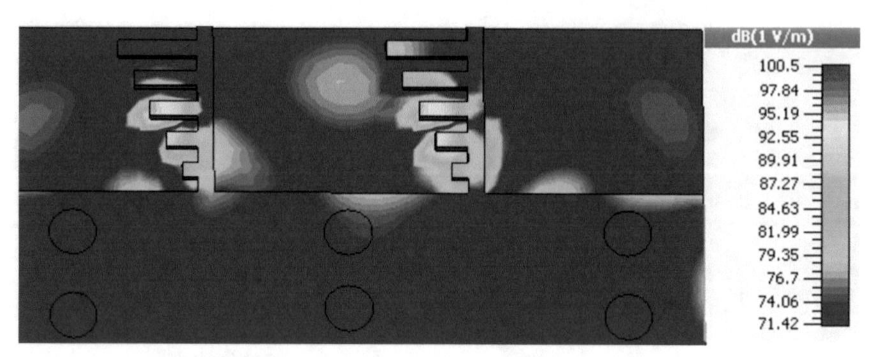

Fig. 8 Electric field distribution for the frequency of 0.172 THz

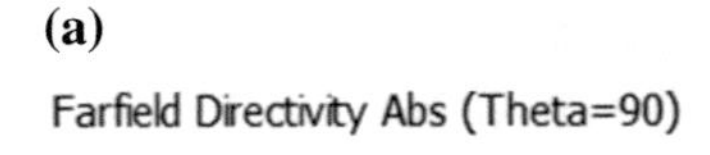

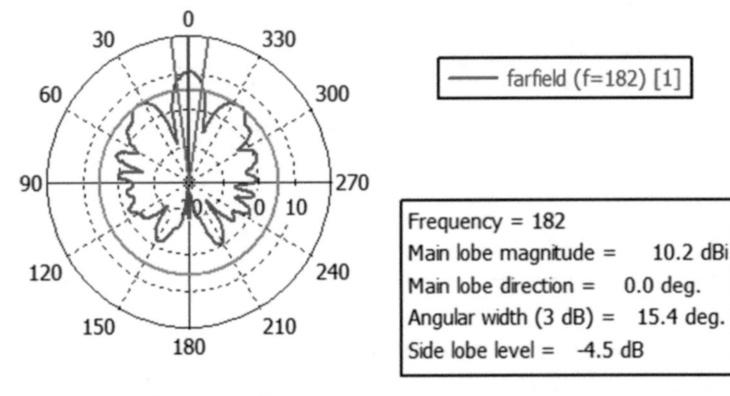

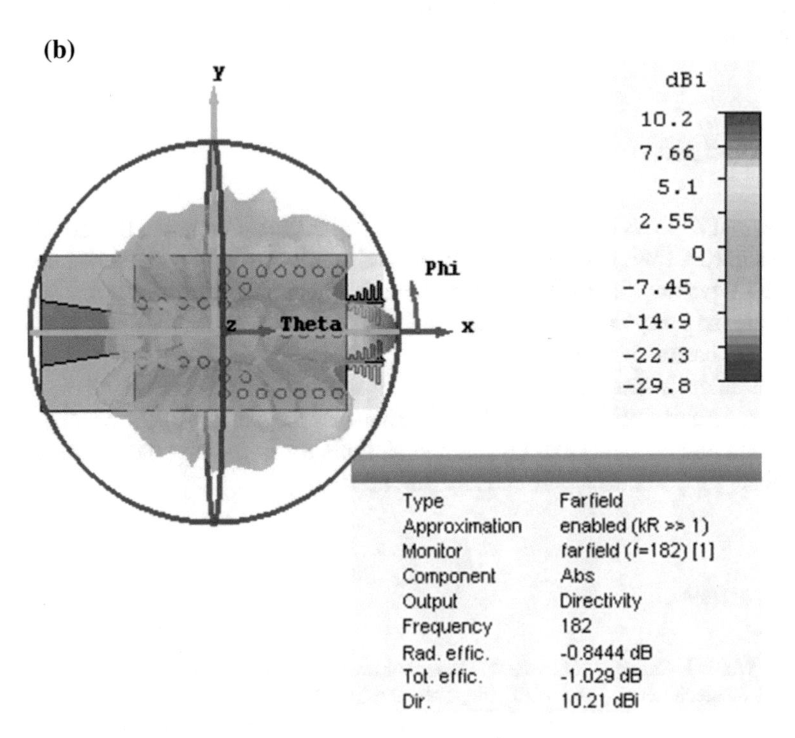

Fig. 9 1 × 2 SIW array antenna pattern in **a** polar coordinates **b** 3D

Table 2 Proposed SIW array antenna compared with previous works

References	Technology	Frequency (GHz)	BW/FBW (GHz/%)	Maximum Gain (dBi)
Li and Chiu [8]	On-chip DRA patch antenna	340	25/7.3	7.9
Vettikalladi et al. [9]	Dipole 1 × 4 array antenna	300	82/27.3	13.6
Kanaya et al. [10]	Multilayer 1 × 4 Slot antenna array	300	50/16.6	7.35
This work	SIW 1 × 2 array antenna	172	35/20.3	10

the longitudinal direction, and the maximum directivity is around the 10.2 dB with an aperture angle of 3 dB of 15.4°.

According to these results, we can use this SIW antenna array for Terahertz applications, and we can consider it as a novel contribution for scientific research (Table 2).

5 Conclusion

A 1 × 2 SIW TWA array for Terahertz applications is presented in this article. First we designed a TWA with a single element, and then we used this antenna for construct 1 × 2 TWA array fed by SIW power divider. The simulated result shows that the 1 × 2 antenna array range from 0.135 to 0.205 THz with an average gain is about of 8 dBi and narrow beamwidth of 15.4°.

This antenna array can be used for ultra wideband radars imaging and biomedical sensors. As a perspective, we will work to increase the gain and directivity by adding more antenna elements and proposing solutions for improving the return loss and eliminate the resonance between adjacent elements.

References

1. K. Wu, Y.J. Cheng, T. Djerafi, W. Hong, Substrate-integrated millimeter-wave and terahertz antenna technology. Proc. IEEE **100**, 2219–2232 (2012)
2. X. Li, J. Xiao, Z. Qi, H. Zhu, Broadband and high-gain millimeter-wave and terahertz antenna arrays. in *2019 International Conference on Microwave and Millimeter Wave Technology (ICMMT) Guangzhou China*, (2019), pp. 19–22
3. F. Benzerga, M. Abri, Design of antipodal linearly tapered slot antennas (ALTSA) arrays in SIW technology for UWB imaging. Recent Adv. Electr. Eng. Control Appl. **411**, 381–389 (2016). Springer

4. X. Li, K. Xu, Y. Li, S. Ye, J. Tao, C. Wang, SIW-fed vivaldi array for mm-wave applications. in *2019 IEEE International Conference on Computational Electromagnetics (ICCEM)* (Shanghai China, 2019)
5. L. Bin, D. Liang, Z. Jiao-cheng, The research of broadband millimeter-Wave vivaldi array antenna using SIW technique. In *IEEE 2010 International Conference on Microwave and Millimeter Wave Technology*, (Chengdu China, 2010), pp. 997–1000
6. N. Cherif, M. Abri, F. Benzerga, H. Badaoui, J. Tao, A compact wideband DGS bandpass filter based on half mode substrate integrated waveguide technology. Int. J. Microw. Opt. Technol. **16**(2), 142–147 (2021)
7. F. Benzerga, M. Abri, H. Badaoui, Optimized bends, corporate 1×4 and 1×8 SIW power dividers junctions analysis for V-Band applications using a rigorous finite element method. Arab. J. Sci. Eng. **41**, 3335–3343 (2016)
8. C.H. Li, T.Y. Chiu, 340-GHz low-cost and high-gain on-chip higher order mode dielectric resonator antenna for THz applications. IEEE Trans. Terahertz Sci. Technol. **7**(3), 284–294 (2017)
9. H. Vettikalladi, W. Sethi, A.F. B. Abas, Ko. Wonsuk, M. Alkanhal, M. Himdi, Sub-thz antenna for high-speed wireless communication systems. Hindawi Int. J. Antennas Propag. 1–9 (2019)
10. H. Kanaya, T. Oda, N. Iizasa, K. Kato, 300 GHz one-sided directional slot array antenna on indium phosphide substrate. in *International Symposium on Antennas and Prpagation (ISAP)* (Hobart TAS Australia, 2015), pp. 1–2

Polarization of THz Signals Using Graphene-Based Metamaterial Structure

Vishal Sorathiya and Sunil Lavadiya

Abstract Graphene-based Terahertz devices have attracted huge attention because of their ultrathin design and tunable property. The graphene-based polarizer can be formed using a single or multilayer of graphene sheet over the dielectric substrate. The different shapes and size of the engraved graphene geometry make possible to design different band and different mode of the polarizer which was ultrathin in design. The graphene-assisted polarizer also has the tunable by various physical parameters such as chemical potential, frequency, scattering rate. The graphene-based polarizer also provided unusual material properties like negative refractive index which makes the overall polarizer structure a metamaterial device. The proposed book chapter provides the fundamentals of graphene-based polarization devices. The chapter includes the mathematical modeling of the graphene-based polarizers devices and numerical investigation techniques used to identify the performance of the graphene-based polarizer structure. The chapter also includes a detailed comparative analysis of the previously published and available polarization devices in the market.

Keywords Graphene · Polarizer · Terahertz · Tunability

1 Introduction

Metamaterials (MMs), a modern form of the artificial substance that was recently investigated in terms of their non-traditional electromagnetic properties. These features are used to achieve numerous results such as negative refractive index [1], perfect lensing [2], bolometer [3], etc. On the other way, Graphene owns exceptional optical, electrical, and mechanical properties, such as large young modules, high

V. Sorathiya (✉) · S. Lavadiya
Department of Information and Communication Technology, Marwadi University, Rajkot, Gujarat 360002, India
e-mail: vishal.sorathiya@marwadieducation.edu.in

S. Lavadiya
e-mail: sunil.lavadiya@marwadieducation.edu.in

S. Das et al. (eds.), *Advances in Terahertz Technology and Its Applications*,
https://doi.org/10.1007/978-981-16-5731-3_13

thermal conductivity, and massive carrier mobility [4]. Graphene material can be realized with the bandgap structure of Dirac tapered. Graphene also offers the properties of linear dispersion. Graphene is a one-atom-thick two-dimensional material that provides electrical and optical tunability over the ultra-wide range of THz and GHz [5] frequency. It is also attracted huge attention for research on preparation methods of graphene and its composition material. Similarly, the research community also identifying the various material property extraction from graphene and its composite devices. The property of the single-layer graphene can be controlled by temperature, scattering rate, chemical potential, and frequency [6]. Several devices were studied previously for Gigahertz to Terahertz region for achieving wide-angle [7], perfect absorption, tunability [8], and polarization-insensitive [9]. These types of properties of graphene provide the scope to design different tunable graphene-based photonics components like absorber [10], THz antenna[11], polarizer [12], grating structure [13, 14], Frequency selective surfaces [8], and bandpass filters. The current metamaterial does not have features of tunability. The conventional metamaterial is realized by the different layers of metal, Si_3N_4, SiO_2, and Al material. It is possible to achieve the tunable feature only with the thicker size of conventional material. The limitation of the size (specially thickens) and tunability can be fulfilled by the graphene-assisted metamaterial structures. Al, silica and Au substrate are used to form the single-layered graphene-based metamaterial devices. There are different shapes like T shaped [15], L shaped [16], rectangular split ring-shaped, C shaped [12] are engraved on graphene sheets to realize metamaterial polarizer. The graphene-based polarizer with composited graphene metallic wire grid array is one of the famous types of the polarizer to fabricate. There is wide use of the graphene-based polarizer in various applications like an attenuator, modulator, photonics sensor, etc.

2 Graphene Conductivity Model

The single layer of the graphene sheet between two different materials has been depicting as shown in Fig. 1. The sheet of graphene is considered as laterally infinite over the x-z plane. Graphene is interfaced as $\mu_1\varepsilon_1$ for $y > 0$ and $\mu_2\varepsilon_2$ for $y \leq 0$ between two mediums as shown in Fig. 1. Here we may consider the all-parameters values consisting of complex terms. The proposed graphene structure can be considered as ultrathin with a two-sided surface conductivity model with surface conductivity $\sigma(\omega, \mu_c, \Gamma, T)$. The conductivity of the graphene is considered by Kubo's Formula [6, 17, 18]. The equation of the surface conductivity model is shown in Eq. 1.

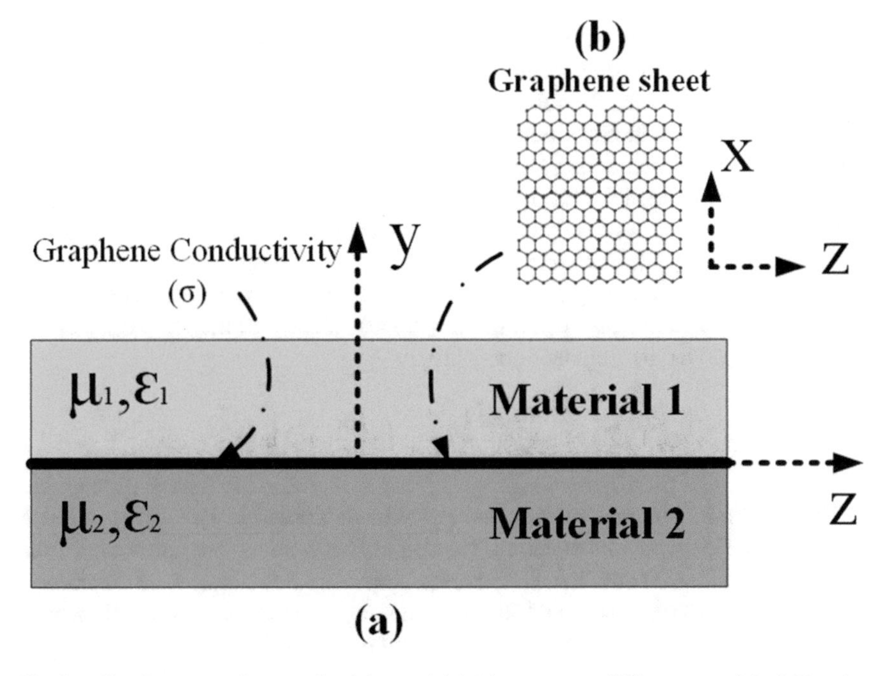

Fig. 1 **a** Graphene as surface conductivity model (σ) between two different materials. **b** Top view of the graphene. Hexagon is denoted as a six atom carbon structure

$$\sigma(\omega, \mu_c, \Gamma, T) = \frac{je^2(\omega - j2\Gamma)}{\pi\hbar^2}\left[\frac{1}{(\omega - j2\Gamma)^2}\int_0^\infty \varepsilon\left(\frac{\partial f_d(\varepsilon)}{\partial\varepsilon} - \frac{\partial f_d(-\varepsilon)}{\partial\varepsilon}\right)d\varepsilon - \int_0^\infty \frac{f_d(-\varepsilon) - f_d(\varepsilon)}{(\omega - j2\Gamma)^2 - 4(\varepsilon/\hbar)^2}d\varepsilon\right] \quad (1)$$

Here, ω is defined as the frequency in radian, μ_c is considered as the chemical potential of the graphene sheet. The graphene's chemical potential can be controlled externally. Γ and T are the scattering rate and temperature respectively. Here Γ is assumed to be not dependent on energy. $\hbar$ and e are the reduced Planck constant and charge of electron respectively. The function $f_d(\varepsilon)$ is defined as $f_d(\varepsilon) = \left(e^{(\varepsilon-\mu_c)/k_BT} + 1\right)^{-1}$. Here the local conductivity is considered isotropic by assuming no magnetic field. Equation 1 is categorized in the two conditions as intraband (first part) and interband (second part) conductivity. The carrier density values n_s is used to identify the value of chemical potential μ_c as shown in Eq. 2.

$$n_s = \frac{2}{\pi\hbar^2 v_F^2}\int_0^\infty \varepsilon[f_d(\varepsilon) - f_d(\varepsilon + 2\mu_c)]d\varepsilon \quad (2)$$

where the value of Fermi velocity v_F is considered as approximately 9.5×10^5 m/s. This carrier density can be controlled using external biasing and/or doping voltage. It also needs to consider the for non-doped conditions T = 0K as well as $n_s = \mu_c = 0$.

The intraband conductivity term can be derived as shown in Eq. 3. This equation is first derived by considering the case $\mu_c = 0$ for graphite (by considering the additive factor included for the interlayer connection between two-layer graphene in graphite) intraband conductivity of single-walled carbon nanotube by considering the infinite radius [19]. The simple conductivity, in this case, is considered as real and imaginary part using $\sigma_{intra} = \sigma' + j\sigma''$. Where the values of both part $\sigma_{intra}^{'} \geq 0$ and $\sigma_{intra}^{''} < 0$. The imaginary part of the graphene sheet will help to propagate the surface wave guided by graphene sheet [20].

$$\sigma_{\text{intra}}(\omega, \mu_c, \Gamma, T) = \frac{-je^2 k_B T}{\pi \hbar^2 (\omega - j2\Gamma)} \left(\frac{\mu_c}{k_B T} + 2 \ln \left(e^{-\frac{\mu_c}{k_B T}} + 1 \right) \right) \tag{3}$$

The interbrand conductivity for the approximate condition $k_B T \ll |\mu_c|, \hbar\omega$ is presented by Eq. 4. In this equation, the conductivity values are considered from scattering rate $\Gamma = 0$ and $2|\mu_c| > \hbar\omega$, $\sigma_{\text{inter}} = j\sigma_{\text{inter}}^{''}$ and $\sigma_{\text{inter}}^{''} > 0$ conditions. For the conditions of $\Gamma = 0$ and $2|\mu_c| < \hbar\omega$, σ_{inter} are the complex values with $\sigma_{\text{inter}}^{'} = \pi e^2/2h = \sigma_{\min} = 6.085 \times 10^{-5} (S)$ and $\sigma_{\text{inter}}^{''} > 0$ for $\mu_c \neq 0$.

$$\sigma_{\text{inter}}(\omega, \mu_c, \Gamma, T) = \frac{-je^2}{4\pi \hbar} \ln \left(\frac{2|\mu_c| - (\omega - j2\Gamma)\hbar}{2|\mu_c| + (\omega - j2\Gamma)\hbar} \right) \tag{4}$$

3 Graphene-Based Polarizer

They are generally using the Drude models of conductivity to form conventional polarizers which are based on the quantum well-based semiconductor structure. While the metal's imaginary conductivity part is greater than 0, which is used to design only TM polarizer [21] devices. Some extensive research is available on how graphene conductivity is to be calculated [22]. The results published in Hanon [18] in 2008 include most scholars from the industry. Graphene was equated with a 2D material, The surface conductivity of the graphene was measured using the Kubo formula. Because of its adjustable conductivity, graphene can achieve the properties of a metal or transparent medium under the control of different external applied voltages. The polarization controller can therefore be implemented in TE or TM mode with graphene material that cannot possibly be obtained by previous materials. In this respect, graphene is increasingly attractive as a new material for the development of polarizers.

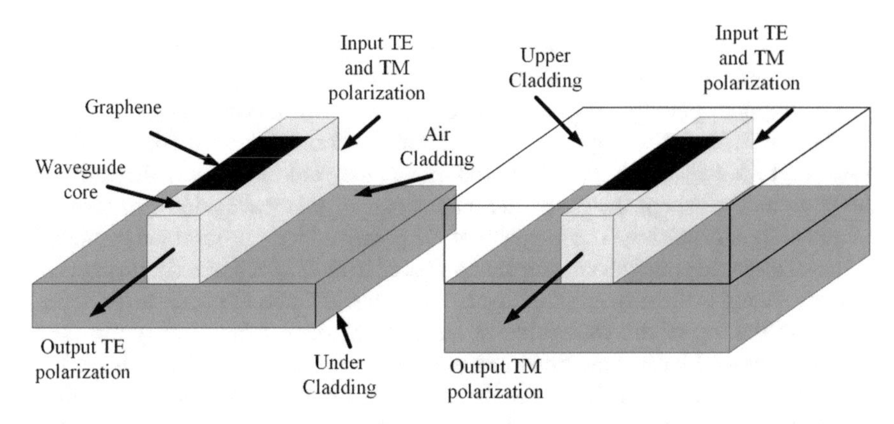

Fig. 2 Schematic structure of polymer-based graphene polarizer device. **a** Growth of the graphene sheet realized. This structure supports the one mode pass (TE mode) polarizer because the air cladding supports a graphene sheet. **b** Modified TM polarizer structure realized by placing the UV curable polymer of CVD grown graphene sheet

There is a huge demand for the creation of compatible photonics integrated circuits with complementary metal-oxide-semiconductor (CMOS) structures. Graphene-based photonics devices are fortunately highly compatible with available conventional CMOS devices. Kim et al. developed a hybrid, optical flat waveguide polarizer. The structure is shown in Fig. 2. In this case, it can be used for selection as a TM or TE pass polarizer to identify whether a top layer cladding is available in the system [23].

The polarizer is based on the graphene-based single-layer sheet. In this structure, both layer (Upper and lower) is formed with per fluorinated polymer acrylate, UV curable, which can adjust graphene's carrier density and conductivity, to allow TM or TE surface waves to be selectively supported. The TE pass polarizer consists of graphene, which consists of a rectangular cross-section corner where the RI is 1.37 and 1.39, as shown in Fig. 2a, respectively. In addition, since in the absence of upper cladding, the waveguide is covered with air. The electrically tunable TM polarizer structure with the graphene monolayer sheet on a guide core as shown in Fig. 2b, The difference between Fig. 2a is only a polymer with a UV curable on top of the layer of the waveguide. The rest of the RI values of the other layer is the same as the TE pass polarizer.

The polymer waveguide with graphene sheet in the polarizer is created first by assembling a lower cladding structure with 20 μm-thick sizes on a silicon slab wafer and then treating it with UV light. The core of the waveguide is transferred to the lower cladding layer, which has a thickness of 5 μm and formed with polymer, which is irradiated with UV light to cure it. The thermochemical vapor deposition (CVD) method has been used to transfer the graphene film on SiO_2/Si substrate. The graphene film for this structure has been grown using thick nickel sputtering with a thickness of 300 nm. The mask is created by photolithography and then modifies using O_2 plasma to form a linear optical fiber waveguide with graphene tape on the top of the core structure in the polarizer waveguide region.

The graphene polarizer's optical characteristics were tested first by coating it with air. Afterward, the upper layer of the polarizer was formed by spin-coated polymer resin and tested again. To get an infrared image, they did the same as they did with the photometer, except they used the output port of this waveguide structure as a polarized light capturing. The intensity of TE polarized light is observed strongly in the air-clad-based polarizer structure as compared to the TM-polarized light intensity. The key factor is the air-coupling layer. While the UV curable resin layer is been placed on the top of the waveguide in the TM polarizer. After placing this layer, the TM-polarized light is modified to a more intense spot. The TE light has a faint intensity, which may be detectable by polarized light. There is a very small RI difference between the upper and lower layers. Although the slab mode can be seen in TM-polarized light. This structure with UV curable resin work as a TM-polarizing waveguide. Also, the insertion loss is measured in the waveguide to investigate the polarizing effect.

Graphene strips have a notable effect on insertion loss in waveguides with cores. For polarizing TE-treated graphene light, generates the 10.9 dB of the extinction ratio 20.7 dB of the insertion loss. On average, the TE-related modified graphene polarizer had a 50 dB of insertion loss, and the TE insertion loss was found to be 19.8 dB. This results in the majority of the optical power being radiating in a radiative mode. As a result, Kim concluded that the proposed waveguide device takes full advantage of the variation in optical and electrical characteristics concerning external biasing connected to graphene for use in on-chip PICs.

4 Graphene-Based Polarizer Using Optical Fiber

In an optical network, an optical fiber polarizer is one of the most important passive components, as it controls the orientation of the phase and polarization. It is especially popular due to its lightweight, quick rise time, long lifetime, and ease of integration with the optical fiber system (High compatibility). At present, the fiber optic sensing system and fiber gyroscope is the primary device for polarized light generation in the fiber. This system will also play an important role in the optical fiber system. An optical attenuator can be split into an optic polarizer. The difference between two modes of polarization, one stronger and one weaker, is to be increased for creative interference. The longer the mode is used, the less loss it produces, so the shorter the other mode gets. The polarization identity of these two endpoints follow is that one mode of polarization end of the line and the polarization mode flows from it. For the typical fiber, polarizers based on metal-clad fiber, toroidal clad fiber [24], and crystal clad fiber are frequently used [25].

Graphene's outstanding optical and electrical properties mean that its chemical potential can be externally manipulated. The "n-doped" state of graphene can be obtained by adding metals to the atom-doped state of graphene, and the "p-doped"

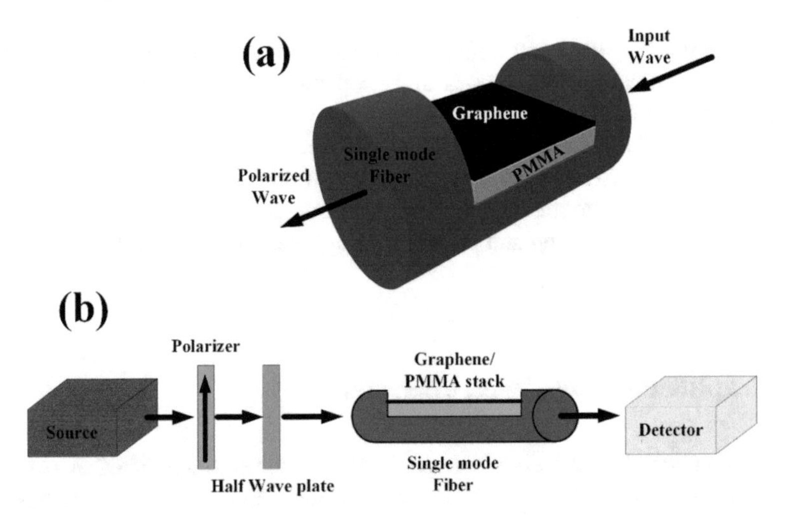

Fig. 3 **a** Schematic of the stacked PMMA/graphene-based fiber-based polarizer structure. **b** Setup for identifying extinction ratio measurement

graphene state can be attained through the introduction of polymer molecules such as F, O, N materials. This can be also attained with other semiconductor elements. Additionally, the surface transfer doping of graphene is containing more stable conditions. While the variationally doping having less unstable conditions than surface transfer doping. The quality of the overall surface coverage of the number of atoms or chemical species or dopant atoms that are mixed into the basal plane is impossible to replicate. We need precise ways to uniformly and precisely dope n-type/p-type graphene for various applications, and then use these methods to control the concentration of dopant. It is possible to effectively suppress the Transverse magnetic and transverse electric modes by controlling the chemical potential of graphene.

Figure 3a shows 3D schematic of the PMMA/Graphene stacked fiber-based polarizer structure and Fig. 3b shows the proposed setup for identifying extinction ratio measurement values. The wheel side-polished method is used to create the polished surface. The fiber polarizer loss is dependent on parameters such as a side-polished facet roughness and the length and surface. The measured fiber polarizer loss was smaller than 1 dB/mm is observed for the polishing depth of 62.5 μm. This loss can be reduced by making a more polished surface. The CVD method is used to grow graphene on copper foil. A spin coater with a very high rotation speed is used to form the 400 nm thick PMMA film.

While this treated copper foil was sitting in ferric nitrate nonahydrate solution for two hours, the rest of the PMMA-graphene surface was flushed with deionized water. The evaporation of any leftover solvent left the fiber materials in contact, which resulted in a single layer of graphene and PMMA. It was coated a second time with a graphene/PMMA sheet and transferred again, yielding a double-sided PMMA stack-based graphene polarizer. Because of a modal evanescent field enhancement associated with the high-index fiber cladding, light can propagate throughout the core of the fiber. With strong light interaction between the modes, the attenuation between orthogonal polarizations is high, and this results in a lower value of the insertion loss because of the device compactness. When the device is only 2.5 mm in length, a double-layer PMMA/graphene stacked fiber polarizer is composed of a 5 dB insertion loss and a 36 dB extinction ratio. A double-layer of bilayer graphene (on both sides) has outperformed single-layer graphene and single-coated graphene (one side only) in designing and manufacturing polarizers with a ratio of 44 dB for 4 mm device length.

5 Graphene Array-Based Polarizer

Several graphene-based THz polarizers with the composition of other materials have been presented previously [26]. Metasurfaces devices are manipulating EM waves using two-dimensional anisotropic array structures. It is possible to create tunable THz polarizers, either using pure graphene metasurfaces structure or through graphene and metal incorporated metamaterial geometries. These structures can control the polarization direction of the incident light and convert the linear state of polarization to circular polarization states. More interestingly, graphene-based THz polarizers are essential to offer THz attenuator devices for switching devices for THz communications [27]. The wire grids-based THz polarizer with the substrate [28] or free-standing [29] structures offer an approximately 40 dB of high extinction ratio and 1 dB of low insertion loss for lower THz range up to 1 THz [30].

This type of polarizer with metallic wire grid structure shows a high transmission amplitude for TM polarizer wave and low transmission for TE polarized wave [29]. The same effect can be observed for wire grids made up of graphene sheets, for the near-infrared and microwave range [31–33] of the EM wave spectrum. In addition to graphene wire grid structure generates the low transmission for THz waves (TM polarized) with the effect of wave plasmon coupling [34, 35]. This graphene wire grid-based THz polarizers having a limitation of low transmission in both transmission modes. Another limitation in graphene wire grid-based polarizer structure is the low modulation (~10%) (TE mode of polarization), which is created by the Drude absorption [34] effect.

This polarizer is formed such that both metal and graphene elements are interleaved. Metal metasurface in the polarizer structure is fabricated from a graphene grid and metal patch array structure. A graphene sheet is used in the metal grid structure. This offers a higher conductivity than the grid structure formed from individual graphene sheets. The graphene wire hybrid polarizer structure, with graphene wire grid placed onto a metal metasurface, is shown in Fig. 4. The graphene filling factor (GFF) responsible for the conductivity modulation. Smaller values of GFF along the wire make a larger conductivity modulation, which will make larger THz transmission modulation.

The metal metasurface along with the graphene patch wires makes a strong TM-polarized plasmon mode, which causes the separated graphene patch geometry to be

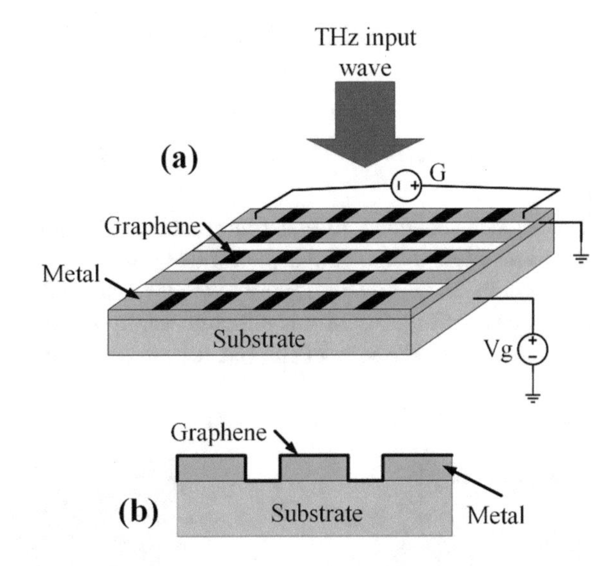

Fig. 4 **a** Three-dimensional view of hybrid wide grid formed with graphene metal-quarts based polarizer structure. **b** Cross-sectional view of the graphene metal quarts based polarizer structure

connected with an individual metal strip. The coupling efficiency of wave plasmon interaction is depending on the dimensions of graphene patch up to a certain width, the TM-polarized coupling can be suppressed by decreasing graphene wire length [36]. The metal metasurface provides the TE polarized plasmon into graphene. The polarized plasmon (TE mode) interacts with the THz wave and it will also tunable by different values of graphene conductivity. The THz transmission modulation variation can be realized with these TE polarized plasmons [37]. The metal graphene structure forms the inductive and capacitive circuit which generates the coupling resonance with TE polarized plasmons. In this circuit, charge accumulation is formed by metal and an inductive channel formed by graphene sheet.

The graphene-wide grid-based polarizer is fabricated on ST-cut SSA quartz substrate with 0.35 mm thickness. DLP laser lithography uses to fabricate the metal patterns. In this lithography, the layer of Ti (7 nm) and Au(70 nm) has been formed with electron beam evaporation. The wet transfer procedure [38, 39] is used to transfer the CVD graphene sample on the metal array. The spin coat of the PMMA is taken as support in this process. The copper foil was removed after this using an ammonium persulfate solution with a 2% concentration. Initially, graphene is covered the whole area of the metal surface. The oxygen plasma at 50 W of power has been applied for 2 min to each graphene wire structure. The photoresist material S1813 is used to protect the graphene wire. This photoresist was removed in acetone after etching. The ion gel was prepared by dissolving PVDF-HFP in acetone with the aid of a magnetic stirrer for an hour, before adding the [EMIM][TFSI] into the solution and stirring for 24 h. The gel was prepared with ionized solvent as suggested in [40]. This gel offers large carrier density and good transparency at lower applied gate voltage [41].

6 Graphene Gold Patch-Based Three-Layered THz Polarizer

6.1 Introduction and Design Modeling

The graphene conductivity equation allows to articulation single-layer graphene sheet in various FEM computational platforms as a surface conductivity model. The single-layer graphene-based polarizer structure is numerically investigated as per the structure shown in Fig. 5a. In this structure width and length of the structure (W) are set as 7.6 μm. The size of the patch is set as L set as 4 μm. The height of the

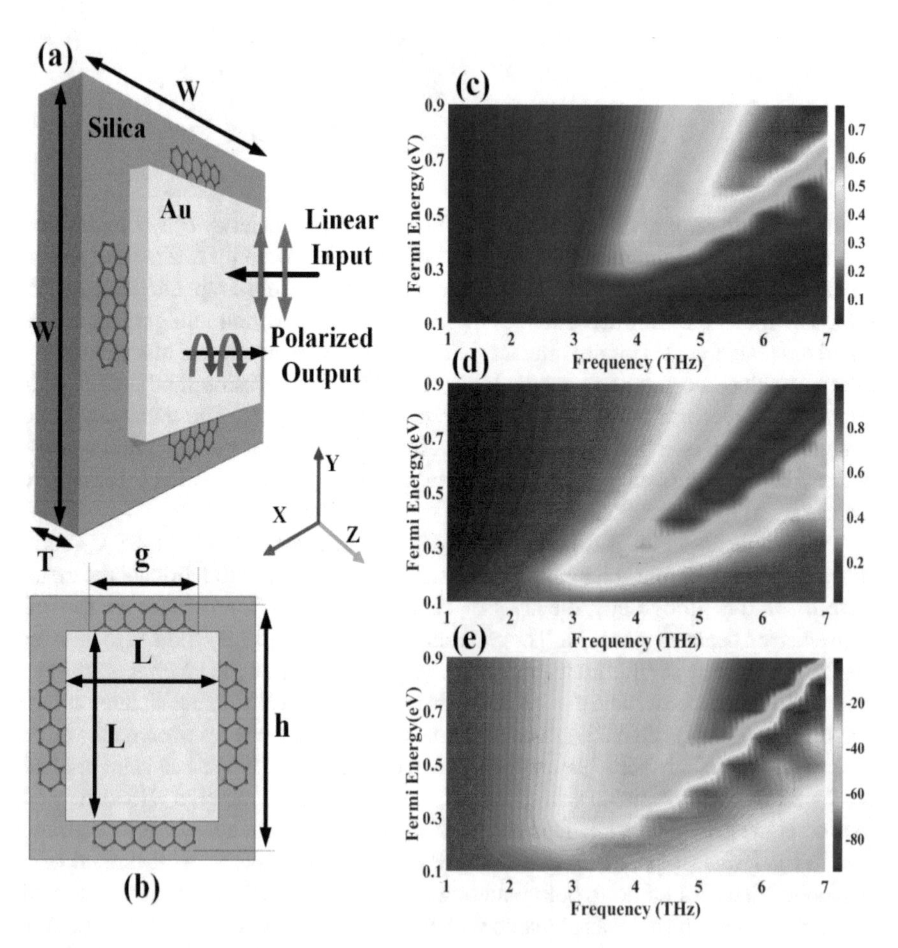

Fig. 5 Graphene-based polarizer structure for THz frequency spectrum. **a** 3D view of the polarizer structure. **b** Top view of the polarizer structure. **c** Reflectance and **d** transmittance amplitude for different fermi energy values of graphene and frequency. **e** Phase variation for the different fermi energy values of graphene and frequency

gold patch is set as 200 nm. The graphene is articulated as a cross-shaped structure with a single-layer structure. The graphene conductivity equation is used as given in Eqs. 1–4. The length and width of the graphene patch are set as ($g \times h$) 3.25 × 6.5 μm. The thickness of the silica is set as 1.5 μm. The wave is excited about the *Z*-axis. *X*-*Y* boundaries are set as periodic boundary conditions. The reflection and transmitted amplitude variation are calculated by considering the two-port system for the proposed structure. The fermi voltage of the graphene is varying from 0.1 to 0.9 eV for the graphene sheet.

6.2 Results and Discussion

The reflectance and transmittance amplitude variations are shown in Fig. 5c, d. the reflected wave phase change is shown in Fig. 5e. This fundamental structure is used to identify the wave polarization for the THz range and different Fermi energy of the graphene sheet. The reflected wave amplitude and its polarization values help to identify the polarization conversion (linear/circular/elliptical). The comparative table of such layered polarizer devices is shown in Table 1.

Table 1 Comparative study with the proposed polarizer structure and previously studied the structure in terms of various physical parameters

References	Material	Dimensions (W, L, H) in μm^3	Max. reflectance (%)	Type of layer	Wide-angle stability	Min. and max. frequency band
Liu et al. [42]	Metal	(200, 200, 90)	55	Single layer	70	1.6–5
Liu et al. [43]	Metal	(100, 100, 26)	70	Multi layer	40	0.76–1.48
Yu et al. [44]	Graphene dielectric gold	(120, 120, 59)	40	Single layer	–	0.4–1
Fardoost et al. [45]	Graphene	(15, 15, 19)	25	Multi layer	–	1.65–4.35
Deng et al. [46]	Tantalum nitride	(70, 70, 25)	60	Single layer	45	1.17–2.99
Zhu et al. [47]	Graphene dielectric Silica	(16, 16, 25)	22	Single layer	40	0.6–2.6
Shi et al. [48]	Metal	(105, 105, 25)	80	Multi layer	60	0.7–2.5
Gao et al. [49]	Graphene silica	(4, 4, 0.25)	75	Single-layer	–	1–4

7 Graphene Metamaterial-Based Polarizer

Metamaterials, which are not found in nature, are frequently the focus of interest over the last two decades, are experiencing increasing levels of public scrutiny in academic research. a Few kinds of materials can be utilized in the application, including semiconductors [50–52], periodic materials containing multiple metals, which are used in optics and electromagnetics metamaterials, [50–52]. particularly, their property of being able to be expanded to make lightweight or deformable items without being of physical substance is noteworthy until that time static does not prove useful, expansion of metals is unviable, we must restrict development.

The properties of grapheme—an active semiconductor that can also act as a plasmonic device—have recently increased in use as metamaterials have started to be focused on its new dynamic properties [53]. Giant gold nanoparticles have the potential for helping with certain optical devices, especially those with light transmission, this is known as plasmon-induced transparency (PIT) The rule is that if the linearly polarized light enters a semi-transparent glass, the model will shift from clear to opaque. Other than that, dark and bright modes based on the structure of the graphene can be additionally configured with a fermi level, there are bright and dark ones that are based on switching to and lowering or increasing the Fermi level within graphene. PITIs are commonly employed, which may make the PIT phenomenon common. Thus, metamaterials used in this way have also begun to be discovered as a research interest.

The metamaterial structure as a single periodic unit comprising four graphene blocks and strips is shown in Fig. 6a. The linearly polarized light is excited from the

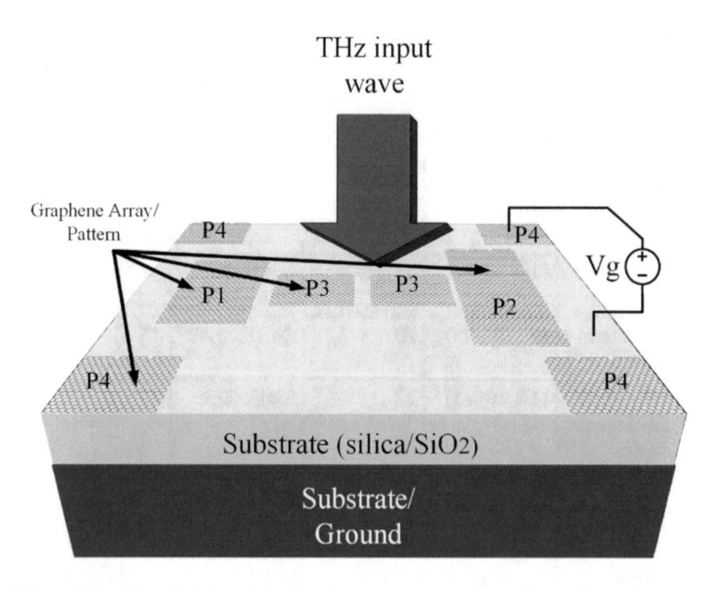

Fig. 6 **a** Three-dimensional view of the graphene patch array-based polarizer structural unit

top of the structure as shown in Fig. 6. The different graphene metamaterial structure generates different transmission spectrum for the different incident source polarization conditions by exciting different graphene patch on the array. It is possible to design such an array-based graphene polarizer structure using different graphene engraved geometries [54, 55]. Many studies show that the polarization angle of the light will have a great influence on the performance of the metamaterial structure. To check the behavior of the variation in the polarization state in this study the different structures of graphene metamaterial will produce individually. The polarizer structure with only the P1 patch in Fig. 6 will generate the red Lorenz lines for *X* polarization incident conditions. It is possible to generate the bright modes as they are interacting directly with the incident light and graphene in presence of patch P1/P3. This interaction generates the red Lorenz line. However, The polarizer structure with a single P4 patch generates the dark mode as it is not interacting with the direct incident light. Surprisingly, In *Y* polarized light the for all the structures the red Lorenz lines disappear. A graphene structure can produce multiple resonances points. Multiple resonant dips appear at different frequencies for this graphene-based metamaterial structure. In this research study the low-frequency first-order resonance shown; complex phenomenon results for the second or higher order resonating modes. This mode is ignored in the studies. Similarly, in many cases of split ring structure, *I* shaped, *L* shaped, cross-shaped graphene engrave geometry it possible to generates the different polarization effects with different input incident angles.

The applications of the graphene-based polarizer device are shown in Table 2. The graphene-based polarizer is used in many fields such as light manipulation devices, Photodetector, polarization sensors, modulators, radiating antenna, etc. The short working methodology along with the advantages as compared to other non-graphene-based devices are shown in Table 2.

8 Conclusion

Graphene-based THz polarizer has attracted the large attention of the scientific community due to its reconfigurable and ultrathin design. The different graphene-based polarizer structure offers to generate tunability, TE to TM mode conversion, ultrathin design. The graphene-based metamaterial polarizer is used for potential applications such as sensor, photodetector, antenna, modulator, polarizer, etc. In this chapter, we presented various designs and structures of the polarizer with its working principles, fabrication techniques, and advantages. We presented useful mathematical equations to articulate the graphene-based conductivity model which helps to define properties of graphene in FEM calculation. A Three-layer silica-graphene-gold-based structure is numerically investigated using the same graphene conductivity formulation and FEM environment. This chapter will give the idea of forming fundamental graphene-based polarizer and other photonics devices for various THz and electro-optical applications.

Table 2 Application of the graphene-based polarizer devices with working methodology and possible advantages of structure

Type of devices	Working methodology	Advantages
Polarization sensor	• Enhance the absorption values • Use the attenuated total internal reflection	• Fast speed of measurement • Real-time processing • Sensitivity is high • Low sample consumption
Polarization Photodetector	• The different shaped structure used with SiO_2 substrate • Different pattern structure created different surface plasmon • Photodetector tunability achieved using an external biasing voltage of graphene	• Fast and responsive speed • High rate of detection • Works on broad-spectrum and tunable
Polarization modulator	• Absorption is changed concerning the applied voltage by refractive index variation • Light absorption rate controlled by an electrical signal which is ultimately modulation variation	• Small loss • Small volume • Modulation efficiency and extinction ratio is high
Polarizer	• Depends on the leak mode created by the graphene surface • When epsilon reaches zero, one mode loses from TE/TM and one remains available at another side of the structure	• Easy to integrate • Small size • High compatibility with another device • Working on a wideband spectrum
Polarizer based antenna	• Different shape of structure generates different polarized radiation signal • Graphene work as a conductive medium for radiation • Tunable with external biasing	• Thin and small size • Tunable

References

1. R.A. Shelby, Experimental verification of a negative index of refraction. Science (80-.) **292**(5514), 77–79 (2001)
2. R. Singh, E. Plum, W. Zhang, N.I. Zheludev, Highly tunable optical activity in planar achiral terahertz metamaterials. Opt. Express **18**(13), 13425 (2010)
3. T. Maier, H. Brückl, Wavelength-tunable microbolometers with metamaterial absorbers. Opt. Lett. **34**(19), 3012 (2009)
4. A.K. Geim, K.S. Novoselov, The rise of graphene. Nat. Mater. **6**(3), 183–191 (2007)
5. F. Ding, Y. Cui, X. Ge, Y. Jin, S. He, Ultra-broadband microwave metamaterial absorber. Appl. Phys. Lett. **100**(10), 103506 (2012)

6. G.W. Hanson, Dyadic green's functions for an anisotropic, non-local model of biased graphene. IEEE Trans. Antennas Propag. **56**(3), 747–757 (2008)
7. Y. Bai, L. Zhao, D. Ju, Y. Jiang, L. Liu, Wide-angle, polarization-independent and dual-band infrared perfect absorber based on L-shaped metamaterial. Opt. Express **23**(7), 8670 (2015)
8. Y. Guo, T. Zhang, W.Y. Yin, X.H. Wang, Improved hybrid FDTD method for studying tunable graphene frequency-selective surfaces (GFSS) for THz-wave applications. IEEE Trans. Terahertz Sci. Technol. **5**(3), 358–367 (2015)
9. V. Sorathiya, V. Dave, Numerical study of a high negative refractive index based tunable metamaterial structure by graphene split ring resonator for far infrared frequency. Opt. Commun. **456**(June 2019), 124581 (2020)
10. L. Thomas, V. Sorathiya, S.K. Patel, T. Guo, Graphene-based tunable near-infrared absorber. Microw. Opt. Technol. Lett. **61**(5), 1161–1165 (2019)
11. S.K. Patel, V. Sorathiya, T. Guo, C. Argyropoulos, Graphene-based directive optical leaky wave antenna. Microw. Opt. Technol. Lett. **61**(1), 153–157 (2019)
12. V. Sorathiya, S.K. Patel, D. Katrodiya, Tunable graphene-silica hybrid metasurface for far-infrared frequency. Opt. Mater. (Amst) **91**, 155–170 (2019)
13. S.K. Patel, M. Ladumor, V. Sorathiya, T. Guo, Graphene based tunable grating structure. Mater. Res. Express **6**(2), 025602 (2019)
14. J. Parmar, S.K. Patel, M. Ladumor, V. Sorathiya, D. Katrodiya, Graphene-silicon hybrid chirped-superstructure bragg gratings for far infrared frequency. Mater. Res. Express **6**(6) (2019)
15. Y. Niu, J. Wang, Z. Hu, F. Zhang, Tunable plasmon-induced transparency with graphene-based T-shaped array metasurfaces. Opt. Commun. **416**(January), 77–83 (2018)
16. T. Guo, C. Argyropoulos, Broadband polarizers based on graphene metasurfaces. Opt. Lett. **41**(23), 5592 (2016)
17. K. Ziegler, Minimal conductivity of graphene: nonuniversal values from the Kubo formula. Phys. Rev. B **75**(23), 233407 (2007)
18. G.W. Hanson, Dyadic green's functions and guided surface waves for a surface conductivity model of grapheme. J. Appl. Phys. **103**(6) (2008)
19. P.R. Wallace, The band theory of graphite. Phys. Rev. **71**(9), 622–634 (1947)
20. S.K. Patel, V. Sorathiya, S. Lavadiya, Y. Luo, T.K. Nguyen, V. Dhasarathan, Numerical analysis of polarization-insensitive squared spiral-shaped graphene metasurface with negative refractive index. Appl. Phys. B Lasers Opt. **126**(5) (2020)
21. Q. Bao, et al., Broadband graphene polarizer. Nat. Photonics **5**(7), 411–415 (July 2011)
22. A. Pandya, V. Sorathiya, S. Lavadiya, Graphene-based nanophotonic devices, in *Recent Advances in Nanophotonics—Fundamentals and Applications*, IntechOpen (2020)
23. S.K. Patel, V. Sorathiya, S. Lavadiya, L. Thomas, T.K. Nguyen, V. Dhasarathan, Multi-layered Graphene silica-based tunable absorber for infrared wavelength based on circuit theory approach. Plasmonics **15**(6), 1767–1779 (2020)
24. Y. Takuma, H. Kajioka, K. Yamada, High performance polarizers and sensing coils with elliptical jacket type single polarization fibers, in *Optical Fiber Sensors* (1988), p. FEE3
25. S.K. Patel, V. Sorathiya, Z. Sbeah, S. Lavadiya, T.K. Nguyen, V. Dhasarathan, Graphene-based tunable infrared multi band absorber. Opt. Commun. **474**, 126109 (2020)
26. Y.V. Bludov, M.I. Vasilevskiy, N.M.R. Peres, Tunable graphene-based polarizer. J. Appl. Phys. **112**(8) (2012)
27. L.-J. Cheng, L. Liu, Optical modulation of continuous terahertz waves towards cost-effective reconfigurable quasi-optical terahertz components. Opt. Express **21**(23), 28657 (2013)
28. K. Imakita, T. Kamada, M. Fujii, K. Aoki, M. Mizuhata, S. Hayashi, Terahertz wire grid polarizer fabricated by imprinting porous silicon. Opt. Lett. **38**(23), 5067 (2013)
29. F. Yan, C. Yu, H. Park, E.P.J. Parrott, E. Pickwell-MacPherson, Advances in polarizer technology for terahertz frequency applications. J. Infrared, Millimeter, Terahertz Waves **34**(9), 489–499 (2013)

30. A. Ferraro, D.C. Zografopoulos, M. Missori, M. Peccianti, R. Caputo, R. Beccherelli, Flexible terahertz wire grid polarizer with high extinction ratio and low loss. Opt. Lett. **41**(9), 2009 (2016)
31. J.-H. Hu et al., Enhanced absorption of graphene strips with a multilayer subwavelength grating structure. Appl. Phys. Lett. **105**(22), 221113 (2014)
32. J. Wu, Enhancement of absorption in graphene strips with cascaded grating structures. IEEE Photonics Technol. Lett. **28**(12), 1332–1335 (2016)
33. M. Grande et al., Optically transparent microwave polarizer based on quasi-metallic grapheme. Sci. Rep. **5**(1), 17083 (2015)
34. L. Ju et al., Graphene plasmonics for tunable terahertz metamaterials. Nat. Nanotechnol. **6**(10), 630–634 (2011)
35. J.W. You, N.C. Panoiu, Polarization control using passive and active crossed graphene gratings. Opt. Express **26**(2), 1882–1894 (2018)
36. Y. Xia et al., Polarization dependent plasmonic modes in elliptical graphene disk arrays. Opt. Express **27**(2), 1080 (2019)
37. M.M. Jadidi et al., Tunable terahertz hybrid metal–graphene plasmons. Nano Lett. **15**(10), 7099–7104 (2015)
38. G.B. Barin, Y. Song, I. de Fátima Gimenez, A.G.S. Filho, L. S. Barreto, J. Kong, Optimized graphene transfer: influence of polymethylmethacrylate (PMMA) layer concentration and baking time on graphene final performance. Carbon N. Y. **84**, 82–90 (Apr 2015)
39. M.P. Lavin-Lopez, J.L. Valverde, A. Garrido, L. Sanchez-Silva, P. Martinez, A. Romero-Izquierdo, Novel etchings to transfer CVD-grown graphene from copper to arbitrary substrates. Chem. Phys. Lett. **614**, 89–94 (2014)
40. K. Chae et al., Electrical properties of ion gels based on PVDF-HFP applicable as gate stacks for flexible devices. Curr. Appl. Phys. **18**(5), 500–504 (2018)
41. B.J. Kim, H. Jang, S.-K. Lee, B.H. Hong, J.-H. Ahn, J.H. Cho, High-performance flexible graphene field effect transistors with ion gel gate dielectrics. Nano Lett. **10**(9), 3464–3466 (2010)
42. X. Liu, Q. Zhang, X. Cui, Ultra-broadband polarization-independent wide-angle THz absorber based on plasmonic resonances in semiconductor square nut-shaped metamaterials. Plasmonics **12**(4), 1137–1144 (2017)
43. S. Liu, H. Chen, T.J. Cui, A broadband terahertz absorber using multi-layer stacked bars. Appl. Phys. Lett. **106**(15), 1–6 (2015)
44. X. Yu, X. Gao, W. Qiao, L. Wen, W. Yang, Broadband tunable polarization converter realized by graphene-based metamaterial. IEEE Photonics Technol. Lett. **28**(21), 2399–2402 (2016)
45. A. Fardoost, F.G. Vanani, A. Amirhosseini, R. Safian, Design of a multilayer graphene-based ultrawideband terahertz absorber. IEEE Trans. Nanotechnol. **16**(1), 68–74 (2017)
46. G. Deng, J. Yang, Z. Yin, Broadband terahertz metamaterial absorber based on tantalum nitride. Appl. Opt. **56**(9), 2449 (2017)
47. J. Zhu, S. Li, L. Deng, C. Zhang, Y. Yang, H. Zhu, Broadband tunable terahertz polarization converter based on a sinusoidally-slotted graphene metamaterial. Opt. Mater. Express **8**(5), 1164 (2018)
48. C. Shi et al., Compact broadband terahertz perfect absorber based on multi-interference and diffraction effects. IEEE Trans. Terahertz Sci. Technol. **6**(1), 40–44 (2016)
49. E. Gao et al., Dynamically tunable dual plasmon-induced transparency and absorption based on a single-layer patterned graphene metamaterial. Opt. Express **27**(10), 13884 (2019)
50. J. Gu et al., A close-ring pair terahertz metamaterial resonating at normal incidence. Opt. Express **17**(22), 20307 (2009)
51. N.I. Landy, S. Sajuyigbe, J.J. Mock, D.R. Smith, W.J. Padilla, Perfect metamaterial absorber. Phys. Rev. Lett. **100**(20), 207402 (2008)
52. J.B. Pendry, Negative refraction makes a perfect lens. Phys. Rev. Lett. **85**(18), 3966–3969 (2000)
53. Y. Zhang et al., A graphene based tunable terahertz sensor with double Fano resonances. Nanoscale **7**(29), 12682–12688 (2015)

54. X. Zhang et al., Polarization-sensitive triple plasmon-induced transparency with synchronous and asynchronous switching based on monolayer graphene metamaterials. Opt. Express **28**(24), 36771 (2020)
55. V. Dave, V. Sorathiya, T. Guo, S.K. Patel, Graphene based tunable broadband far-infrared absorber. Superlattices Microstruct. **124**, 113–120 (2018)

Terahertz Frequency, Heisenberg's Uncertainty Principle, Einstein Relation, Dimensional Quantization, and Opto-Electronic Materials

K. Bagchi, P. K. Bose, P. K. Das, S. D. Biswas, T. Basu, K. P. Ghatak, R. Paul, M. Mitra, and J. Pal

Abstract In this chapter, the Einstein relation (ER) in quantum wells (QWs) and nano wires (NWs) of opto-electronic materials (taking QWs and NWs $n - Hg_{1-x}Cd_xTe and\, n - In_{1-x}Ga_xAs_yP_{1-y}\ lattice\, matched\, to\, InP$ as examples) under terahertz frequency on the basis of Heisenberg's Uncertainty Principle (HUP) has been investigated. It has been found that the ER exhibits quantum steps variations with electron statistics, film thickness, intensity, and wave length for all the cases. A relationship where the ER is inversely proportional to the thermo electric power in the presence of a strong magnetic field (L_0), which is an easily measurable experimental quantity, has been suggested in this context.

K. Bagchi
237, Canal Street, Sreebhumi, Kolkata 700048, India

P. K. Bose
Department of Mechanical Engineering, Swami Vivekananda Institute of Science and Technology, DakshinGobindapur, Sonarpur, Kolkata, West Bengal 700145, India

P. K. Das · S. D. Biswas · T. Basu · K. P. Ghatak (✉)
Department of Basic Science and Humanities, Institute of Engineering and Management, 1, Management House, Salt Lake, Sector–V, Kolkata, West Bengal 700091, India

P. K. Das
e-mail: prabir.das@iemcal.com

S. D. Biswas
e-mail: samapika.dasbiswas@iemcal.com

R. Paul
Department of Computer Science and Engineering, University of Engineering and Management, Kolkata 700156, India
e-mail: rajashree.paul@uem.edu.in

M. Mitra
Department of Electronic and Telecommunication Engineering, Indian Institute of Engineering Science and Technology, Shibpur, Howrah 711103, India

J. Pal
Department of Physics, MeghnadSaha Institute of Technology, Nazirabad, P.O. Uchepota, Anandapur, Kolkata 700150, India

S. Das et al. (eds.), *Advances in Terahertz Technology and Its Applications*,
https://doi.org/10.1007/978-981-16-5731-3_14

Keywords Terahertz frequency · Einstein relation · Quantum wells · Nano wires · Experimental determination

1 Introduction

The importance of ER is well-known in the literature and has been investigated widely for the last forty years mainly by the group of Ghatak et al. [1–35] and few others [36–40]. In this chapter, an attempt is made to explore the ER in QWs and NWs of opto-electronic materials [31–39] under terahertz frequency using the HUP. We shall study the dependence of ER on electron statistics, film thickness, intensity, and wave length, respectively, taking QWs and NWs of the said compounds. Besides, a relationship where the ER is inversely proportional to L_0 which is an easily measurable experimental quantity has also been suggested in this context.

2 Theoretical Background

The use of HUP leads to the expression of the electron statistics in the present case in QWs of opto-electronic materials in accordance with three and two band models of Kane together with parabolic energy bands as [35]

$$n_{2D} = \frac{m^* g_v}{\pi \hbar^2} \sum_{n_z=1}^{n_{z\max}} \left[\beta_0(E_{F2DL}, \lambda) - \frac{\hbar^2}{2m^*} \left(\frac{n_z \pi}{d_z} \right)^2 \right] \tag{1}$$

$$n_{2D} = \frac{m^* g_v}{\pi \hbar^2} \sum_{n_z=1}^{n_{z\max}} []\tau_0(E_{F2DL}, \lambda) - \frac{\hbar^2}{2m^*} \left(\frac{n_z \pi}{d_z} \right)^2 \tag{2}$$

$$n_{2D} = \frac{m^* g_v}{\pi \hbar^2} \sum_{n_z=1}^{n_{z\max}} \left[\rho_0(E_{F2DL}, \lambda) - \frac{\hbar^2}{2m^*} \left(\frac{n_z \pi}{d_z} \right)^2 \right] \tag{3}$$

where the notations are defined in [35].

The ER, in general, assumes the form

$$\frac{D}{\mu} = \theta_0(|e|)^{-1} n_0 \left[\frac{\partial n_0}{\partial E_F} \right]^{-1} \tag{4}$$

where θ_0 is the correction factor for Terahertz frequency.

Thus, using the above mentioned equations, we can write

$$\frac{D}{\mu} = \theta_0(|e|)^{-1}\left[\sum_{n_z=1}^{n_{z\max}}\left[\beta_0(E_{F2DL},\lambda) - \frac{\hbar^2}{2m^*}(\frac{n_z\pi}{d_z})^2\right]\right]$$
$$\left[\sum_{n_z=1}^{n_{z\max}}[\beta_0'(E_{F2DL},\lambda)]\right]^{-1} \tag{5}$$

$$\frac{D}{\mu} = \theta_0(|e|)^{-1}\left[\sum_{n_z=1}^{n_{z\max}}\left[\tau_0(E_{F2DL},\lambda) - \frac{\hbar^2}{2m^*}\left(\frac{n_z\pi}{d_z}\right)^2\right]\right]$$
$$\left[\sum_{n_z=1}^{n_{z\max}}[\tau_0'(E_{F2DL},\lambda)]\right]^{-1} \tag{6}$$

$$\frac{D}{\mu} = \theta_0(|e|)^{-1}\left[\sum_{n_z=1}^{n_{z\max}}\left[\rho_0(E_{F2DL},\lambda) - \frac{\hbar^2}{2m^*}\left(\frac{n_z\pi}{d_z}\right)^2\right]\right]$$
$$\left[\sum_{n_z=1}^{n_{z\max}}[\rho_0'(E_{F2DL},\lambda)]\right]^{-1} \tag{7}$$

For NWs, the electron statistics in the present case for the said three types of band models can be expressed as [35]

$$n_{1D} = \frac{2g_v}{\pi}\frac{\sqrt{2m^*}}{\hbar}K_1\left[\beta_0(E_{F1DL},\lambda) - \phi(n_y,n_z)\right]^{1/2} \tag{8}$$

$$n_{1D} = \frac{2g_v}{\pi}\frac{\sqrt{2m^*}}{\hbar}K_1\left[\tau_0(E_{F1DL},\lambda) - \phi(n_y,n_z)\right]^{1/2} \tag{9}$$

$$n_{1D} = \frac{2g_v}{\pi}\frac{\sqrt{2m^*}}{\hbar}K_1\left[\rho_0(E_{F1DL},\lambda) - \phi(n_y,n_z)\right]^{1/2} \tag{10}$$

where $K_1 = \sum_{n_y=1}^{n_{y\max}}\sum_{n_z=1}^{n_{z\max}}$ and $\phi(n_y,n_z) = \frac{\hbar^2\pi^2}{2m_c}\left[\left(\frac{n_y}{d_y}\right)^2 + \left(\frac{n_z}{d_z}\right)^2\right]$ and $\phi(n_y,n_z) = \frac{\hbar^2\pi^2}{2m_c}\left[\left(\frac{n_y}{d_y}\right)^2 + \left(\frac{n_z}{d_z}\right)^2\right]$.

The use of the above-mentioned equations leads to the expression of the respective ER for NW as

$$\frac{D}{\mu} = \theta_0(2/e)\left[K_1\left[\beta_0(E_{F1DL},\lambda) - \phi(n_y,n_z)\right]^{1/2}\right]$$
$$\left[K_1\{\beta_0(E,\lambda)\}'\left[\beta_0(E,\lambda) - \phi(n_y,n_z)\right]^{-1/2}\right]^{-1} \tag{11}$$

$$\frac{D}{\mu} = \theta_0(2/e)\Big[K_1\big[\tau_0(E_{F1DL}, \lambda) - \phi(n_y, n_z)\big]^{1/2}\Big]$$
$$\Big[K_1\{\tau_0(E, \lambda)\}'\big[\tau_0(E, \lambda) - \phi(n_y, n_z)\big]^{-1/2}\Big]^{-1} \quad (12)$$

$$\frac{D}{\mu} = \theta_0(2/e)\big[K_1\big[\rho_0(E_{F1DL}, \lambda) - \phi(n_y, n_z)\big]\big]^{1/2}$$
$$\Big[K_1\{\rho_0(E, \lambda)\}'\big[\rho_0(E, \lambda) - \phi(n_y, n_z)\big]^{-1/2}\Big]^{-1} \quad (13)$$

3 Suggestion for Experimental Determination of ER

The L_0 assumes the form

$$L_0 = \left(\frac{\pi^2 k_B^2 T}{3|e|n_0}\right)\left(\frac{\partial n_0}{\partial E_F}\right) \quad (14)$$

Using (4) and (14), we can write

$$\frac{D}{\mu} = \theta_0\left(\frac{\pi^2 k_B^2 T}{3|e|^2 L_0}\right) \quad (15)$$

Thus, we can experimentally determine the ER.

4 Result and Discussion

Using of Eqs. (1) to (3) and (5) to (7) and the values of E_g, Δ *and* m^*[35], the plots of the normalized ER (denoted by ϕ) versus d_z, n_0, I, λ *and* x for the QWs and NWs of the said materials have been drawn in Figs. 1, 2, 3, 4, 5, 6, 7, 8, 9, 10, 11, 12, 13, 14, 15, 16, 17, 18, 19 and 20.

From the said graphs, we note the followings:

a. The ER decreases with increasing d_z, decreasing n_0, decreasing I, and decreasing λ, respectively, in quantized steps, and the numerical values depend on the particular dispersion relation.
b. With the decrement of x, the ER assumes increasing values.
c. The points a and b are valid for both QWs and NWs, respectively.

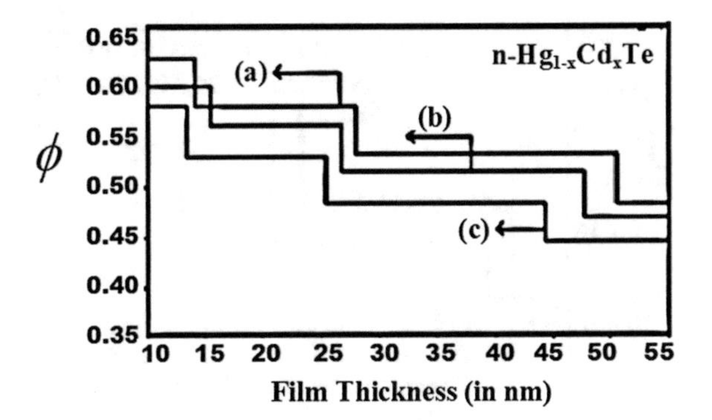

Fig. 1 Plot of ϕ versus d_z for QWs under terahertz frequency by using Eqs. (1) and (5) (plot a), Eqs. (2) and (6) (plot b), and Eqs. (3) and (7) (plot c), respectively

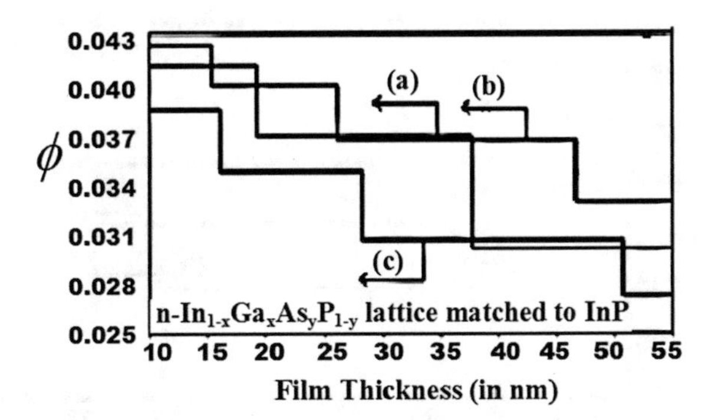

Fig. 2 Plot of ϕ versus d_z for QWs under terahertz frequency by using Eqs. (1) and (5) (plot a), Eqs. (2) and (6) (plot b), and Eqs. (3) and (7) (plot c), respectively

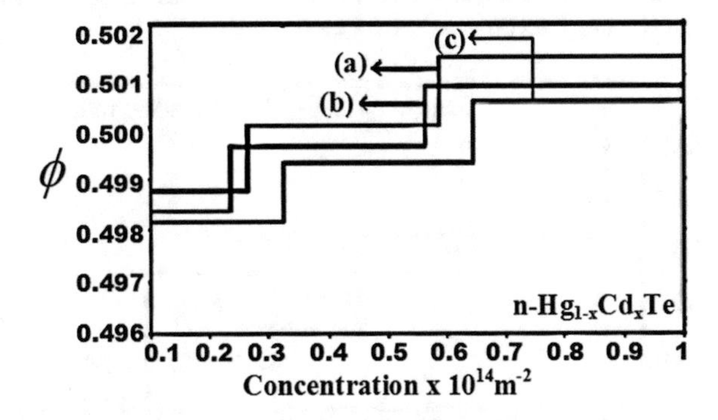

Fig. 3 Plot of ϕ versus n_0 for QWs under terahertz frequency by using Eqs. (1) and (5) (plot a), Eqs. (2) and (6) (plot b), and Eqs. (3) and (7) (plot c), respectively

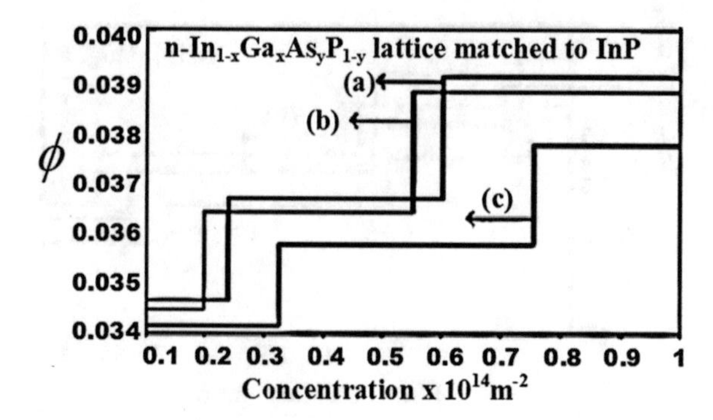

Fig. 4 Plot of ϕ versus n_0 for QWs under terahertz frequency by using Eqs. (1) and (5) (plot a), Eqs. (2) and (6) (plot b), and Eqs. (3) and (7) (plot c), respectively

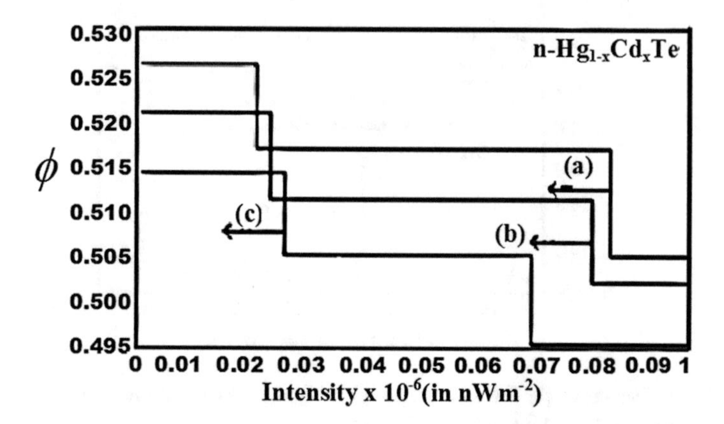

Fig. 5 Plot of ϕ versus I for QWs under terahertz frequency by using Eqs. (1) and (5) (plot a), Eqs. (2) and (6) (plot b), and Eqs. (3) and (7) (plot c), respectively

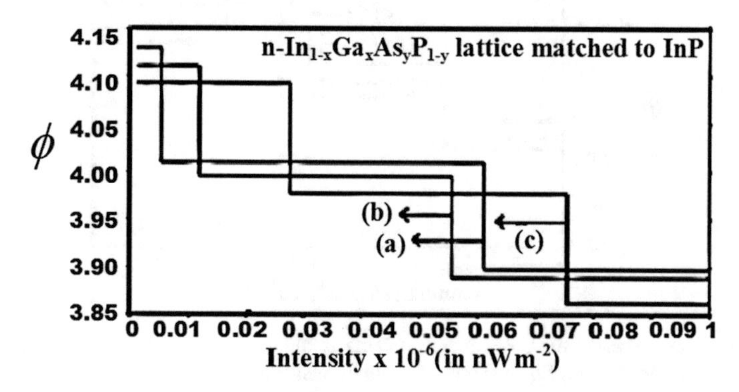

Fig. 6 Plot of ϕ versus I for QWs under terahertz frequency by using Eqs. (1) and (5) (plot a), Eqs. (2) and (6) (plot b), and Eqs. (3) and (7) (plot c), respectively

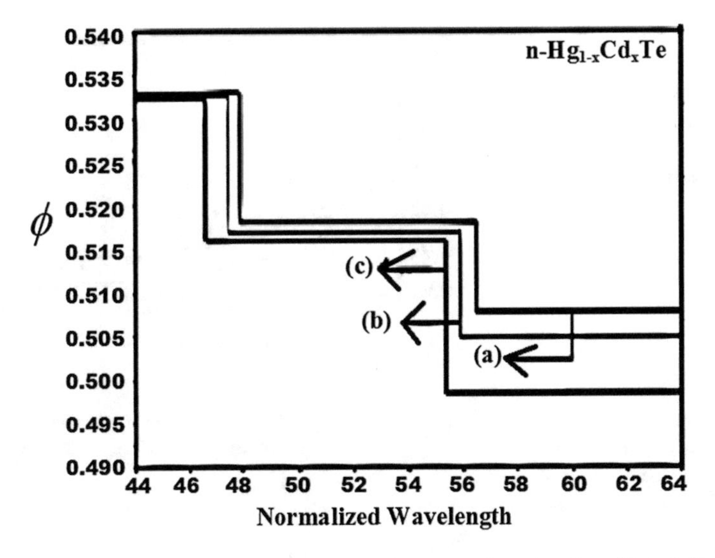

Fig. 7 Plot of ϕ versus λ for QWs under terahertz frequency by using Eqs. (1) and (5) (plot a), Eqs. (2) and (6) (plot b), and Eqs. (3) and (7) (plot c), respectively

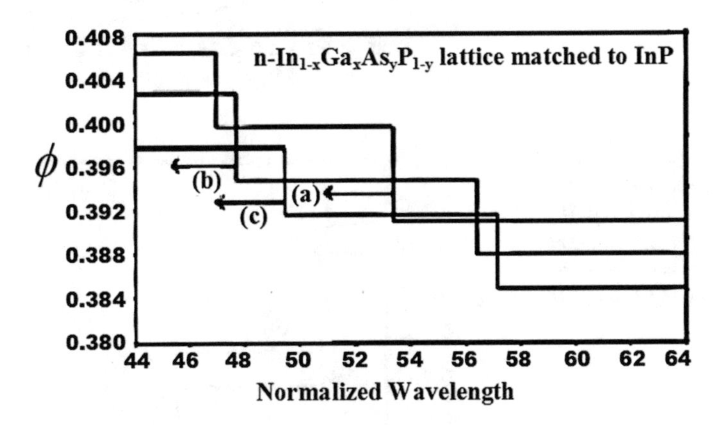

Fig. 8 Plot of ϕ versus λ for QWs under terahertz frequency by using Eqs. (1) and (5) (plot a), Eqs. (2) and (6) (plot b), and Eqs. (3) and (7) (plot c), respectively

5 Conclusion

In this chapter, the Einstein relation (ER) in quantum wells (QWs) and nano wires (NWs) of opto-electronic materials (taking QWs and NWs $n-Hg_{1-x}Cd_xTe$and $n-In_{1-x}Ga_xAs_yP_{1-y}$lattice matched to InP as examples) under terahertz frequency on the basis of Heisenberg 's Uncertainty Principle (HUP) has been investigated. It has

Fig. 9 Plot of ϕ versus x for QWs under terahertz frequency by using Eqs. (1) and (5) (plot a), Eqs. (2) and (6) (plot b), and Eqs. (3) and (7) (plot c), respectively

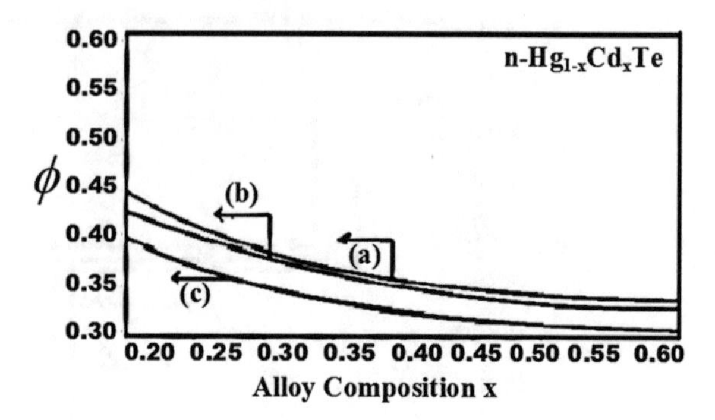

Fig. 10 Plot of ϕ versus x for QWs under terahertz frequency by using Eqs. (1) and (5) (plot a), Eqs. (2) and (6) (plot b), and Eqs. (3) and (7) (plot c), respectively

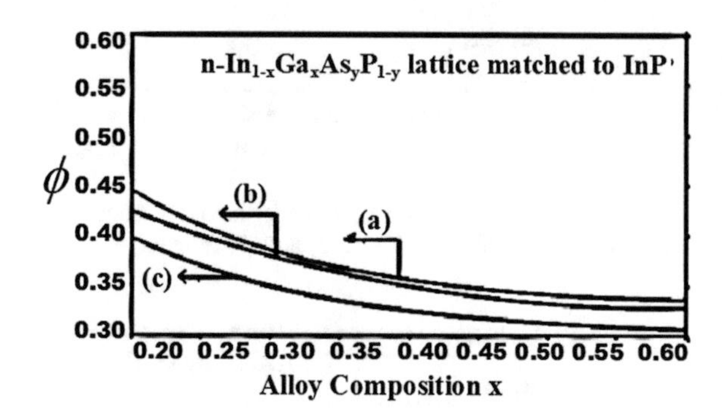

Fig. 11 Plot of ϕ versus d_z for NWs under terahertz frequency by using Eqs. (8) and (11) (plot a), Eqs. (9) and (12) (plot b), and Eqs. (10) and (13) (plot c), respectively

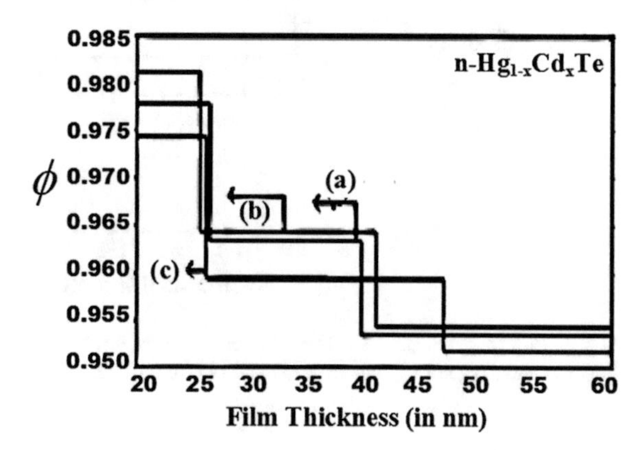

been found that the ER exhibits quantum steps variations with electron statistics, film thickness, intensity, and wave length for all the cases. A relationship where the

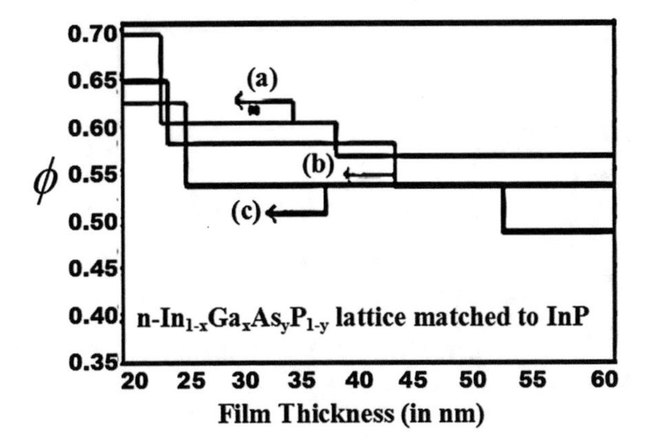

Fig. 12 Plot of ϕ versus d_z for NWs under terahertz frequency by using Eqs. (8) and (11) (plot a) Eqs. (9), and (12) (plot b) and Eqs. (10) and (13) (plot c), respectively

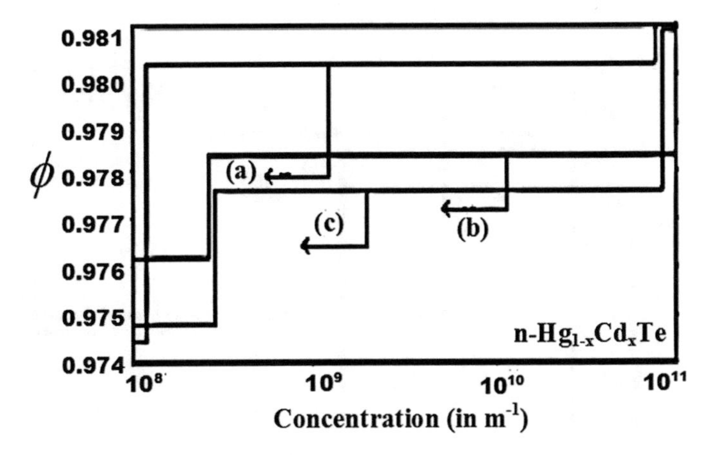

Fig. 13 Plot of ϕ versus n_0 for NWs under terahertz frequency by using Eqs. (8) and (11) (plot a), Eqs. (9) and (12) (plot b), and Eqs. (10) and (13) (plot c), respectively

ER is inversely proportional to the thermo electric power in the presence of a strong magnetic field (L_0), which is an easily measurable experimental quantity, has been suggested in this context.

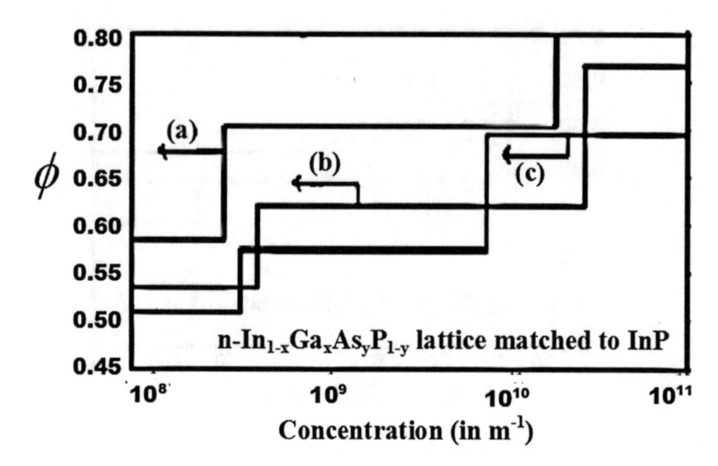

Fig. 14 Plot of ϕ versus n_0 for NWs under terahertz frequency by using Eqs. (8) and (11) (plot a), Eqs. (9) and (12) (plot b), and Eqs. (10) and (13) (plot c), respectively

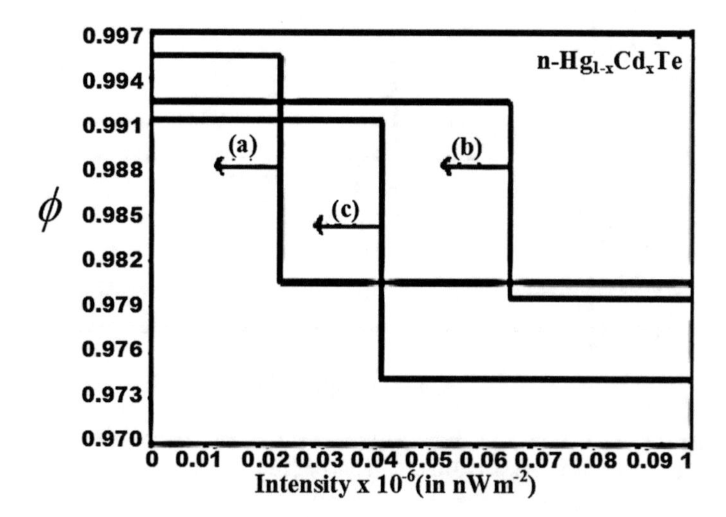

Fig. 15 Plot of ϕ versus I for NWs under terahertz frequency by using Eqs. (8) and (11) (plot a), Eqs. (9) and (12) (plot b), and Eqs. (10) and (13) (plot c), respectively

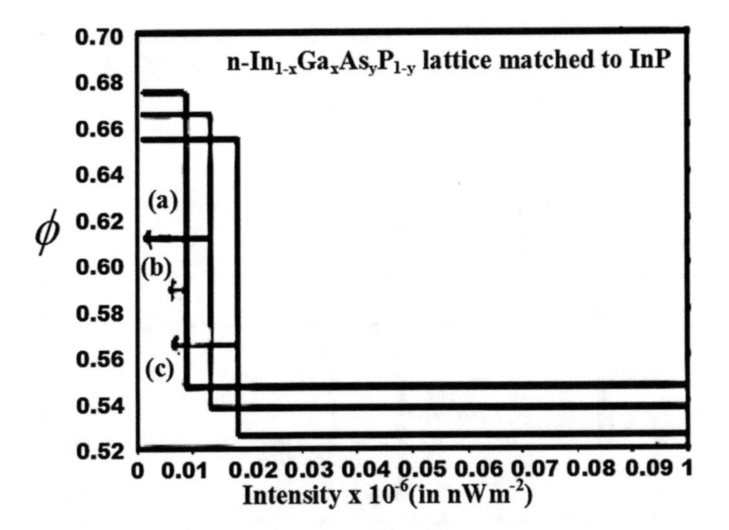

Fig. 16 The plot of ϕ versus I for NWs under terahertz frequency by using Eqs. (8) and (11) (plot a), Eqs. (9) and (12) (plot b) and Eqs. (10) and (13) (plot c), respectively

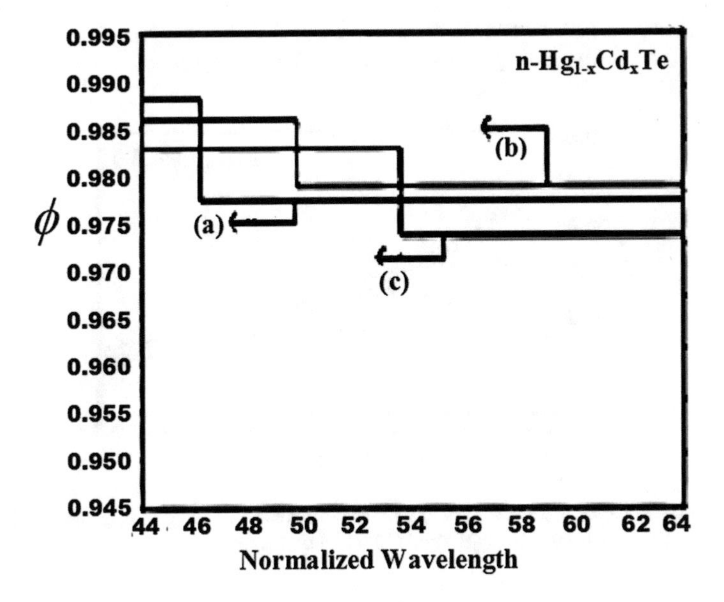

Fig. 17 Plot of ϕ versus λ for NWs under terahertz frequency by using Eqs. (8) and (11) (plot a), Eqs. (9) and (12) (plot b), and Eqs. (10) and (13) (plot c), respectively

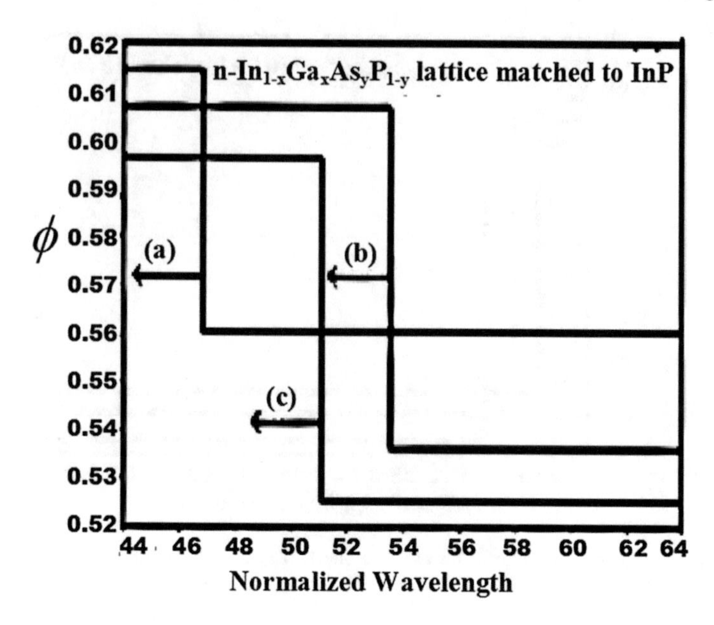

Fig. 18 Plot of ϕ versus λ for NWs under terahertz frequency by using Eqs. (8) and (11) (plot a), Eqs. (9) and (12) (plot b), and Eqs. (10) and (13) (plot c), respectively

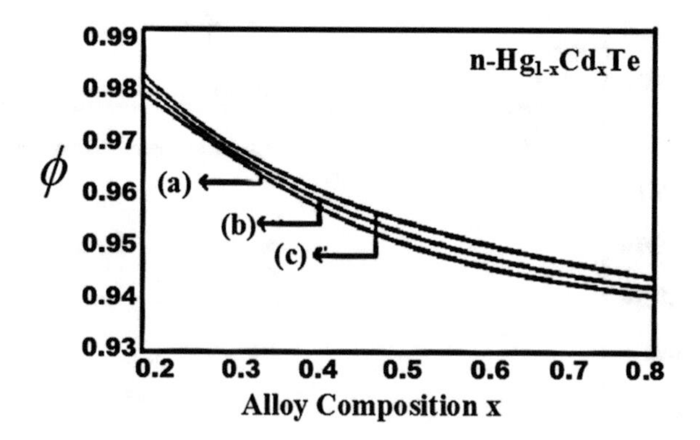

Fig. 19 Plot of ϕ versus x for NWs under terahertz frequency by using Eqs. (8) and (11) (plot a), Eqs. (9) and (12) (plot b), and Eqs. (10) and (13) (plot c), respectively

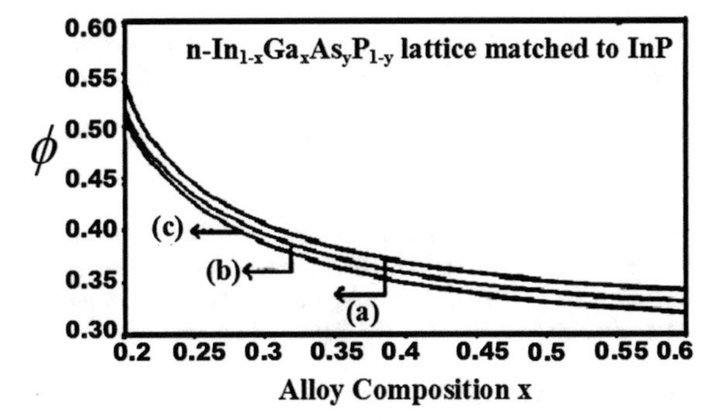

Fig. 20 Plot of ϕ versus x for NWs under terahertz frequency by using Eqs. (8) and (11) (plot a), Eqs. (9) and (12) (plot b), and Eqs. (10) and (13) (plot c), respectively

Acknowledgements The authors are grateful to Prof. Dr. S. Chakrabarti, Director of the Institute of Engineering and Management, Kolkata, India her constant encouragement and inspiration.

References

1. S. Choudhury, D. De, S. Mukherjee, A. Neogi, A. Sinha, M. Pal, S.K. Biswas, S. Pahari, S. Bhattacharya, K.P. Ghatak, J. Comp. Theo. Nanosc. **5**, 375–400 (2008)
2. K.P. Ghatak, S.N. Biswas, J. Appl. Phys. **70**, 4309–4316 (1991)
3. B. Mitra, K.P. Ghatak, Phys. Letts. A **135**, 397–400 (1989)
4. K.P. Ghatak, D. Bhattacharyya, Phys. Letts. A **184**, 366–369 (1994)
5. K.P. Ghatak, B. Nag, D. Bhattacharyya, J. Low Temp. Phys. **14**, 1–22 (1995)
6. K.P. Ghatak, *Influence of Band Structure on Some Quantum Processes in Tetragonal Semiconductors, D. Eng. Thesis*, (Jadavpur University, Kolkata, India, 1991)
7. S.N. Biswas, K.P. Ghatak, *Proceedings of the Society of Photo-Optical and Instrumentation Engineers (SPIE), Quantum Well and Superlattice Physics*, vol. 792 (USA, 1987), pp. 239–252
8. K.P. Ghatak, B. De, *Defect Engineering in Semiconductor Growth, Processing and Device technology Materials Research Society Proceedings*, (MRS) Spring meeting, vol. 262 (1992) pp. 911–914
9. M. Mondal, K.P. Ghatak, J. Phys. C, Sol. State **20**, 1671–1681 (1987)
10. M. Mitra, B. Chatterjee, K.P. Ghatak, J. Comput. Theor. Nanosci. **12**, 1527–1539 (2015)
11. K.P. Ghatak, L.S. Singh, K. Sarkar, N. Debbarma, M. Debbarma, Mater. Focus **4**, 85–110 (2015)
12. S. Mukherjee, S.N. Mitra, P.K. Bose, A.R. Ghatak, A. Neogi, J.P. Banerjee, A. Sinha, M. Pal, S. Bhattacharya, K.P. Ghatak, J. Comp. Theo. Nanosc. **4**, 550–573 (2007)
13. K.P. Ghatak, A. Ghoshal, S.N. Biswas, Nouvo. Cimento. **15D**, 39–63 (1993)
14. K.P. Ghatak, D. Bhattacharyya, Phys. Scr. **52**, 343–348 (1995)
15. K.P. Ghatak, N. Chattropadhyay, M. Mondal, Appl. Phys. A **44**, 305–309 (1987)
16. K.P. Ghatak, S. Bhattacharya, D. De, *Einstein Relation in Compound Semiconductors and Their Heterostructures*, Springer Series in Materials Science, vol. 116 (Springer, 2009)
17. K.P. Ghatak, D. De, S. Bhattacharya, *Photoemission from Optoelectronic Materials and Their Nanostructures,* Springer Series in Nanostructure Science and Technology, (Springer Berlin Heidelberg, Germany 2009)

18. K.P. Ghatak, S. Bhattacharya, *Thermo Electric Power In Nano structured Materials Strong Magnetic Fields*, Springer Series in Materials Science, vol. **137** (Springer, Berlin Helderberg, Germany, 2010)
19. S. Bhattacharya, K.P. Ghatak, *Fowler-Nordheim Field Emission: Effects in Semiconductor Nanostructures*, Springer Series in Solid state Sciences, vol. 170 (Springer, Berlin Helderberg, Germany, 2012)
20. S. Bhattacharya, K.P. Ghatak, *Effective Electron Mass in Low Dimensional Semiconductors*, Springer Series in Materials Sciences vol. 167, (Springer, Berlin, Helderberg, 2013)
21. K.P. Ghatak, S. Bhattacharya, *Debye Screening Length: Effects of Nanostructured Materials;* Springer Tracts in Modern Physics, vol. 255 (Springer International Publishing, Switzerland, 2014)
22. K.P. Ghatak, S. Bhattacharya, *Heavily Doped 2D Quantized Structures and the Einstein Relation;* Springer Tracts in Modern Physics, vol. 260 (Springer International Publishing, Switzerland, 2015)
23. K.P. Ghatak, *Einstein's Photo-Emission: Emission from Heavily Doped Quantized Structures; Springer Tracts in Modern Physics*, vol. 262 (Springer International Publishing, Switzerland, 2015)
24. K.P. Ghatak, *Dispersion Relations in Heavily—Doped Nanostructures;* Springer Tracts in Modern Physics, vol. 265 (Springer cham Helderberg, New York, USA, 2015)
25. K.P. Ghatak, *Magneto Thermoelectric Power in Heavily Doped Quantized Structures;* Series on the Foundations of Natural Science and Technology. vol. 7 (World Scientific Publishing Co. Pte. Ltd. Singapore, 2016)
26. K.P. Ghatak, *Quantum Effects, Heavy Doping, and The Effective Mass; Series on the Foundations of Natural Science and Technology*, vol. 8 (World Scientific Publishing Co., Ltd., Singapore, USA, 2017)
27. K.P. Ghatak, M. Mitra, *Electronic Properties, Series in Nanomaterials*, vol. 1 (De Guyter, Germany, 2020)
28. K.P. Ghatak, M. Mitra, *Quantization and Entropy, Series in Nanomaterials*, vol. 2 (De Guyter, Germany, 2020)
29. K.P. Ghatak, M. Mitra, *Elastic Constants in Heavily Doped Low Dimensional Materials,* Series on the Foundations of Natural Science and Technology. vol. 17 (World Scientific Publishing Co. Ltd., Singapore, USA, 2021)
30. B. Chatterjee, N. Debbarma, S. Debbarma, S. Chakrabarti, K.P. Ghatak, J. Advan. Sci. Eng. Med. **6**, 1177–1190 (2014)
31. S. Chakrabarti, S.K. Sen, S. Chakraborty, L.S. Singh, K.P. Ghatak, J. Advan. Sci. Eng. Med. **6**, 1042–1057 (2014)
32. S.M. Adhikari, K.P. Ghatak, J. Quantum Matter **2**, 296–306 (2013)
33. S. Bhattacharya, N. Paitya, A. Sharma, D. De, A. Kumar, S. Choudhary, P.K. Bose, K.P. Ghatak, *Bismuth Characteristics, Production and Application*, vol. 2, ed. by K.P. Ghatak, S. Bhattacharya (Materials Science and Technologies Series), (Nova Science Publishers, Inc., New York, USA, 2012) pp. 1–56
34. K.P. Ghatak, S. Bhattacharya, S. Pahari, D. De, R. Benedictus, Superlattices Microstruct. **46**, 387–414 (2009)
35. K.P. Ghatak, S. Bhattacharya, D. De, P.K. Bose, S.N. Mitra, S. Pahari, J. Phys. B: Condens. Matter **403**, 2930–2948 (2008)
36. A.N. Chakravarti, B.R. Nag, Int. J. Elect. **37**, 281–284 (1974)
37. B.R. Nag, A.N. Chakravarti, Solid State Electron. **18**, 109–110 (1975)
38. B.R. Nag, A.N. Chakravarti, P.K. Basu, Phys. Stat. Sol. (A) **68**, K75–K80 (1981)
39. B.R. Nag, A.N. Chakravarti, Phys. Stal Sol. (A) **67**, K113–K118 (1981)
40. A.N. Chakravarti, D.P. Parui, Phys. Letts, **40A**, 113–116 (1972)

Design and Characterization of Novel Reconfigurable Graphene Terahertz Antenna Using Metamaterials

T. Sathiyapriya, T. Poornima, and R. Sudhakar

Abstract The desire for wide bandwidth persists in this era of low latency, which drives research into the high frequency band of the electromagnetic spectrum. There are several hundred gigahertz of bandwidth available above 100 GHz for applications such as super-fast wireless communication devices, material investigation, security, medical scanners, and wireless imaging. The Recent exploration of Terahertz (THz) band (0.1–10 THz) paves the way for significant advancements in Terahertz devices, Signal sources, and measuring instruments. The primary performance metrics imposed on Terahertz (THz) antennas are high gain and a reduction in size. The unique properties of graphene, such as its high conductivity, make it perfect for constructing THz-enabled antennas. This chapter proposes the design and analysis of Novel Terahertz (THz) antenna loaded with different metamaterial techniques on reconfigurable graphene layer. The THz antenna's design complexity, material selection problems, and performance specifications are discussed. With simulation results at 4.6 THz, the influence of metasurface on the graphene patch antenna at THz band has been investigated. Throughout the design and investigation process, antenna miniaturization is considered as a crucial point and the same is obtained with metasurface. The antenna has a dimension of 30×30 μm with a gold conductor on a Teflon substrate. Proposed antenna achieves reasonable return loss of −40 dB in the range of 4.3–5.3 THz with bandwidth of 1 THz.

Keywords Terahertz · Graphene · Metasurface · Miniaturization · High gain

1 Introduction

Today's world is evolving to go wireless anywhere, at any time, for any service. Many wireless devices have been developed to meet the needs of consumers. All of these applications have overburdened the microwave band, resulting in significant

T. Sathiyapriya (✉) · R. Sudhakar
Department of ECE, Dr. Mahalingam College of Engineering and Technology, Pollachi, Coimbatore, India

T. Poornima
Department of ECE, Amrita Vishwa Vidyapeetham, Coimbatore, India

S. Das et al. (eds.), *Advances in Terahertz Technology and Its Applications*,
https://doi.org/10.1007/978-981-16-5731-3_15

bandwidth scarcity [1]. Competitions and challenges to provide ultra-fast services in the blink of an eye drive researchers to look beyond the microwave spectrum [2]. Millimeter-wave band has been identified as having potential applications in 5G communication [3], Autonomous Vehicle Systems [4], and other areas [5]. Without pausing, researchers and wireless service providers have gone to the upper band of the electromagnetic spectrum, Terahertz band, which ranges from (0.1–10 THz) between microwave band and infrared light [6].

The digital revolution continues to transform the behavior and frequencies of Internet use around the world. For millions of people, mobile devices have become an integral part of everyday life. World Wide Web-enabled hand-held devices like smartphones and tablets have developed into key communication, information and entertainment tools. In 2020, there were 4.28 billion single mobile Internet users, which indicated that over 90% of the world's Internet population used mobile devices for online use. Mobility and Internet use are forecast to increase in the future, because the affordability and availability of mobile technology is growing ever greater. This revolution induces the huge data traffic which is supposed to handle the reliable communication at data rates of Gbps or Tbps. THz communication presently has a data rate of Gigabits per second (Gbps), while the Terabit per second (Tbps) data rate is still in the early stages of research [7].

1.1 Traits and Properties of THz Band

The phrase Terahertz was coined in the microwave community in the 1970s to delineate Michelson interferometer's spectrum frequency coverage, operating frequency range of diode detector, and resonance frequency of water laser [8, 9]. Terahertz is widely applied to submillimeter-wave radiation with frequencies ranging from 100 to 10 THz during the year 2000 [10]. The distinction between far infrared and submillimeter is still somewhat unclear, and the categorization is likely to be resolved at that time [11]. Terahertz electromagnetic waves is specified as 0.3–10 THz per the IEEE standard and are commonly classified as having a wavelength of 0.03–3 mm [12]. All THz technology characteristics that are expected to transform telecommunications landscape and change how people connect and receive information. The features of THz technology expected are unbounded capacity, continuous data transmission, ultra-low latency and super-fast download [13]. As seen in Fig. 1, Terahertz waves are found on the electromagnetic spectrum between optical waves and microwaves.

THz radiation has a number of appealing characteristics: It can, for example, produce extraordinarily high-resolution photographs and swiftly transport large amounts of data. Despite this, it is nonionizing, which means that its photons aren't powerful enough to knock electrons off atoms and molecules in human flesh, potentially causing hazardous chemical reactions. The wave also excites molecular and electronic motions in many materials, which reflect off some, propagate through others, and are absorbed by the remainder. The THz wave has the following excellent features [14]:

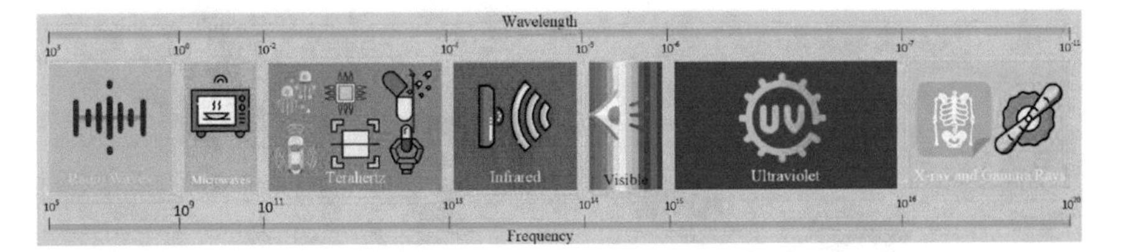

Fig. 1 THz wave location in the electromagnetic spectrum and respective applications [13]

(1) **The ability to penetrate nonmetallic objects**

Terahertz radiations have the capacity to permeate nonmetallic things and perceive molecular movements. Because of the strong transmission of terahertz radiation through metallic and dielectric materials, the ability to see through the objects is the most beneficial for object detection imagers. THz waves' penetrating characteristic makes them ideal for concealed object detection, which is used in high-end surveillance systems, the military sector, and security robots to identify suspicious things.

(2) **The attenuation loss**

The attenuation loss of electromagnetic waves over 275 GHz is significant and it can be exploited for detection and diagnosis in a variety of medical sectors. Because THz radiation is non-destructive, it may be used to detect organisms in live time without ionizing their individual components and damaging their structure. This characteristic also distinguishes cancer tissues from normal tissues on the basis of their water content.

(3) THz waves have a **strong directivity** due to substantial reflection and absorption losses at high frequencies, which shrinks the communication range to a directive line-of-sight (LoS) situation due to spreading losses in the NLoS state. THz waves are potential for ultra-speed seamless wireless communications due to strong directivity features.
(4) **The molecular movements** (vibrational, rotational, torsional, and translational) of biological molecules, allowing high-sensitivity spectral probing of molecular events in areas of practical importance.
5) The **High spectral resolution** characteristics of Terra Hertz have been used in lab demonstrations to detect explosives, uncover hidden weapons, inspect space shuttle tiles for faults, and screen for skin cancer and teeth cavities.

1.2 Evolution of THz Technology

The generation of THz signals with usable power levels is a serious challenge that must be overcome. Conventional transistor oscillator and diode multiplier technologies can only reach the lower end of the spectrum. Micromachining methods were

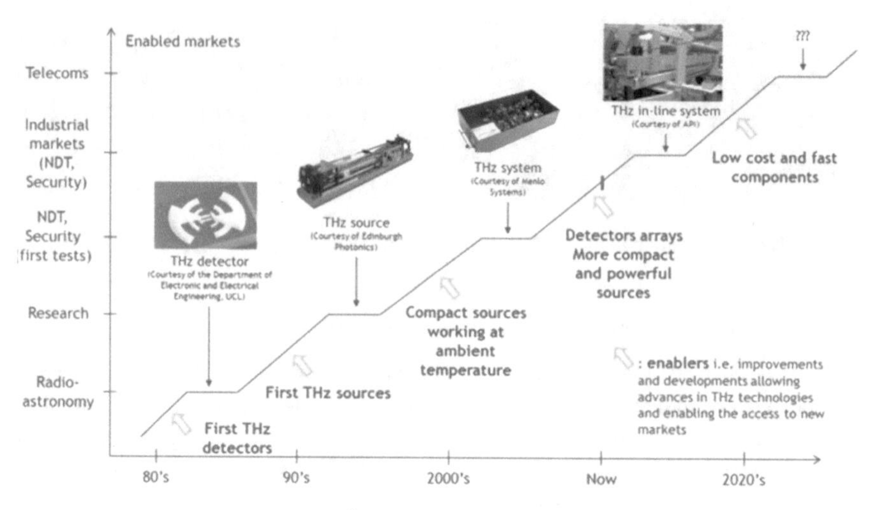

Fig. 2 Evolution of THz technology [©Tematys 2013]

used to create miniature backward-wave oscillators (BWOs). Quantum cascade lasers have produced many mW of power in the 5 THz region; however, they must be cooled to liquid nitrogen or lower temperatures. THz sources, detectors, sensors, and other devices evolved gradually throughout time, as shown in Fig. 2.

1.3 Applications of THz Band

Terahertz bands, like mm wave communications, can be utilized for mobile backhaul to carry enormous capacity signals between base stations. Point-to-point linkages in rural areas and macro-cell communications are two further areas where fiber or copper can be replaced. More notably, terahertz bands may be used in whisper radio applications, which are close-in communications. This comprises circuit board and vehicle wiring harnesses, nano-sensors, and wireless personal area networks (PANs) [15]. Then there are applications that demand short-range communications in the form of enormous bandwidth channels with zero error, such as high-resolution spectroscopy and imaging, as well as high speed communication. Health monitoring, illness detection, and medicine administration are just a few of the nano-scale applications of molecular communication possible by THz band [16, 17]. Biological sensing and threat avoidance, agricultural monitoring and surveillance defense monitoring are the main applications for THz nano-sensors in outdoor situations. In agriculture, the nano-sensors enables the interaction between the parts of plants and trees like leaves to roots. Nano-sensors operating in the THz spectrum can be used to better evolve and develop speedy communication, perform biomedical diagnosis and capture space images [18, 19].

2 Antenna Technologies for Terahertz Communications

This section briefly describes the evolution of THz antennas in order to aid readers' understanding. The design concepts and thorough examination of materials appropriate for making such antennas are then addressed. THz was first researched in the nineteenth century, although it was not considered a separate field at the time. The majority of THz research is in the far-infrared range. Researchers did not begin to expand millimeter-wave research into the THz band and create specialist research on THz technologies until the middle and late twentieth century. During the 1980s, the development of THz radiation sources allowed THz waves to be used in practical systems. Figure 3 shows a timeline of THz communication technology advancements, illustrating how THz research has progressed from an emerging to a more established area in the recent decade, with a clear technical jump [20].

2.1 Consequences for Antenna Design

THz-enabled communication offers several prospects in both micro and macro scale applications. From the standpoint of antenna design for different applications, it imposes various consequences and obstacles to meet out system performance. The most essential problem is to have a thorough knowledge of the physics of THz materials, and challenges pertaining to manufacturing and high throughput must be addressed effectively. Recent advancements in semiconductor component development and manufacturing technologies have made THz systems both possible and

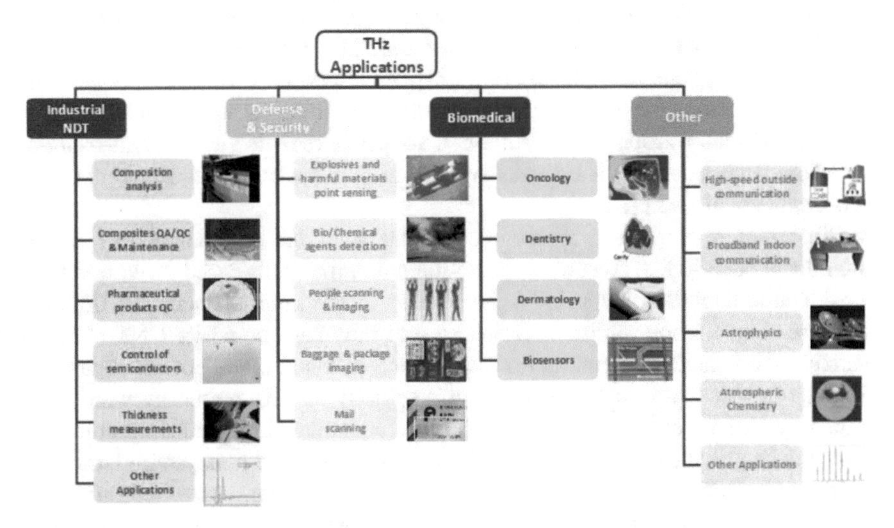

Fig. 3 Applications of THz band [©Tematys 2013]

economical, resulting in small systems. In reality, technological constraints have been overcome through architectural advancements as well as new smart architectures of new device structures [13].

2.2 *Different Types of THz Antenna*

The design of Terra Hertz antennas differs significantly from that of the RF and microwave domain. Furthermore, for such nanoscale antennas, antenna theory necessitates new modeling techniques that account for the shift from RF and microwave range to plasmonic nano-antennas. Furthermore, various efforts to develop appropriate antennas for the THz frequency range have been described in the literature. This section majorly classifies THz antennas based on materials and antenna structure. THz antennas are categorized into three types based on their production material: Lens-based antennas [21], Novel material antennas [22], Photo Conductive Antenna (PCA) [23] and In-chip Antennas [24]. Photoconductive antennas (PCAs) are utilized for THz wave synthesis and detection. The invention of photoconductive antennas has an improvement on the THz communication system and related disciplines.

3 Material Selection at THz Band

Another challenging part of developing an efficient antenna is selecting a good material with minimal propagation losses. Copper's properties make it a good material for antenna construction. Metals such as copper have a high refractive index at THz frequencies, reducing conductivity and skin depth which renders it unfit for efficient THz antenna construction [25]. At lower THz range of frequencies, the sheet resistance plays a dominant role in contributing to metal conductivity as they are indirectly proportional. Although copper is regarded as an ideal material for Microwave antennas, it imposes significant drawbacks to the development of THz-enabled antennas [26].

Other factors have been investigated by the scientific community in order to circumvent such limits. According to the literature, materials based on carbon have considerable attention due to its various forms of the same element that differ in structure. The Carbon allotrope Graphene and polymer composite CNT (Carbon nanotubes) are the best alternative to Copper for the manufacturing of THz antennas. Graphite is black and good conductor of electricity at THz frequency which was coined by Novoselov et al. in 2004 [27]. The fundamental repeating structure is a graphene layer which are structured in hexagonal ring called honeycomb lattice.

In the terahertz electromagnetic spectrum, CNT-based composites exhibit favorable conditions for making active and passive components. CNTs have a variety of properties and topologies, and some of their physical impacts include surface plasmon excitation, EM surface wave guiding, nano-antenna construction, and optical imaging. When single-walled Carbon Nano-Tubes and multi-walled CNTs are treated with various oxidizing agents, they perform well [28]. Due to the high resistivity factor, the performance of single-walled CNTs (SWCNT) is restricted. Bundled Carbon Nano-Tubes are made using graphene layers, which reduce the skin effect to a negligible level at THz frequencies, increasing antenna efficiency [29, 30].

3.1 Graphene for THz Devices

Graphene is a two-dimensional material made up of carbon sheets of one atom thickness organized in a honeycomb lattice [31]. It has been a popular research topic due to its appealing characteristics and potential new uses [32, 33], particularly in the terahertz frequency ranges. Currently, experimental research of terahertz graphene-based devices is being conducted [34]. The fascinating feature of graphene is that its complicated surface conductivity can be adjusted using the Kubo formalism [35]. By providing a bias voltage to the graphene sheet, μ_c can switch over to different values [36]. The operating frequency and electromagnetic behavior of graphene-based Terahertz antenna can be dynamically adjusted by this technique [37]. The properties of electromagnetic structures created by graphene material are unprecedentedly changeable by manipulating Fermi energy [38]. Graphene-based antennas can also have a directed beam and those are made up of huge number of unit elements. The reflect array proposed in [39] included almost 25,000 elements.

3.2 Graphene Modeling

The Kubo formula may be used to determine the electric conductivity of a graphene sheet [40].

$$\sigma(\omega) = \frac{e^2(\omega + i\gamma_c)}{i\pi h^2}\left[\int_{-\infty}^{+\infty} \frac{|\int|}{(\omega + i\gamma_c)^2}\frac{df_0(\int)}{d\int} d\int - \int_{0}^{+\infty} \frac{f_0(-\int) - f_0(\int)}{(\omega + i\gamma_c)^2 - 4(\int/h)^2} d\int\right] \quad (1)$$

$$\sigma(\omega) = \sigma_{intra}(\omega) + \sigma_{inter}(\omega)$$

where e denotes elementary charge, γ_c is carrier scattering rate with $\boldsymbol{\tau}$ as the charge carrier relaxation time, $\hbar$ is the reduced Plank constant, and $\boldsymbol{f_0(\epsilon)}$ validates the Fermi energy [41] which is expanded as

$$f_0\left(\int\right) = \left\{\exp\left[\left(\int - \mu_c\right)/k_B T\right] + 1\right\}^{-1} \quad (2)$$

where T temperature in K and k_B is the Boltzmann's constant respectively. A range of variables, including frequency response, temperature and chemical potential, impact the surface conductivity of a graphene sheet [42].

The conductivity of graphene is split into two parts: First one is intraband conductivity (σ_{intra}) then next is interband (σ_{inter}) where intraband conductivity is dominating at THz frequencies [42]. The graphene chemical

$$n_0 = \frac{2}{\pi h^2 v^2 F} \int_0^{\infty} \int \left[f_0\left(\int\right) - f_0\left(\int +2\mu_c\right) \right] d \int \tag{3}$$

potential and charging carrier densities are related by following equation.

The Fermi velocity is denoted by vF. The intraband term of graphene conductivity may be approximated in a simple media as

$$\sigma^{\text{intra}}(\omega) = \frac{ie^2|\mu_c|}{\pi h^2(\omega + i\gamma_c)} \tag{4}$$

which is under the supervision of μ_c Furthermore, the electric field Intensity (E_ω) and surface current density on the graphene layer (J_ω) are related as follows:

$$J_\omega = \sigma_\omega E_\omega \tag{5}$$

As a result, E_ω and σ_ω have an effect on the performance of graphene-based devices.

4 Proposed Antenna Design

4.1 Design of Circular Patch Antenna

The method of cavity model analysis is proven as accurate for microstrip patch antenna analysis. In the cavity-model approach, resonant frequency and the quality factor of antenna were proven to depend on the solution of a patch-related transcendental equation. Other expressions relating to the patch geometry, for instance, the trigonometric functions in the case of the rectangle and Bessel functions for the circular patch, are also required to complete antenna design. As per the terminology of Carver, [39] the radius of the patch is represented by a refer Fig. 4.

The equation to determine the deign parameters of antenna is usually written in the form of and it can be represented as like,

$$k_0 J_0(k) = (1 - \alpha) J_1(k) \tag{6}$$

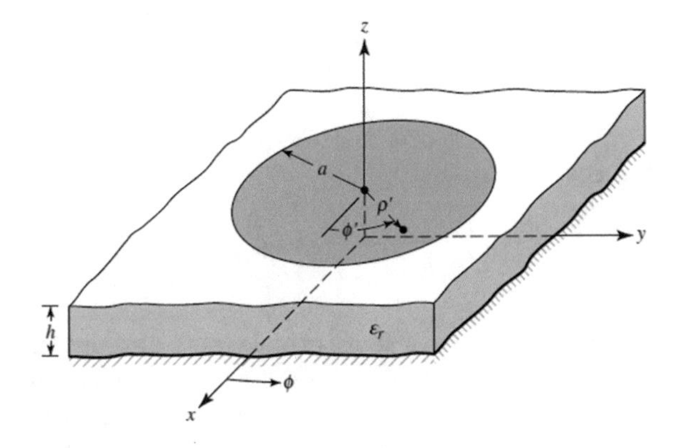

Fig. 4 Geometry of circular microstrip patch antenna

$$k\frac{J_1'(k)}{J_1(k)} = -\alpha \tag{7}$$

Here the term k is denoted as complex wave number and α is said to be input quantity as given below

$$\alpha = j\left(\frac{\eta_0}{\lambda_0}\right)h(G + jB) \tag{8}$$

Here G and B are given as,

$$G = 0.01254(\pi a/\lambda_0) \tag{9}$$

$$B = 0.00834(\pi a/\lambda_0)\varepsilon_r \tag{10}$$

The wave number k is a complex term hence we are denoting real and imaginary part as x and the quantities of attention, f_r and Quality factor are given by:

$$f_r = (x/z_0) f_n \tag{11}$$

Here the first derivative of Bessel function is given as $Z_0 = 1.84118$, for order unity. The normalizing magnetic wall frequency is given by

$$f_{n=}[(cz_0)/(2\pi a)](\varepsilon_r)^{-0.5} \tag{12}$$

Another term of interest the quality factor is given as,

$$Q = 0.5(x/y) \tag{13}$$

The design equation of circular patch geometry is given as,

$$a = \frac{F}{\left\{1 + \frac{2h}{\pi \varepsilon_r F}\left[\ln\left(\frac{\pi F}{2h}\right) + 1.7726\right]\right\}^{1/2}} \tag{14}$$

where

$$F = \frac{8.791 \times 10^9}{f_r \sqrt{\varepsilon_r}} \tag{15}$$

The dimensions of above equations are measured in cm.

This section proposes the design of Circular Patch antenna using Graphene conducting material which is operating at THz frequency range. As the increase in conduction loss of the copper at high frequency regimes finds, it is unsuitable for THz applications [54–60]. The replacements suggested are Gold, Graphene, Carbon Nano-Tube (CNT) and so on. The admirable properties of graphene nanostructure motivate to choose graphene for our design. The simple circular patch antenna on graphene material is shown in Fig. 5.

Here the circular patch antenna is designed on a Teflon Substrate with thickness of 2.8 μm and microstrip line feeding technique is chosen to supply the circular patch.

Dimensions	Symbol	Value(μm)
Length of substrate	Ls	30
Width of substrate	Ws	30
Height of substrate	h	2.8
Circular patch radius	a	6.7
Feed line length	L_f	8

(continued)

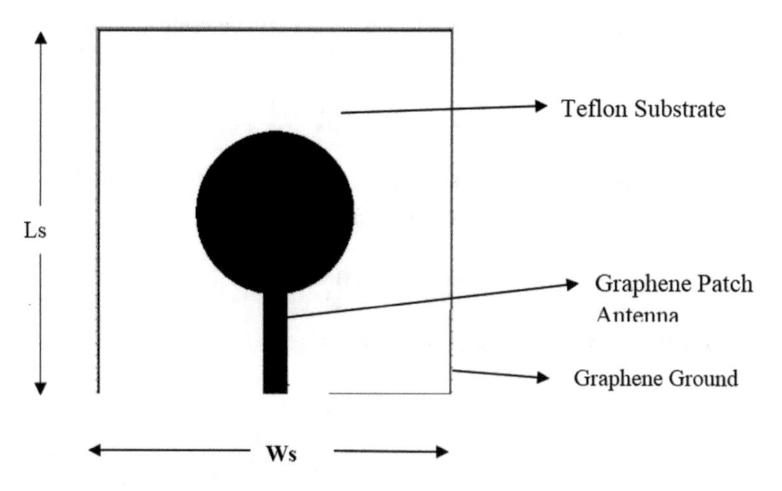

Fig. 5 Geometry of circular patch antenna

(continued)

Dimensions	Symbol	Value(μm)
Feed line width	W_f	2

The approximate model of graphene material is provided in Computer Simulation Software (CST) Microwave Studio for terahertz band and optical applications. The graphene antenna placed upon Teflon substrate with $\varepsilon_r = 2.1$ is shown in Fig. 6.

The intended antenna design and simulation techniques are carried out using CST Microwave Studio's time-domain solver, which employs the FIT (Finite Integration Method) for transient analysis. As per [43], FIT was presented for the first time in 1977, and its comprehensive time-domain formulation is published in [44, 45]. In contrast to the Finite-Difference Time-Domain (FDTD) method, which use the differential version of Maxwell's curl equations, FIT employs integral balances to discretize Maxwell's integral equations using finite volume space grids, allowing the conservation and stability features of the discrete field to be demonstrated numerically. Discrete port feeding with a 50 input impedance excites the microstrip feed-line. Figure 7 depict the proposed antenna's reflection coefficient (S_{11}) and Voltage Standing Wave ratio (VSWR) curves, respectively. The designed antenna resonates at 8.8 THz, 9.9 THz and 10.5 THz respectively without a superstrate layer, with reasonable return loss of 21, 28, and 16 dB.

The simulated antenna shows the better radiation performance over three different frequency bands 8.8 THz, 9.9 THz and 10.5 THz respectively as reported in Fig. 8.

The graphene patch antenna has a gain of 6.23, 4.75 and 5.47 dBi correspondingly at three different resonance band which shows the better performance of antenna as depicted in Fig. 9.

The two-dimensional *E* field and *H* field pattern of proposes antenna are displayed in Fig. 10 which indicates the directional radiation pattern in elevation plane.

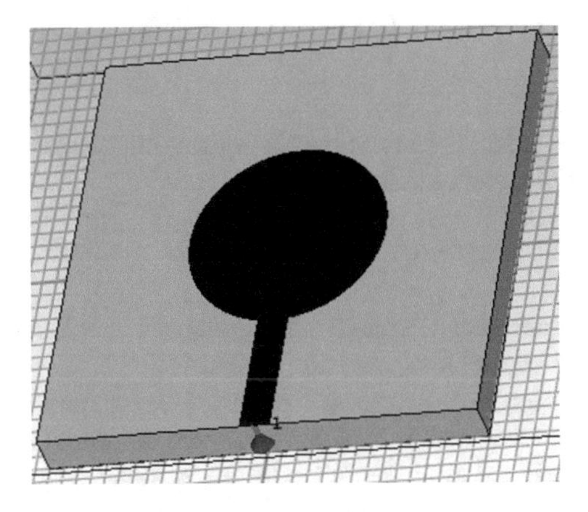

Fig. 6 Layout of circular patch antenna on Teflon substrate

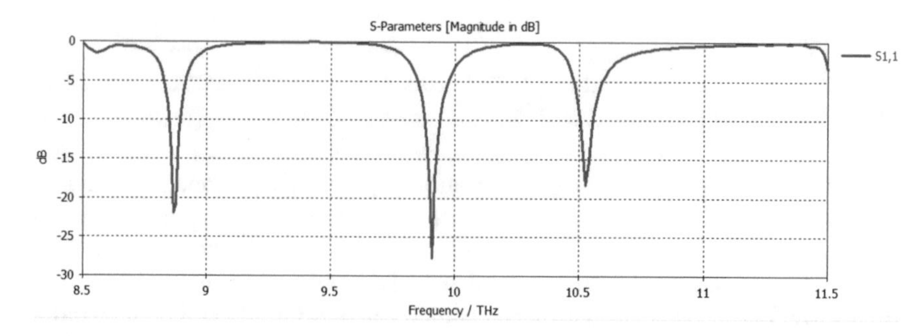

Fig. 7 Simulation results of circular patch antenna

4.2 Metasurface Loaded Circular Patch Antenna

The concentric double ring resonators with split gap are chosen as a unit cell for metasurface design [46–48]. The equivalent circuit of single resonator consists of Capacitor and Inductor in parallel to each other as depicted in Fig. 11.

The resonant frequency ($\boldsymbol{f_0}$) of the square Split Ring Resonator is given by [49]

$$f_0 = \frac{1}{2\pi\sqrt{L_{SRR}C_{SRR}}} \tag{16}$$

$\boldsymbol{L_{SRR}}$ and $\boldsymbol{C_{SRR}}$ are the net inductance and net capacitance of the unit cell structure which can be calculated as follows.

$$L_{SRR} = (0.0002)l\left(2.303log_{10}\frac{4l}{\omega} - \gamma\right) \tag{17}$$

$$C_{SRR} = \left(2a_e - \frac{g}{2}\right)c_{\text{pul}} + \left(\frac{\int_0 \omega t}{g}\right) \tag{18}$$

where the term $\gamma = 2.853$ is fixed value for square unit cell resonator [48] and the length of the conductor can be calculated using

$$l = 8a_e - g \tag{19}$$

The resonant frequency of a defined unit cell ring can be calculated by substituting total inductance and capacitance value in (f_0) Eq. 16.

As illustrated in Fig. 12, the suggested metasurface unit cell is made up of a pair of concentric square resonators in a planar arrangement. Here Gold is a chosen as a conductor for unit cell and multiple unit cells are located over the Teflon Substrate. An outer unit cell has a length and width of 10 μm (l), inner unit cell has a length of 8.5 μm and they are separated by a distance of 0.5 μm. The gap coupling of each unit

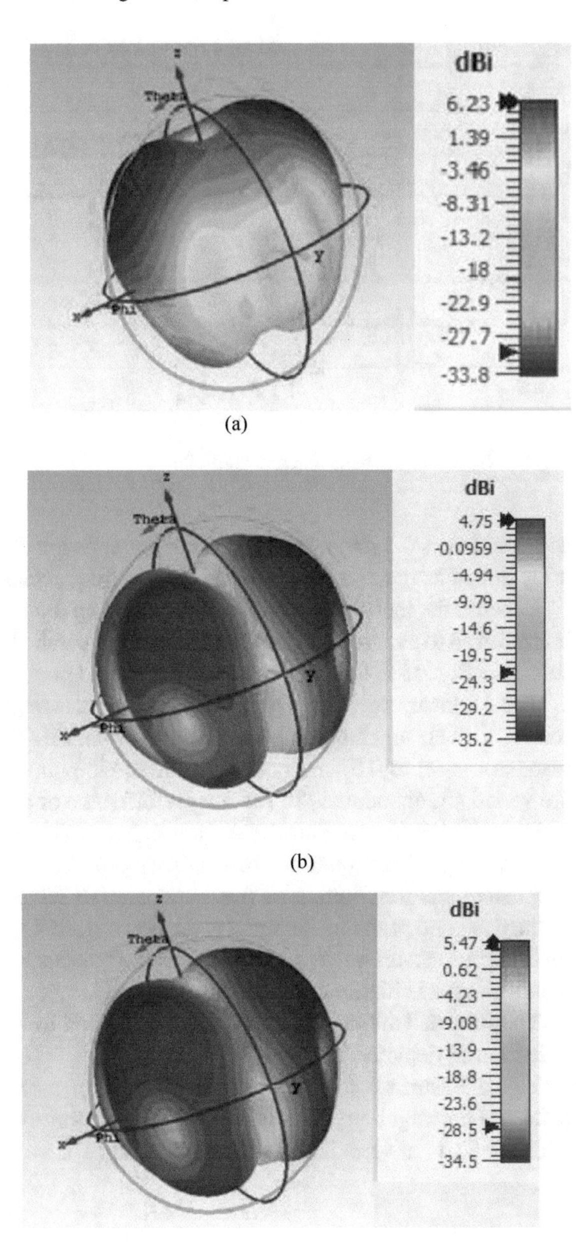

Fig. 8 Simulated 3D radiation pattern of metasurface loaded graphene antenna **a** 8.8 THz **b** 9.9 THz **c** 10.5 THz

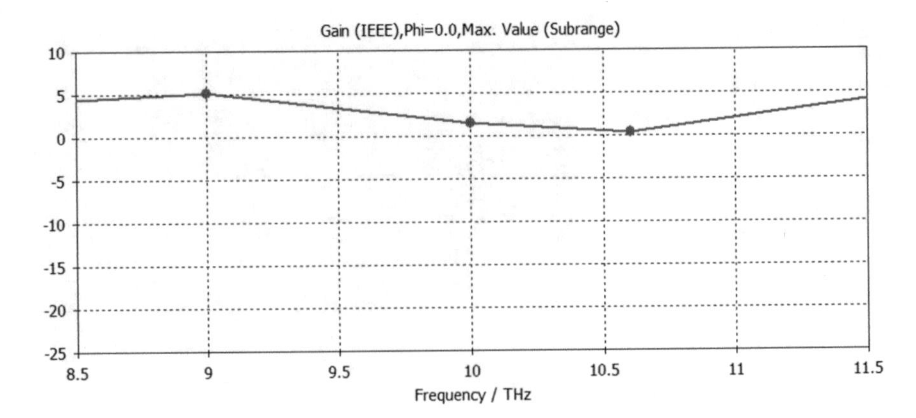

Fig. 9 Simulated graphene antenna gain

cell c and c_1 is 0.5 μm by default. The unit cells are repeated to make metamaterial array called as metasurface [50–53] at THz frequencies.

Because the metamaterial array is mounted on topside of a circular patch antenna, it is referred to as a metasurface loaded circular patch antenna. The proposed antenna shown in Fig. 13 is then simulated with discrete port.

The Metasurface loaded graphene Antenna resonates at 4.68 GHz with a bandwidth of 1 THz and has a good return loss of -40 dB. The gap (c) of square split ring resonator is set to 0.5 μm by default. Hence the inner ring gap and outer ring gaps are varied simultaneously to see the performance of proposed antenna.

Figure 14 indicates that the default value of 0.5 m keeps a suitable return loss value among all permutations of split ring gap. Again, it is a property proof of split ring resonators functioning on the metamaterials concept. The resonance frequency of the designed graphene antenna is pushed toward 4.68 GHz when it is loaded with metasurface, resulting in a size decrease. Thus the meat surface loaded graphene patch antenna is highly compact at 4.68 THz.

The intended antenna gain has been improved to 7 dBi in conjunction with size reduction as depicted in Fig. 15.

The metasurface loading over the graphene patch has a directional radiation pattern at boresight direction with main lobe magnitude of 20.5 dB (v/m) and angular width of 83.1° at 4.68 THz as depicted in Fig. 16.

5 Conclusion

Because of the development trend of new generation wireless communication networks (up to 100 Gbit/s), future spectrum resources are transferring to the THz band. THz antennas are highly promising for high data rate transmission between electronic devices. The efficiency of THz antennas also has a considerable impact

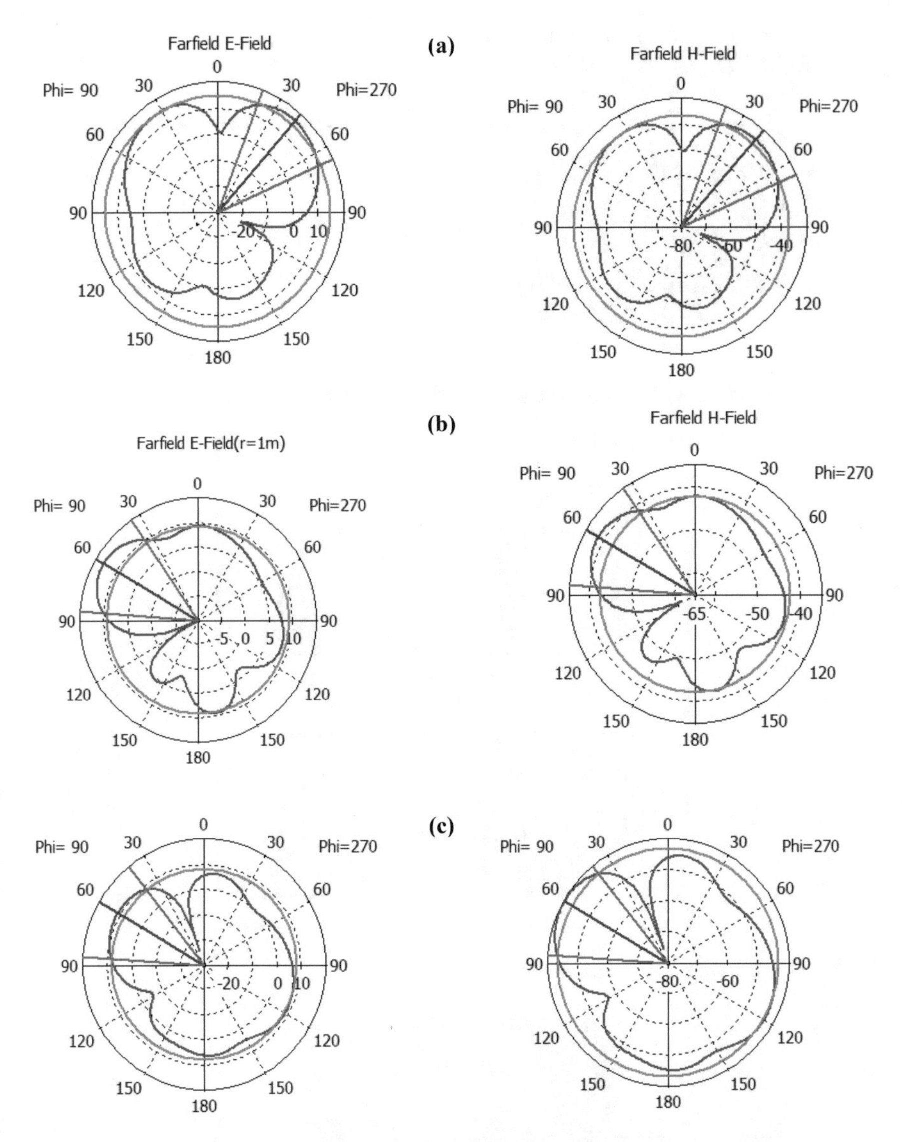

Fig. 10 *E* field and *H* field pattern **a** 8.8 THz **b** 9.9 THz **c** 10.5 THz

on the development of completely secured wireless networks and communication channels. An overview of THz antennas, material suitability in the THz band, the usage of reconfigurable metamaterials, and a comprehensive analysis of graphene characteristics have all been discussed in this chapter. This chapter proposes a metasurface loaded graphene patch antenna 4.68 THz. The suggested antenna improves gain while reducing size at the same time is considered as novelty of the proposed

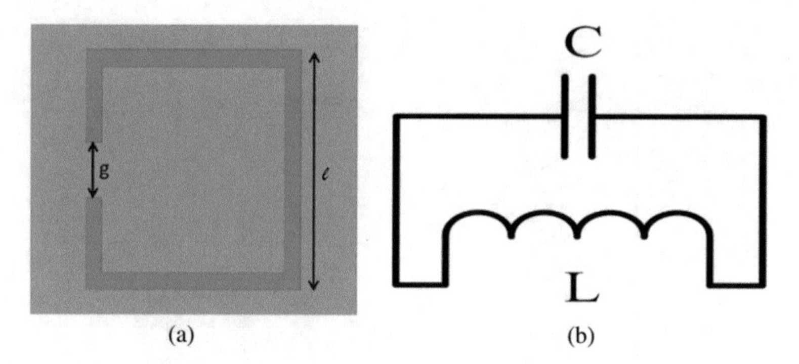

Fig. 11 Complementary SRR unit cell. **a** Layout diagram **b** CSRR Equivalent circuit

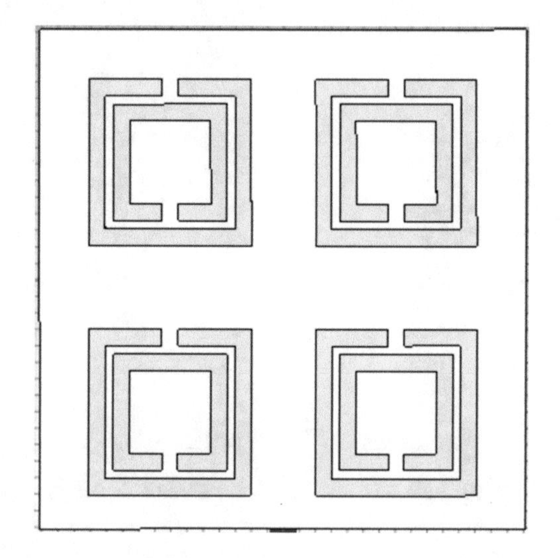

Fig. 12 Unit cell square split ring resonator

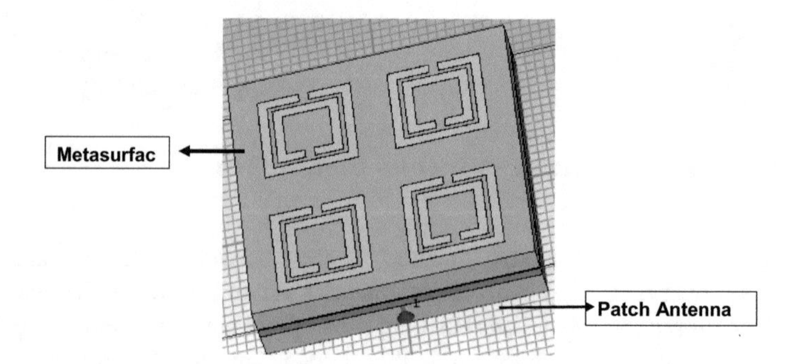

Fig. 13 Layout of metasurface top loaded on graphene antenna

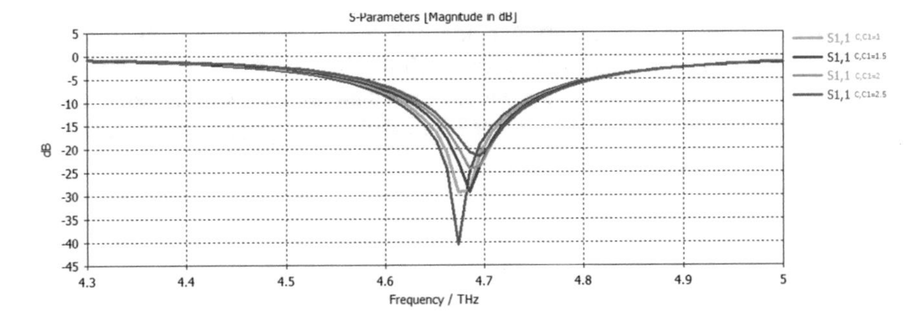

Fig. 14 Return loss of Metasurface loaded graphene antenna

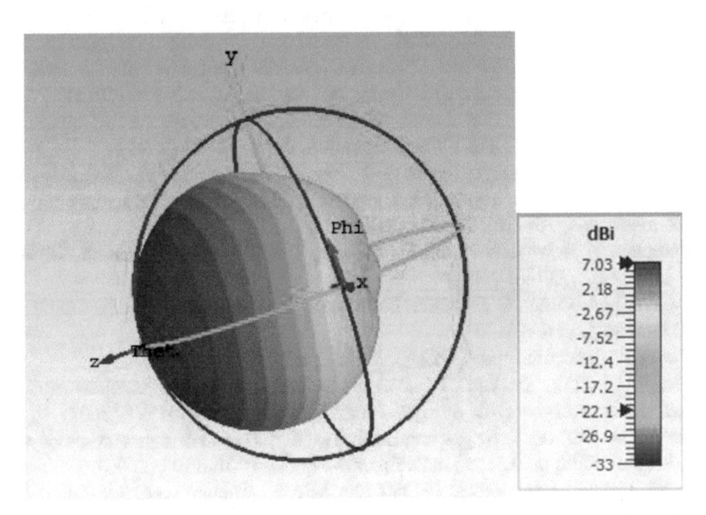

Fig. 15 Simulated far-field radiation pattern and antenna gain

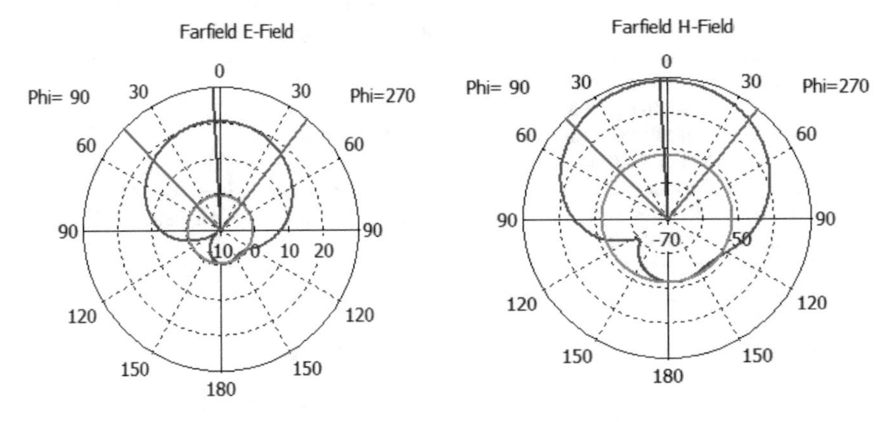

Fig. 16 Simulated 2D radiation pattern of metasurface loaded graphene antenna

work. When compared to a graphene antenna without a metasurface, the miniaturization achieved here is nearly 60%. The performance parameters were examined and determined to be satisfactory in all respects. The research will be expanded in the future to include the development of beam switching compact antennas for the THz frequency band.

References

1. M.A. Jamshed, F. Heliot, T.W.C. Brown, A survey on electromagnetic risk assessment and evaluation mechanism for future wireless communication systems. IEEE J. Electromagn. RF Microw. Med. Biol. **4**(1), 24–36 (2020)
2. D. Wu, J. Wang, Y. Cai, M. Guizani, Millimeter-wave multimedia communications: challenges, methodology, and applications. IEEE Commun. Mag. **53**(1), 232–238 (2015)
3. I.P. Roberts, J.G. Andrews, H.B. Jain, S. Vishwanath, Millimeter-wave full duplex radios: new challenges and techniques. IEEE Wirel. Commun. **28**(1), 36–43 (2021)
4. F.J. Martin-Vega, M.C. Aguayo-Torres, G. Gomez, J.T. Entrambasaguas, T.Q. Duong, Key technologies, modeling approaches, and challenges for millimeter-wave vehicular communications. IEEE Commun. Mag. **56**(10), 28–35 (2018)
5. P. Christopher, A new view of millimeter-wave satellite communication. IEEE Antennas Propag. Mag. **44**(2), 59–61 (2002)
6. I.F. Akyildiz, J.M. Jornet, C. Han, Terahertz band: next frontier for wireless communications. Phys. Commun. **12**, 16–32 (2014)
7. https://www.statista.com/topics/779/mobile-internet/
8. J. Grade, P. Haydon, D. Van Der Weide, Electronic terahertz antennas and probes for spectroscopic detection and diagnostics. Proc. IEEE **95**(8), 1583–1591 (2007)
9. T. Minotani, A. Hirata, T. Nagatsuma, A broadband 120-GHz Schottky-diode receiver for 10-Gbit/s wireless links. IEICE Trans. Electron. **86**(8), 1501–1505 (2003)
10. P.H. Siegel, Terahertz technology. IEEE Trans. Microw. Theory Tech. **50**(3), 910–928 (2002). https://doi.org/10.1109/22.989974
11. https://spectrum.ieee.org/aerospace/military/the-truth-about-terahertz
12. T. Nagatsuma, G. Carpintero, Recent progress and future prospect of photonics-enabled terahertz communications research. IEICE Trans. Electron. **98**(12), 1060–1070 (2015)
13. M.A. Jamshed, et al., Antenna selection and designing for THz applications: Suitability and performance evaluation: a survey IEEE Access **8** (2020)
14. Z. Chen, X. Ma, B. Zhang, Y.X. Zhang, Z. Niu, N. Kuang, W. Chen, L. Li, S. Li, A survey on terahertz communications. China Commun. **16**(2), 1–35 (2019)
15. W. Haiming, H. Wei, C. Jixin, S. Bo, P. Xiaoming, IEEE 802.11aj (45GHz): a new very high throughput millimeter-wave WLAN system. China Commun. **11**(6), 51–62 (2014)
16. H. Tabata, Application of terahertz wave technology in the biomedical field. IEEE Trans. Terahertz Sci. Technol. **5**(6), 1146–1153 (2015)
17. Y.A. Qadri, A. Nauman, Y.B. Zikria, A.V. Vasilakos, S.W. Kim, The future of healthcare Internet of things: a survey of emerging technologies. IEEE Commun. Surv. Tutorials **22**(2), 1121–1167 (2nd Quart., 2020)
18. https://spectrum.ieee.org/telecom/internet/wireless-industrys-newest-gambit-terahertz-communication-bands
19. N.V. Petrov, M.S. Kulya, A.N. Tsypkin, V.G. Bespalov, A. Gorodetsky, Application of terahertz pulse time-domain holography for phase imaging. IEEE Trans. Terahertz Sci. Technol. **6**(3), 464–472 (2016)
20. T. Nagatsuma, Antenna technologies for terahertz communications, in ISAP 2018—2018 International symposium antennas propagation (2019)

21. M. Liang, H. Xin, Microwave to terahertz: characterization of carbon-based nanomaterials. IEEE Microw. Mag. **15**(1), 40–51 (2014)
22. S. Dash, A. Patnaik, Material selection for THz antennas. Microw. Opt. Technol. Lett. **60**(5), 1183–1187 (2018)
23. E. Moreno, M.F. Pantoja, A.R. Bretones, M. Ruiz-Cabello, S.G. Garcia, A comparison of the performance of THz photoconductive antennas. IEEE Antennas Wirel. Propag. Lett. **13**, 682–685 (2014)
24. H.M. Cheema, The last barrier. IEEE Microw. Mag. 79–91 (2013)
25. S. Choi, K. Sarabandi, Design of efficient terahertz antennas: carbon nanotube versus gold. 2010 IEEE international symposium antennas propagation CNC-USNC/URSI radio science meeting—leading the wave, AP-S/URSI 2010 (2010)
26. P.J. Burke, An RF circuit model for carbon nanotubes. IEEE Trans. Nanotechnol. **2**, 55–58 (2003)
27. K.S. Novoselov, Electric field effect in atomically thin carbon films. Science **306**(5696), 666–669 (2004)
28. G.W. Hanson, Fundamental transmitting properties of carbon nanotube antennas. IEEE Trans. Antennas Propag. **53**(11), 3426–3435 (2005)
29. Y. Huang, W.-Y. Yin, Q.H. Liu, Performance prediction of carbon nanotube bundle dipole antennas. IEEE Trans. Nanotechnol. **7**(3), 331–337 (2008)
30. F.H.L. Koppens, D.E. Chang, F.J.G. de Abajo, 'Graphene plasmonics: a platform for strong light–matter interactions.' Nano Lett. **11**(8), 3370–3377 (2011)
31. G.W. Hanson, Dyadic Green's functions and guided surface waves for a surface conductivity model of graphene. J. Appl. Phys. **103**(6), Article No. 064302 (2008)
32. J. Murdock, E. Ben-Dor, F. Gutierrez, T.S. Rappaport, Challenges and approaches to on-chip millimeter wave antenna pattern measurements, in Proceedings of the IEEE International Microwave Symposium (June 2011), pp. 1–4
33. S. Choi, K. Sarabandi, Performance assessment of bundled carbon nanotube for antenna applications at terahertz frequencies and higher. IEEE Trans. Antennas Propag. **59**(3), 802–809 (2011)
34. Z. Xu, X. Dong, J. Bornemann, Design of a reconfigurable MIMO system for THz communications based on graphene antennas. IEEE Trans. THz Sci. Technol. **4**(5), 609–617 (2014)
35. J.S. Gomez-Diaz, C. Moldovan, S. Capdevila, J. Romeu, L.S. Bernard, A. Magrez, A.M. Ionescu, J. Perruisseau-Carrier, Self-biased reconfigurable graphene stacks for terahertz plasmonics. Nat. Commun. **6**, 6334 (2015)
36. K. Li, X. Ma, Z. Zhang, J. Song, Y. Xu, G. Song, Sensitive refractive index sensing with tunable sensing range and good operation angle-polarization tolerance using graphene concentric ring arrays. J. Phys. D, Appl. Phys. **47**(40), 1–7 (2014)
37. M. Tamagnone, J.S. Gomez-Diaz, J.R. Mosig, J. Perruisseau-Carrier, Reconfigurable terahertz plasmonic antenna concept using a graphene stack. Appl. Phys. Lett. **101**(21), 1–4 (2012)
38. D. Correas-Serrano, J.S. Gomez-Diaz, A. Alù, A.Á. Melcón, Electrically and magnetically biased graphene-based cylindrical waveguides: analysis and applications as reconfigurable antennas. IEEE Trans. THz Sci. Technol. **5**(6), 951–960 (2015)
39. E. Carrasco, J. Perruisseau-Carrier, Reflectarray antenna at terahertz using graphene. IEEE Antennas Wirel. Propag. Lett. **12**, 253–256 (2013)
40. M. Aldrigo, M. Dragoman, D. Dragoman, Smart antennas based on graphene. J. Appl. Phys. **116**(11), 1–5 (2014)
41. R. Depine, *Graphene Optics: Electromagnetic Solution of Canonical Problems* (Morgan & Claypool, San Rafael, 2017)
42. P. Tassin, T. Koschny, C.M. Soukoulis, Graphene for terahertz applications. Science (80) **341**(6146), 620–621 (2013)
43. Weiland, Discretization method for the solution of Maxwell's equations for six-component fields. AEU-Archiv fur Elektron. und Ubertragungstechnik **31**(3), 116–120 (1977)

44. M. Clemens, T. Weiland, Discrete electromagnetism with the finite integration technique. Prog. Electromagn. Res. **32**, 65–87 (2001)
45. T. Weiland, Time domain electromagnetic field computation with finite difference methods. Int. J. Numer. Model. Electron. Networks Device Fields **9**(4), 295–319 (1996)
46. J.B. Pendry, A.J. Holden, D.J. Robbins, W.J. Stewart, Magnetism from conductors, and enhanced non-linear phenomena. IEEE Trans. Microw. Theory Tech. **47**(11), 2075–2084 (1999)
47. F. Falcone, T. Lopetegi, J. Baena, Effective negative—ε stopband microstrip lines based on complementary split ring resonators. Microw. **14**(6), 280–282 (2004)
48. F. Capolino, *Theory and Phenomena of Metamaterials* (2017)
49. G. Oliveri, D.H. Werner, A. Massa, Reconfigurable electromagnetics through metamaterials—a review. Proc. IEEE **103**(7), 1034–1056 (2015)
50. C.L. Holloway, E.F. Kuester, J.A. Gordon, J. O'Hara, J. Booth, D.R. Smith, An overview of the theory and applications of metasurfaces: the two-dimensional equivalents of metamaterials. IEEE Antennas Propag. Mag. **54**(2), 10–35 (2012)
51. B. Vasić, M.M. Jakovljević, G. Isić, R. Gajić, Tunable metamaterials based on split ring resonators and doped graphene. Appl. Phys. Lett. **103**(1), 011102 (2013)
52. D. Kumar et al., Bandwidth enhancement of planar terahertz metasurfaces via overlapping of dipolar modes. Plasmonics **15**(6), 1925–1934 (2020). https://doi.org/10.1007/s11468-020-01222-7
53. S.S. Karthikeyan, R.S. Kshetrimayum, Slot split ring resonators and its applications in performance enhancement of microwave filter. IEEE applied electromagnetics conference and URSI commission B meeting (2009)
54. Y. Dong, P. Liu, D. Yu, G. Li, F. Tao, Dual-band reconfigurable terahertz patch antenna with graphene-stack-based backing cavity. IEEE Antennas Wirel. Propag. Lett. **15**, 1541–1544 (2016)
55. Y. Huang, L.S. Wu, M. Tang, J. Mao, Design of a beam reconfigurable THz antenna with graphene-based switchable high-impedance surface. IEEE Trans. Nanotechnol. **11**(4), 836–842 (2012)
56. S. Venkatachalam, K. Zeranska-Chudek, M. Zdrojek, D. Hourlier, Carbon-based terahertz absorbers: materials, applications, and perspectives. Nano Sel. **1**(5), 471–490 (2020)
57. K.R. Carver, J.W. Mink, Microstrip antenna technology. IEEE Trans. Antennas Propag. **29**(1), 2–24 (1981)
58. I.J. Bahl, P. Bhartia, *Microwave Solid State Circuit Design* (Wiley, 2003)
59. J.D. Mahony, Hal Schrank, Similarities in the design equations for rectangular and circular patches. IEEE Antennas Propag. Mag. **33**(5) (1991)
60. J.M. Jornet, I.F. Akyildiz, Graphene-based plasmonic nano-antenna for terahertz band communication in nanonetworks. IEEE J. Sel. Areas Commun. **31**(12), 685–694 (2013)

Monopole Patch Antenna to Generate and Detect THz Pulses

Mohamed Lamiri, Mohammed El Ghzaoui, and Bilal Aghoutane

Abstract The use of electromagnetic waves is a powerful tool for observing and understanding the world around us. From the infinitely large and distant, electromagnetic waves allow us to "see" objects and analyze them through spectroscopy. The tremendous scope of the electromagnetic spectrum multiplies the fields of observation and applications. Far infrared is a specific region of the electromagnetic spectrum with frequencies typically between 0.3 and 20 terahertz (THz). The wavelengths associated with these frequencies range from 1 mm to 15 μm. This domain of the so-called terahertz waves is still little exploited at the technological level. The antenna is one of the most frequently used components for THz generation and detection. It generates and detects THz pulses by transient photocurrents induced by ultra-short excitation laser pulses. In this paper, a rectangular patch antenna having $47.916 \times 37.693\mu m^2$ dimension printed on Rogers RT/duriod 6010 substrate is presented and analyzed. The proposed MIMO antenna offers high bandwidth with less return loss. Other antenna parameters were studied in this work such as radiation patterns and current distribution. To have a good picture about the designed antenna we analyzed the isolation of the MIMO antenna. Based on simulation, the proposed antenna can be used efficiently in THz applications.

Keywords Terahertz technologies · THz antenna · MIMO antenna · MIMO diversity · Antenna gain

M. Lamiri
Moulay Ismail University of Meknes, Meknes, Morocco

M. El Ghzaoui (✉)
Faculty of Sciences, Sidi Mohamed Ben Abdellah University, Fez, Morocco

B. Aghoutane
Ibn Toufail University, Kenitra, Morocco

S. Das et al. (eds.), *Advances in Terahertz Technology and Its Applications*,
https://doi.org/10.1007/978-981-16-5731-3_16

1 Introduction

The 5G frequency bands, currently being tested, will not be able to manage all applications when the industry is at full capacity. It is therefore necessary to study solutions to prepare for the evolution of the current infrastructure. The proposed solution is the terahertz band [1]. The TeraHertz (THz) waves define the range of the electromagnetic wave spectrum between 20 and 0.1 THz, which corresponds to wavelengths between 15 and 3000 μm. The THz domain is bounded by the violet square, which extends from the radio waves of the electrical domain (0.1 THz) to the infrared waves of the optical domain (>20 THz).

The THz pulses produced by these photoconductive antennas also allow time-resolved measurements. Moreover, the real and imaginary parts of the dielectric constant of the materials can be determined without recourse to the relations of Kramers Kronig, since in the field of time-resolved spectroscopy, THz field can be measured directly in amplitude and phase [1–3]. Photoconductive antennas are available in various forms, for example, coplanar micro-ribbon [4, 5], butterfly node [6, 7], periodic logarithmic [8] and logarithmic spirals [9].

In a single-antenna system, increasing the size of the modulation or the frequency band used is the only solution to increase the data rate, with all the complexity or congestion problems that this causes. From the point of view of information theory, the capacity of multi-antenna systems increases linearly with the number of transmitting antennas, significantly exceeding the theoretical limit of Shannon [10, 11]. The ability of multi-antenna systems to resist fainting and interference is an undeniable additional benefit [12]. These discoveries made MIMO systems. In this paper, a MIMO antenna is designed and analyzed for THz bands.

2 MIMO Principle

MIMO technologies use transmit and receive antenna arrays (Fig. 1) to improve the quality of the signal-to-noise ratio and transmission rate. This then allows the radio signal emission level to be reduced in order to reduce the surrounding electromagnetic pollution, but also to extend the battery life in the case of a telephone [13].

3 MIMO Skill Categories

Three categories of MIMO can be considered:

- ***The MIMO spatial diversity***: we simultaneously transmit the same message on different antennas to the emission. The signals received on each of the receiving antennas are then brought back into phase and summed consistently. This increases the signal-to-noise ratio (S/B) of the transmission [14].

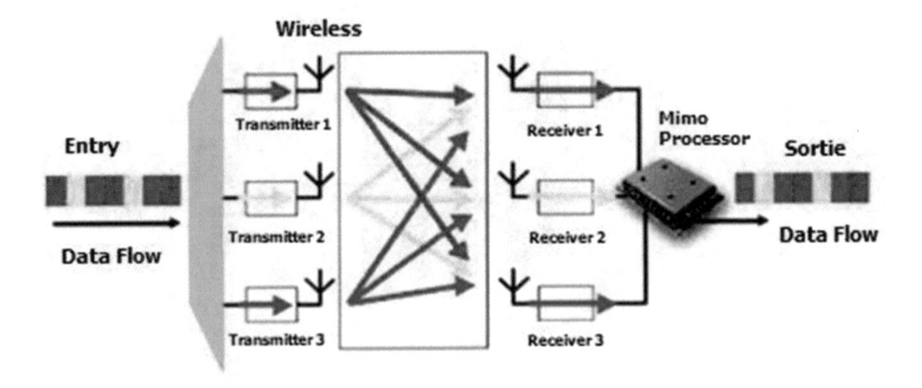

Fig. 1 MIMO principle

- ***MIMO spatial multiplexing***: each message is divided into sub-messages. The different sub-messages are simultaneously transmitted on each of the transmitting antennae. The signals received on the receiving antennas are reassembled to re-form the entire original message. MIMO multiplexing increases the transmission rate [14].
- ***MIMO—beamforming***: The MIMO antenna array is used to orient and control the radio beam. Beamforming techniques allow both to extend radio coverage (from a base station or access point for example) and to limit interference between users and surrounding electromagnetic pollution (by targeting the target receiver) [14].

4 MIMO Channel and Signal Patterns

The channel we are interested in for wireless links is the radio-mobile channel. It can be characterized by a propagation medium: air, by moving obstacles (people, vehicles) or immobile (terrain, buildings). The electromagnetic wave, vector of information on this medium, is created by the transmitter for transmission in the air. Unfortunately, the channel is not transparent to the EM wave and will interact with it. Knowledge of the propagation channel is essential for the dimensioning of a communication system and the optimization of algorithms for processing digital communications signals. For this, a number of propagation channel models were developed to facilitate simulation and a number of measurements were performed in different propagation configurations [15, 16].

The EM wave diffuses in a medium and loses part of its energy as it spreads until it disappears completely. The reception of the signal can only be done if part of the wave can be received by the antenna of the receiving terminal (Fig. 2).

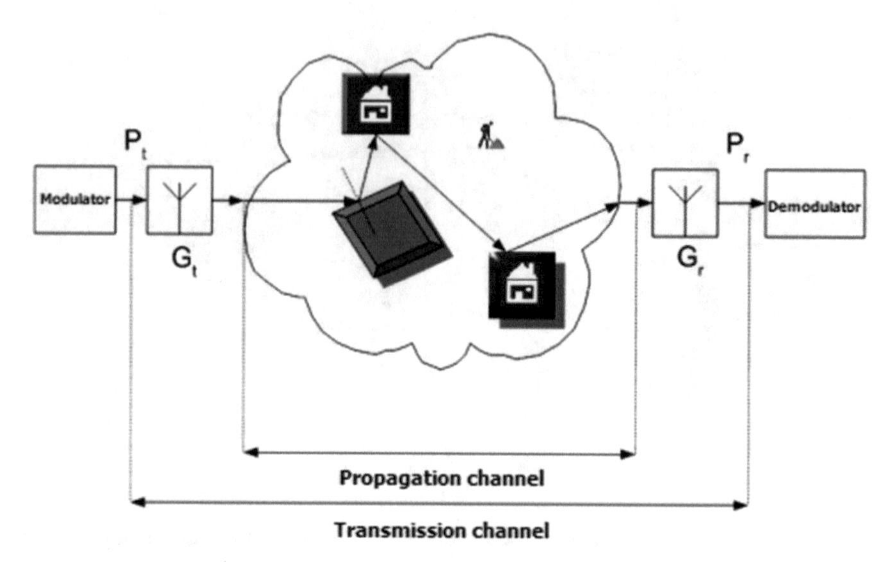

Fig. 2 MIMO channel

The phenomena occurring in the propagation channel and causing variations in the received power are commonly distributed into phenomena of two kinds: large-scale phenomena and small-scale phenomena.

4.1 Large-Scale Phenomena

Large-scale phenomena correspond to two very special phenomena.

- **Free space attenuation**: Free space is defined by the line of sight and by the clearance of the Fresnel ellipsoid. In these propagation conditions, the power losses between the transmitted power and the received power are expressed by the relationship of the link balance. This relationship shows that the power received is inversely proportional to the square of the distance between transmitter and receiver.
- **The masking effect**: The masking effect is a loss of power at the receiving terminal, and occurs when one of the two terminals moves and, in doing so, causes the line of sight to break between them (the terminal can be masked by a building). Part of the transmitted wave still arrives at the receiving terminal, but the power loss of the wave is systematic. This phenomenon is related to the topology of the channel (obstacles).

4.2 Small-Scale Phenomena

By observing the power received by a terminal during a small shift, when the power loss due to masking and free space attenuation is constant, we can still notice a fluctuation in power, a consequence of small-scale phenomena. Large-scale phenomena take into account the existence of a single path, direct or not, to assess the power received at the receiver level. In reality, a multitude of paths is created by the channel according to the phenomena of reflection, refraction, diffraction and diffusion of the wave. Some of the multiple paths recombine at the receiving terminal and the paths with their own amplitude and phases can recombine constructively or destructively, improving or degrading the link balance. It is this phenomenon that is called fainting on a small scale. In addition, if no motion is present in the channel (the channel is time-invariant), then recombination generates a constant attenuation or gain over time. If, conversely, movements exist (the channel evolves over time by the movement of the transmitter, receiver, or moving obstacles), then the attenuation (or gain) varies according to the new multi-path gives. This phenomenon is therefore related to the topology of the channel and varies with the movements of the terminals or channel elements.

4.3 Noise

Noise is any signal that interferes with a communication. The surrounding communications are therefore noise, even if the term interference is preferred. Apart from these interferences, the noise can come from atmospheric disturbances (rain for certain frequency ranges) or cosmic disturbances (solar flares). These noises, which can be described as external to the communication system, are complemented by internal noises, such as the thermal noise of the receiver, or the sound of shot.

4.4 Channel Characterization

The characterization of the channel consists in determining its ability to vary in different domains (temporal, frequency, spatial) [17, 18].

The expression of the impulse response of channel $\boldsymbol{h(\tau,t)}$ is defined as the response of the channel at the moment (t) following a pulse occurring at the moment $\boldsymbol{(t-\tau)}$. This response at time t is composed of P paths arriving each with an amplitude $\beta_p(t)$, a phase $\theta_p(t)$and a delay $\boldsymbol{\tau_p}$

$$\boldsymbol{h(\tau \cdot t)} = \sum_{P=1}^{P} \boldsymbol{\beta_p(t)} \cdot e^{j \cdot \theta_p(t) \cdot \delta(\tau - \tau_p(t))}$$

4.4.1 Dispersion of Delays

Delay spread is defined as the difference $\tau = \boldsymbol{\tau}_p - \boldsymbol{\tau}_1$. This spread can cause inter-symbol interference (IES) if the symbol time of a system is too short and is therefore an important element of channel characterization. The outputs of the channel produce different spreads, which defines the average delay $\boldsymbol{\tau} = \boldsymbol{E}\{\boldsymbol{\tau}\}$, and the variance of delays $\sigma_\tau^2 = E\left\{(\tau - E\{\tau\})^2\right\}$($\boldsymbol{\sigma}_\tau$ is called the delay dispersion). The multi-path channel and its delay dispersion $\boldsymbol{\sigma}_\tau$ allow to define the coherence band B_c, frequency band where attenuations are correlated (above a correlation threshold).

4.4.2 Frequency Dispersion

The notion of temporal variation of the channel is an integral part of its characterization. This temporal variation results from displacements (Tx and Rx terminals, obstacles) which modify the properties of the wave. A displacement of the terminals in the channel at a speed υ creating an angle of incidence α with the direct path of the wave between the two terminals causes the appearance of the Doppler effect, that is to say a frequency shift of the received wave. The resulting wave has a frequency: $\boldsymbol{f}_r = \boldsymbol{f}_0 + \boldsymbol{f}_d$ with $\boldsymbol{f}_0$ the original carrier frequency, and $f_d = \frac{v \cdot f_0}{c} \cdot \cos(\alpha)$ the offset Doppler frequency. The Doppler frequency f_d is included in $[-\boldsymbol{f}_{\text{dmax}};+\boldsymbol{f}_{\text{dmax}}]$, avec $f_{\text{dmax}} = \frac{v \cdot f_0}{c}$. The coherence time τ_c is defined as the time during which the temporal fainting of the channel is correlated (above a correlation threshold). It is inversely proportional to the maximum Doppler frequency f_{dmax}. The coherence time is given by: $\tau_c = \frac{1}{f_{\text{dmax}}}$.

To consider the constant channel, the system symbol time must be much less than the channel coherence time ($\boldsymbol{T}_s \ll \boldsymbol{\tau}_c$).

4.4.3 Angular Dispersion

Signals are emitted by antennas with starting angles. At reception, the 586 angles of arrival determine the incidence of rays on the coil. The study of the dispersion of these angles shows a correlation between the placement of the transmitting antennae and the correlation of the receiving signals [14].

4.5 *Propagation Channel Modeling*

To integrate the model into simulations, the impulse response ***h(τ, t)*** is used as that of a FIR filter (finite impulse response). Thus, provided that the channel response is linear and time-invariant (or at least over a sufficiently long time interval in relation to the data to be sent), then the $y(t)$ output of the channel filter is the convolution product between the $x(t)$ input and the impulsional response $h(t)$:

$$y(t) = [x * h] = \int_{-\infty}^{\infty} x(\tau) \cdot h(t - \tau, t) d\tau$$

4.5.1 The Gaussian Channel Model

The Gaussian channel for wireless communications is the simplest statistical channel from an implementation perspective, but not necessarily the most realistic. It only models the thermal noise of the receiver as a random Gaussian probability density variable $pX(x)$ and adds to the useful signal. The random variable is called the Gaussian additive white noise (BBAG). The noise is called white because it disturbs in an identical way the entire spectrum with a spectral density of constant monolateral power N_0 (W/Hz). The noise is then fully defined statistically by its zero μ_b mean and its variance σ_b^2.

$$P_X(x) = \frac{1}{\sqrt{2\pi\sigma_b^2}} \cdot \exp\left[-\frac{(x - \mu_b)^2}{2\sigma_b^2}\right]$$

4.5.2 The Rayleigh Model

The Rayleigh model is used in the simulation of indoor-type systems because it takes into account multiple paths and thus allows to model the phenomena on a small scale. The Rayleigh model represents multiple paths as a single complex coefficient (attenuation and phase shift), varying over time. The attenuation of the channel $\beta = |h|$ is then represented as a random variable following a Rayleigh law defined by its probability density $P_B(\beta)$ of parameter σ:

$$P_B(\beta) = \frac{\beta}{\sigma^2} \cdot \exp\left[\frac{-\beta^2}{2\sigma^2}\right], \quad \forall \beta \geq 0$$

5 Typical Coil Parameters

An antenna is a device for converting electrical energy into energy electromagnetic emission or reception. Many parameters are used to describe the characteristics and performance of the antennas such as input impedance, reflectance, directivity, gain, efficiency and radiation.

The antenna has several roles, the main ones being:

- Allow proper adaptation between radio equipment and the propagation.
- Transmit or receive energy in preferred directions.
- Convey information as accurately as possible.

In addition, in order to describe the characteristics and performance of the antennas, various parameters are used [13–15]. These parameters are divided into two groups. The first group characterizes the antenna as an electrical circuit element (Zin and S11) and the second group is interested in its radiation properties, such as the radiation diagram, directivity and gain. Finally, the notion of power (absorbed or radiated) plays a role important in the study of antennae.

5.1 Input Impedance

The impedance seen at the input of this component is called the antenna input impedance. It is represented by: $\mathrm{Ze}(f) = \mathrm{Re}(f) + j \cdot \mathrm{Xe}(f)$ (Fig. 3). The input resistance $\mathrm{Re}(f)$ is a dissipation term. It is linked, on the one hand, to radiated power and, on the other, to the power lost by the Joule effect. The latter is generally small compared to the power radiated to ensure the optimal operation of the antenna [19]. However, losses due to Joule can represent significant values depending on the geometry of the antenna. Losses in the mass plane are also to be taken into account. The reactance $\mathrm{Xe}(f)$ is related to the reactive power stored and concentrated in the vicinity of the antenna. The impedance of the antenna is influenced by surrounding objects, in particular by objects or close metal planes or by other antennas. In the latter case, we are talking about of mutual impedances between radiating elements. We will deal in this paragraph only with the antenna's own impedance, that is, that of the antenna placed alone and radiating in the infinite empty space [13].

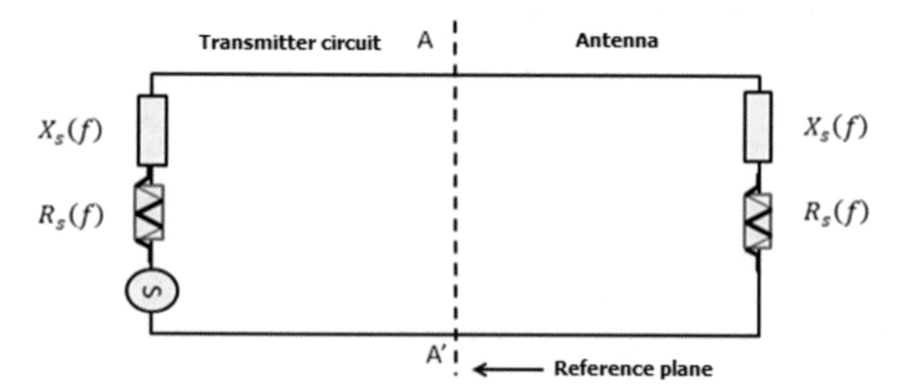

Fig. 3 Input impedance

5.2 Coefficient of Reflection and ROS

In the previously introduced reference plane, the discontinuity presented by the antenna can be characterized by the coefficient of reflection ρ (in voltage or field). This is related to the input impedance of the antenna by a classical relation [10–12]:

$$\rho = \frac{Z_e + R_0}{Z_e - R_0}$$

With R_0 the normalization impedance (equal to 50Ω in microwave technology), this parameter allows to characterize the adaptation of the antenna which is realized ideally for ρ zero (no reflected wave).

In practice, adaptation is characterized by the module of the coefficient of reflection or, more often than not, by the "Standing Wave Ratio" (R.O.S.) defined by:

$$ROS = \frac{1 + |\rho|}{1 - |\rho|}$$

The term TOS (Standing Wave Rate) can be used instead of ROS.

5.3 Radiation Pattern

The graphic representation of the characteristic function of the antenna is called radiation diagram. The direction of the maximum radiation is called the radiation of the antenna. The representation of this function gives the characteristics of the radiation in space. Traditionally, it has become customary to represent the radiation in two perpendicular planes which are: plane E and plane H. Plane E is defined as the plane containing the antenna axis and the electric field. The plane H is defined as the plane containing the axis of the antenna and the magnetic field. The radiation, usually in logarithmic coordinates, is presented either in coordinates rectangular, either in polar coordinates, in the two perpendicular planes (E and H). Some three-dimensional representations have the advantage of showing all the directions of radiation in space, but hardly allow an appreciation quantitative.

The antennas are rarely omnidirectional and emit or receive in preferred directions. The radiation pattern represents the variations in power radiated by the antenna in different directions of space. It indicates the directions of space (θ_0, φ_0) in which the radiated power is maximum. It is important to note that the radiation pattern only makes sense if the wave is spherical. The radiation pattern of an antenna is mainly related to its geometry but can also vary with frequency. Apart from omnidirectional antennas, antennas do not radiate power uniformly in space. In this case, the radiation characteristic is equal to 1 regardless of the direction considered. In general, the power is concentrated in one or more "lobes". The main lobe corresponds to the preferred direction of radiation. Secondary lobes are usually lobes parasites. In these directions, the radiated energy is lost so efforts are made to mitigate them.

5.4 Antenna Directivity, Gain and Performance

La capacité d'une antenne à diriger l'énergie se traduit par la directivité. Le gain d'une antenne est directement proportionnel à sa directivité. Le rapport de proportionnalité entre ces deux grandeurs est le 'rendement' de l'antenne. Finalement, ces trois grandeurs permettent de caractériser la façon dont une antenne convertit la puissance électrique incidente en puissance électromagnétique rayonnée dans une direction particulière. Le gain et la directivité permettent de comparer les performances d'une antenne par rapport à l'antenne de référence qu'est l'antenne isotrope en général.

5.4.1 Directivity

Directivity characterizes the ability of an antenna to concentrate energy in one or several privileged directions. It is an intrinsic parameter of the antenna, without dimension. Directivity $D(\theta, \varphi)$ is a relative measure of the power radiated in one direction $P(\theta, \varphi)$ relative to total radiated power rated PR:

$$D(\theta, \varphi) = \frac{P(\theta, \varphi)}{\frac{P_R}{4\pi}} = 4\pi \cdot \frac{P(\theta, \varphi)}{P_R}$$

5.4.2 Gain of an Antenna

In general, an antenna radiates power that varies with direction considered. We call gain $G(\theta, \varphi)$of an antenna in one direction(θ, φ), the ratio of the $P(\theta, \varphi)$ power in this direction at power $P_0(\theta_0, \varphi_0)$ of an isotropic source of reference per solid angle unit with the same power supply.

The gain is proportional to the directivity. It carries the same information on the directions of radiation.

The gain is expressed in decibel (dB). dB_i is sometimes used to specify the reference to isotropic radiation. Sometimes the gain expressed in dB_d, when a dipole antenna is used as a reference. If the antenna is omnidirectional and without losses, its gain is 1 or 0 dB. According to the previous definitions the gain G can be written as follows:

$$G(\theta, \varphi) = 4\pi \cdot \frac{P(\theta, \varphi)}{P_A}$$

5.4.3 Performance

Let P_A be the power supply of an antenna. This power is transformed into a radiated power P_R. In the direction of transmission, the radiated power is less than the power supply. The antenna is an imperfect transformer. There are losses during the energy transformation, as in any system. The efficiency of the antenna is defined by:

$$\eta = \frac{P_R}{P_A}$$

It measures the rate of transformation. It's a return in the sense of thermodynamics of the term: η 1 Performance is related to losses in the polarization network and in the radiant. we see that the efficiency links the gain and the directivity:

$$P_R = \eta \cdot P_A \Rightarrow G = \eta \cdot D$$

5.5 *Bandwidth and Quality Factor*

The bandwidth, also known as bandwidth, of an antenna defines the frequency domain in which the antenna radiation has the required characteristics. It is also the frequency band where the energy transfer from the power supply to the antenna (or from the antenna to the receiver) is maximum. The bandwidth can be defined according to the coefficient of reflection, provided that the radiation pattern does not change on this band. To determine the bandwidth of an antenna with respect to radiation, trace the reflection parameter S11 as a function of frequency. It is generally accepted that if this parameter is less than -10 dB or -15 dB, the radiation power is sufficient. It's enough then locate on the curve the values of the frequency corresponding to this value as on Fig. 4 (The bandwidth recorded for telecommunications antennas is Coil with ROS or TOS 2 9.5 dB; T0S = 1.4 S11 = 15 dB).

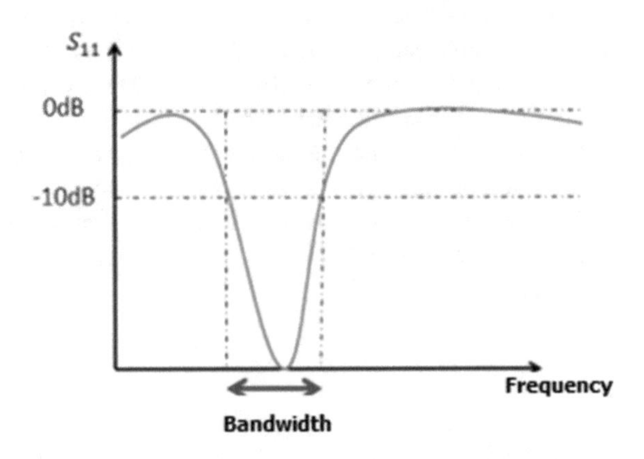

Fig. 4 Bandwidth

From an electrical point of view, an antenna can be seen as a RLC resonant circuit. The BW bandwidth (3 dB bandwidth of the field value) is related to the Q quality factor of the RLC circuit at the resonance frequency $\boldsymbol{f}_{\boldsymbol{Res}}$. The quality factor represents the quantity of resistance present during resonance (for a series resonant circuit) as shown in equation.

$$Q = \frac{\boldsymbol{f}_{\boldsymbol{Res}}}{\boldsymbol{BW}} \Rightarrow \frac{1}{Q} = \frac{R_{ant}}{2\pi f_{Res} \cdot L_{ant}}$$

An antenna with a high quality factor radiates very efficiently at the frequency of radiation on a narrow frequency band, which may limit out-of-band interference. However, if the bandwidth is too narrow, any signal sent or received near the operating frequency band will be attenuated.

An antenna with a low quality factor is considered broadband if the higher frequency (f_2) is at least equal to about twice the lower frequency (f_1).

6 Design and Optimization of a MIMO Antenna

The antenna designed on Rogers RT/duriod 6010 substrate with a thickness with of $0.5\mu m$. The antenna element size where as a ground plan comprise $47.916 \times 37.693\mu m^2$ of a size. With relative dielectric permittivity equal 10.2 and tangent loss 0.0023 to operate on Terahertz. Patch has a much better impedance matching. This antenna is simulated before implementing a two-element MIMO antenna which is our goal, as shown in Fig. 5. In the design of MIMO antenna, one of the more significant element is the monopole. This operated under the λ/2 mode. Figure 5 depicted the basic element. Figure 6 shows the configuration of the proposed MIMO antenna. Two identical patch antennas of rectangular with feed line are used to feed the two patch antennas also $82.386 \times 84.916\mu m^2$ is the dimension for the ground plane. Table 1 shows the antenna's parameters.

Figure 7 compares between the S11 of the proposed antenna and the single element. It is clear from this figure that the MIMO antenna offers wide bandwidth with less return loss. Indeed, the basic element had a bandwidth of about 64 GHz with a minimum return loss of −52 at 1.47 THz. However, the proposed antenna presents a bandwidth of 117 GHz at 1.47 THz. Besides, the MIMO antenna gain is higher than the basic element gain (Table 2).

In order to study the mutual coupling between the two elements, we will plot the isolation of the MIMO antenna. Figure 8 shows the S12 which the same as S21 of the proposed MIMO antenna. As can be seen in this figure the isolation is less than −15 dB for the whole band. In addition, we obtain high isolation when the inter-elements distance is about 5.7777 um.

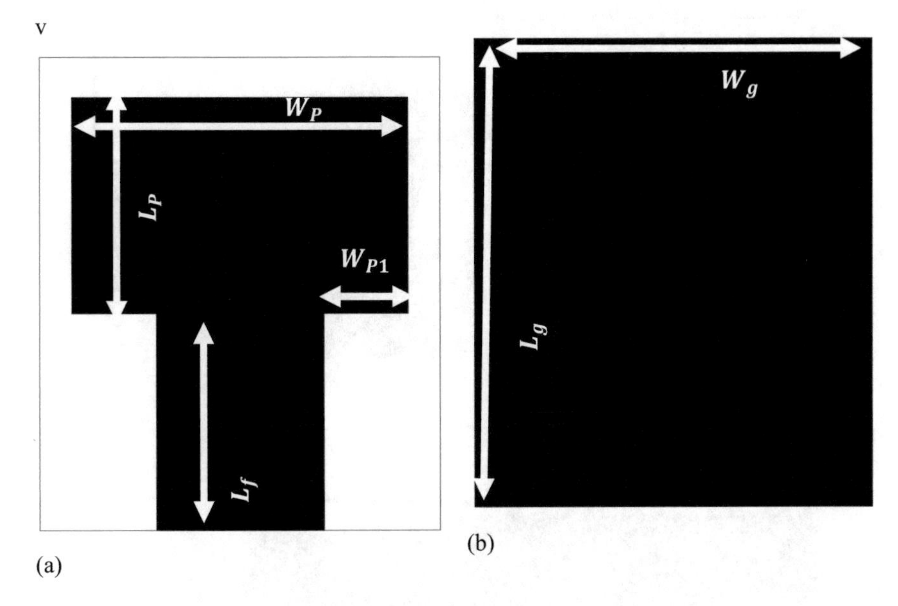

Fig. 5 Antenna geometry for Terahertz. **a** Top view, **b** Bottom view

Figure 9 shows the radiation patterns in E-plane and H-plane of the proposed antenna. We observe from the figure that antenna is omnidirectional at 1.2 and 1.4 THz. However, when we increase the frequency, the antenna becomes directive like for the case when the frequency is 1.6 or 1.8 THz.

Figure 10 shows the current distribution of the proposed antenna. From this, we figure we saw that the horizontal antenna gives the best performance in terms of radiation.

In order to validate our results and to certify better MIMO performance, the proposed antenna is evaluated by examining numerous MIMO metrics such as the Mean Effective Gain (MEG), the Diversity Gain (DG) and the Envelope Correlation Coefficient (ECC), the Channel Capacity Loss (CCL). The MIMO performance metrics are depicted in Fig. 11. From this figure, we can conclude that our antenna present high performance with high diversity at the operating band (1.3–1.5 THz). The ECC value is less than 0.1 at the operating band which make the antenna very attractive for diversity need. The value of MEG and CCL show clearly that the presented antenna present high isolation which mean less mutual coupling.

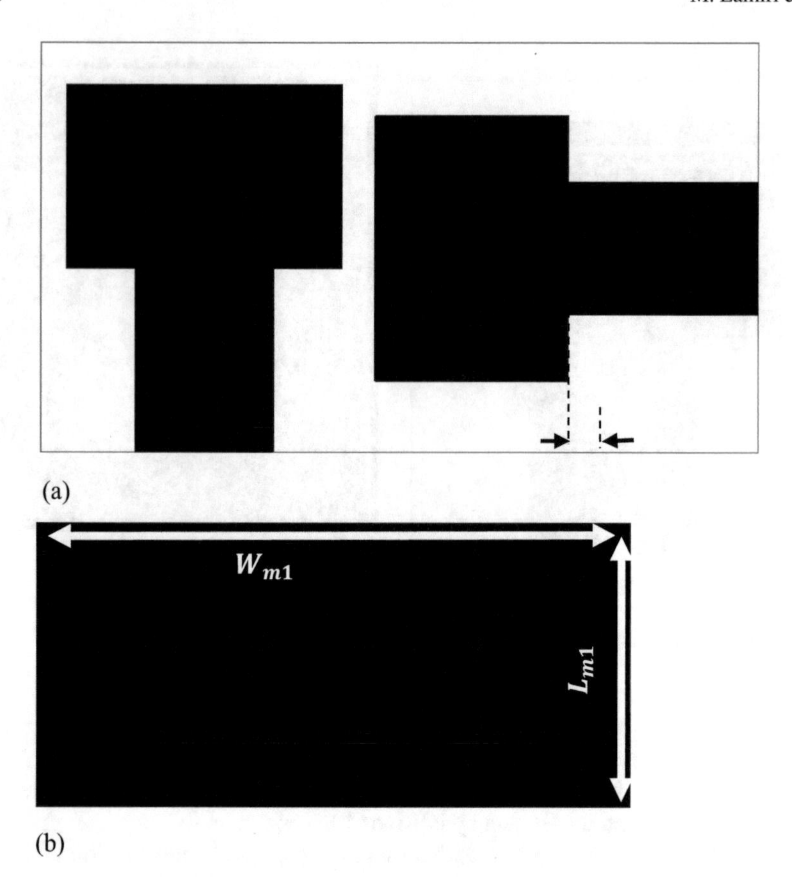

Fig. 6 Two-element antennas for Terahertz, **a** Top view, **b** Bottom view

Table 1 Optimized dimensions of the single- and two-element antennas

Parameters	Dimensions (μm)
W_p	31.693
L_p	21.958
W_f	15.846
L_f	21.958
W_g	37.693
L_g	47.916
W_{p1}	7.923
L_{m1}	84.916
W_{m1}	82.386
d	3.777

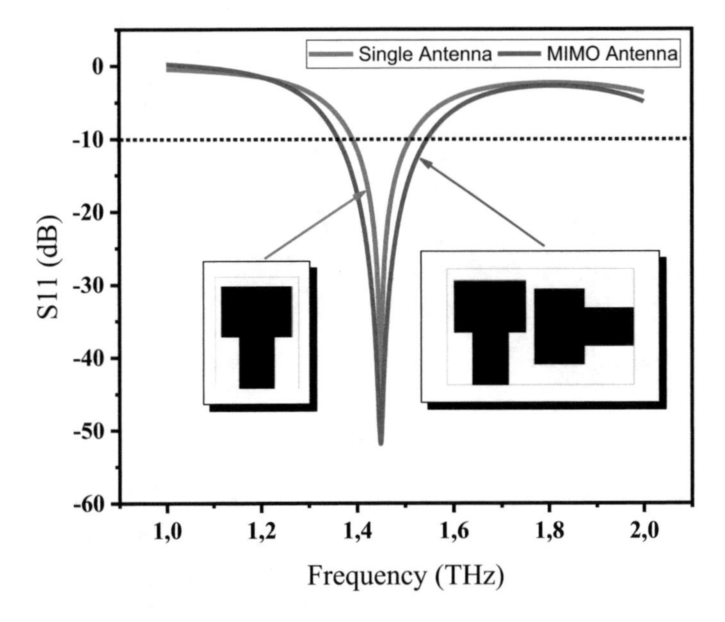

Fig. 7 S11 for the single-element antenna and two-element MIMO antenna

Table 2 Comparison between single-element, two-element MIMO antennas

Antenna	Bandwidth (THz)	Peak gain (dB)	Directivity (dB)
Single element	[1.4169–1.4807]	1.20 (0.97 Eff)	1.23
Two element	[1.3908–1.5078]	3.14 (0.96 Eff)	3.27

Fig. 8 Comparison of S21 and effect of the distance between two-element antenna

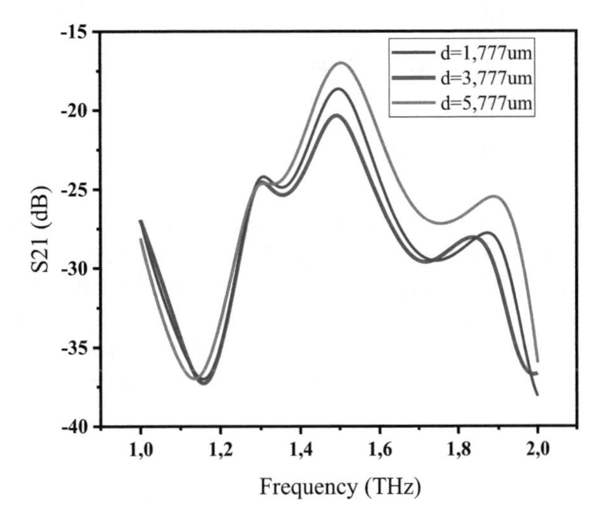

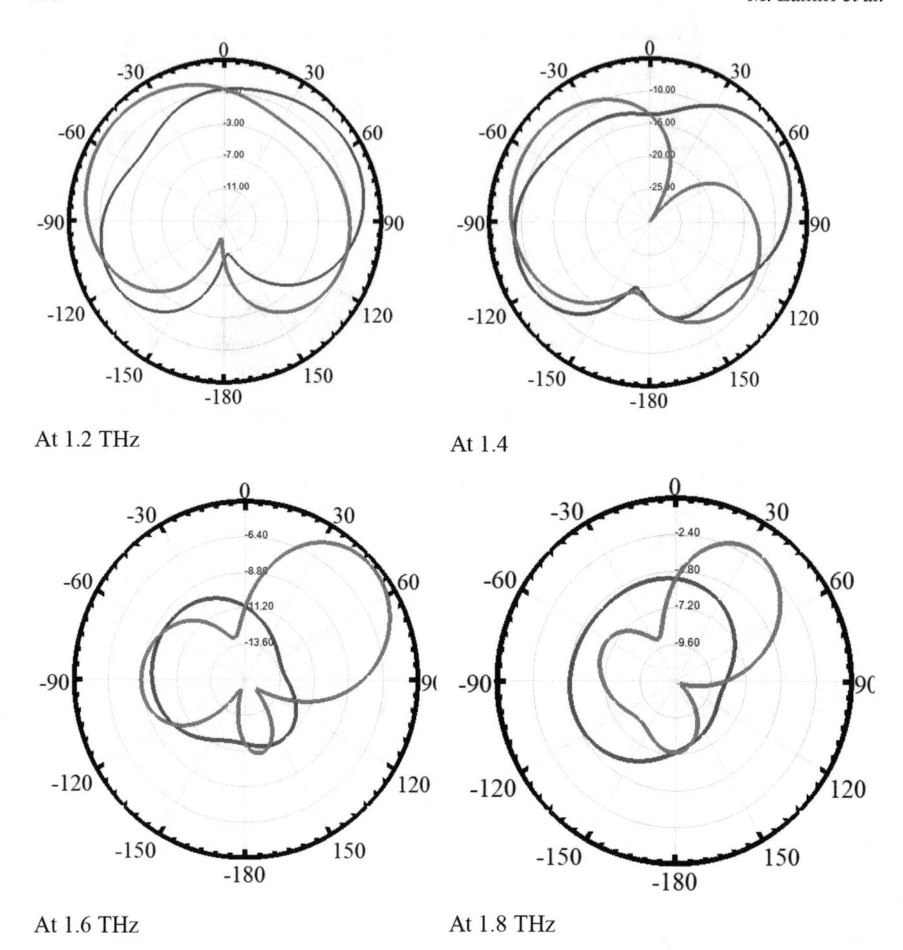

Fig. 9 E-plane (red color) and H-plane (green color) for the proposed MIMO antenna

7 Conclusion

In this chapter, we proposed a MIMO antenna for detection and generation of THz pulse. The proposed antenna offers high performance which allows it to be a suitable choice of THz applications. The proposed antenna has a bandwidth of 117 GHZ with a peak gain more than 3 dB and return loss of about −52 dB. Besides radiation pattern and current distribution are also studied. An isolation less than −20 dB is obtained when the distance between the two elements is about 5777 um.

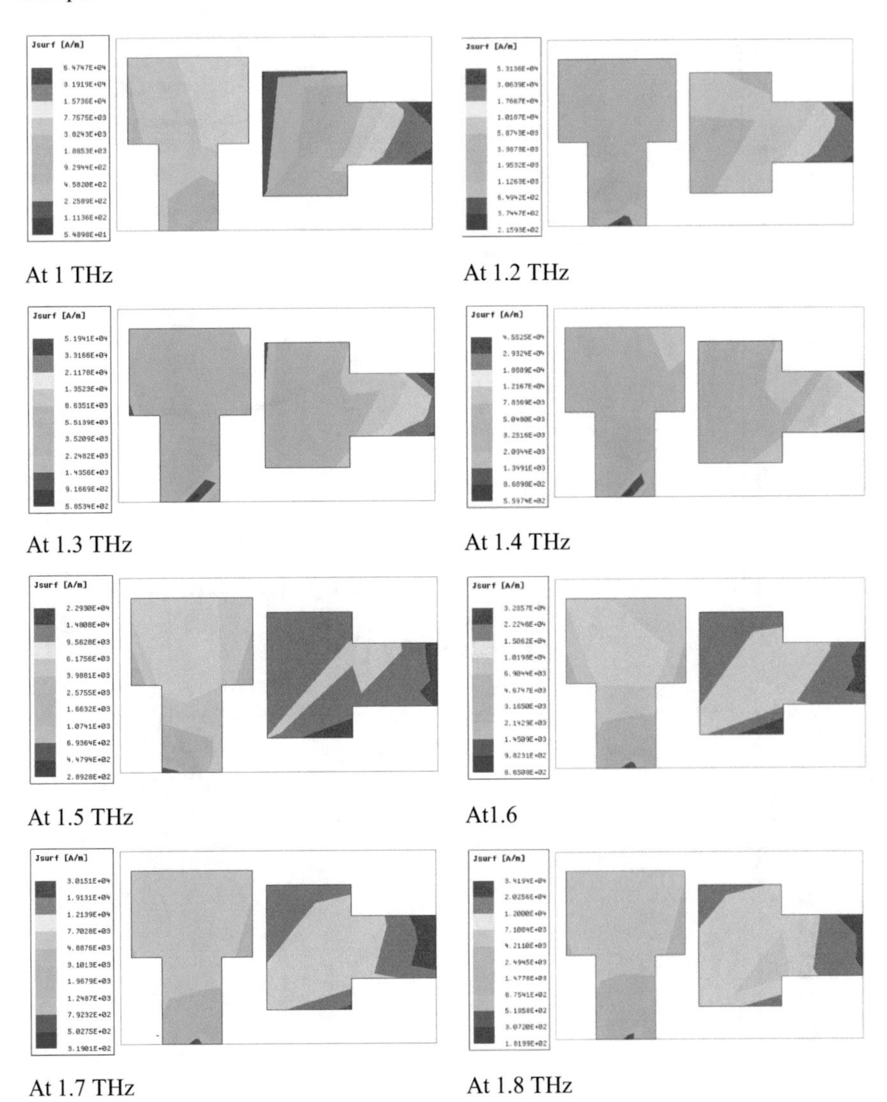

Fig. 10 Current distribution for the proposed MIMO antenna

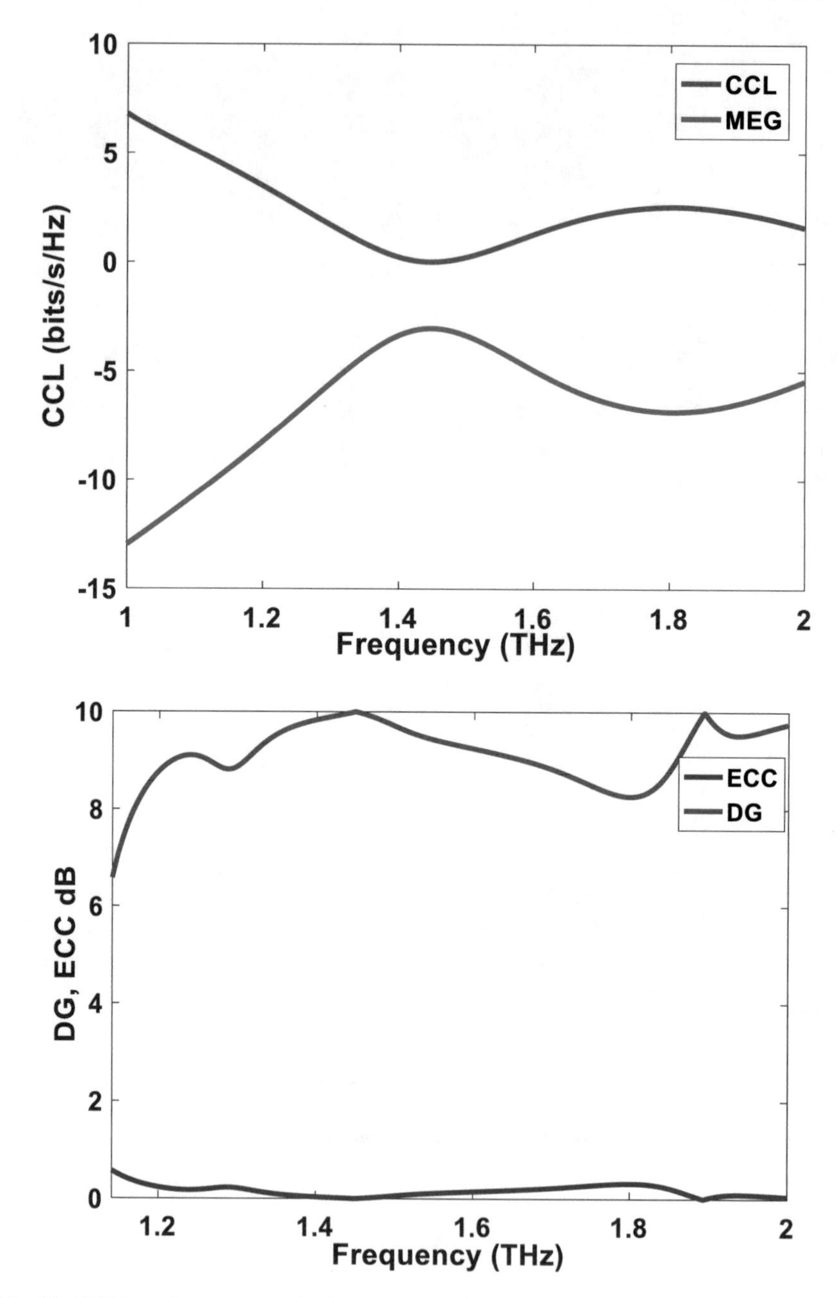

Fig. 11 MIMO performance metrics for the proposed antenna

References

1. B. Aghoutane, M. El Ghzaoui, H. El Faylali, Spatial characterization of propagation channels for terahertz band. SN Appl. Sci. **3**, 233 (2021). https://doi.org/10.1007/s42452-021-04262-8
2. Winnerl, Scalable microstructured photoconductive terahertz emitters, J. Infrared Millimeter Terahertz Waves **33**(4), 431–454 (2012)
3. M. Van Exter, D. Grischkowsky, Optical and electronic properties of doped silicon from 0.1 to 2 THz, Appl. Phys. Lett. **56**(17), 1694–1696 (1990)
4. C. Fattinger, D. Grischkowsky, Point source THz optics, Appl. Phys. Lett. **53**, 1480–1482 (1988); THz beams, Appl. Phys. Lett. **54**, 490–492 (1989)
5. M. Tani, S. Matsuura, K. Sakai, Nakashima, Study of terahertz radiation from InAs and InSb, Appl. Opt. **36**(30), 7853–7859 (1997)
6. R. Yano, H. Gotoh, Y. Hirayama, S. Miyashita, Y. Kadoya, T.Hattori, Terahertz wave detection performance of photoconductive antennas: Role of antenna tructure and gate pulse intensity. J. Appl. Phys. **97**(10), 103103 (2005)
7. H. Harde, D. Grischkowsky, Coherent transients excited by sub- picosecond pulses of terahertz radiation. JOSA B **8**(8), 1642–1651 (1991)
8. D.R. Dykaar, B.I. Greene, J.F. Federici, A.F.J. Levi, L.N. Pfeiffer, R.F. Kopf, Log-periodic antennas for pulsed terahertz radiation, Appl. Phys. Lett. **59**(3), 262–264
9. K.A. McIntosh, E.R. Brown, K.B. Nichols, O.B. McMahon, W.F. DiNatale, T.M. Lyszczarz, Terahertz measurements of resonant planar antennas coupled to low- temperature-grown GaAs photomixers. Appl. Phys. Lett. **69**, 3632–3634 (1996)
10. S. Park, H. Lee, J. Yook, Orthogonal code-based transmitted radiation pattern measurement method for 5G massive MIMO antenna systems, IEEE Trans. Antennas Propag. 68(5), 4007–4013 (May 2020).https://doi.org/10.1109/TAP.2019.2963207
11. J. Guo, F. Liu, G. Jing et al., Mutual coupling reduction of multiple antenna systems. Front. Inform. Technol. Electron. Eng. **21**, 366–376 (2020). https://doi.org/10.1631/FITEE.1900490
12. M. El Ghzaoui, A. Hmamou, J. Foshi, J. Mestoui, Compensation of non-linear distortion effects in MIMO-OFDM systems using constant envelope OFDM for 5G applications. J. Circuits Syst. Comput. **29**(16), 2050257 (2020). https://doi.org/10.1142/S0218126620502576
13. S. Salous, Multiple input multiple output: capacity and channel measurement, Department of Electronic Engineering and Electronics, UMIST, Manchester, M60, 1QD, UK, (August 2003)
14. J. Paulraj et al. An overview of MIMO communications-A key to gigabit wireless, Proc. IEEE 92(2), (Feb, 2004)
15. B. Benadda, F.T. Bendimerad, Quadratic error optimization algorithm applied to 3D space distributed array sensors. J. Appl. Sci. Res. **5**(10), 1320–1324 (2009)
16. K. Mabrouk, « Conception et réalisation d'un système de Télécommunications MIMO avec formation numérique de Faisceaux en réception ; Calibrage aveugle du Démodulateur triphasé Zéro IF et comparaison au démodulateur classique à 2 voies I et Q ». Ecole nationale supérieur des télécommunications, 12 Décembre 2008
17. L. Ros, «Spécificité du canal de propagation en communication radiomobile et choix d'une modulation », Séminaire du LIS, Grenoble, le 12 Mai 2005
18. B. A. Bjerke, J.G. Proakis, Multiple antenna diversity techniques for transmission over fading channels, in IEEE Proceedings of Wireless Communications and Networking Conference (WCNC), vol 3, (New Orleans (LA), USA, September 21–24 1999), pp. 1038–1042
19. B. Aghoutane, S. Das, H. El Faylali, B.T.P. Madhav, M. El Ghzaoui, A. El Alami, Analysis, design and fabrication of a square slot loaded (SSL) millimeter-wave patch antenna array for 5G applications. J. Circuits Syst. Comput. **30**(05), 2150086 (2021). https://doi.org/10.1142/S0218126621500869

Artificial Neuron Based on Tera Hertz Optical Asymmetric Demultiplexer Using Quantum Dot Semiconductor Optical Amplifier

Kousik Mukherjee

Abstract This chapter deals with the basic concepts of artificial neural networks (ANN) and design of an artificial photonic neuron using Tera hertz Optical Asymmetric Demultiplexer. The proposed neuron has sigmoid curve like transfer function. TOAD using quantum dot semiconductor optical amplifier is used to build this neuron for the first time to the best of author's knowledge. ANN is briefly introduced along with optical neural network (ONN). QDSOA and TOAD are also explained in brief. Dynamic response of the neuron for different input powers is also included.

Keywords Artificial neuron · Quantum dot SOA · TOAD · Dynamic response

1 Introduction

Artificial neural networks (ANN) have potential to be applicable to almost all branches of science technology and mathematics [1]. The concept started in 1943 by Walters Pitt [2], designed a mathematical model. Our brain has approximately 100 billion neurons (biological unit of brain). Through synapses, neurons communicate electromagnetic signals by a network that consists of axons, dendrites, etc. Figure 1 shows a simplified diagram of a biological neuron. Dendrites are inputs and the extensions of the axon on the other side of cell body behave like output regions. Two neurons connect through each other through synapses and generate a neural network.

Neural computations take place in the cell body and are transmitted through axon, synapse, and dendrites to the next neuron. In the synaptic junction, some neurotransmitter diffusion gives rise to electrochemical communication. Figure 2 shows a simplified diagram of biological neural network. A specific neuron has several dendrites and synaptic junctions and thus can receive and transmits signals from many other neurons. In this way, the neural network builds in.

K. Mukherjee (✉)
Physics Department, Banwarilal Bhalotia College, Asansol, India

Centre of Organic Spintronics and Optoelectronic Devices, Kazi Nazrul University, Asansol, India

S. Das et al. (eds.), *Advances in Terahertz Technology and Its Applications*,
https://doi.org/10.1007/978-981-16-5731-3_17

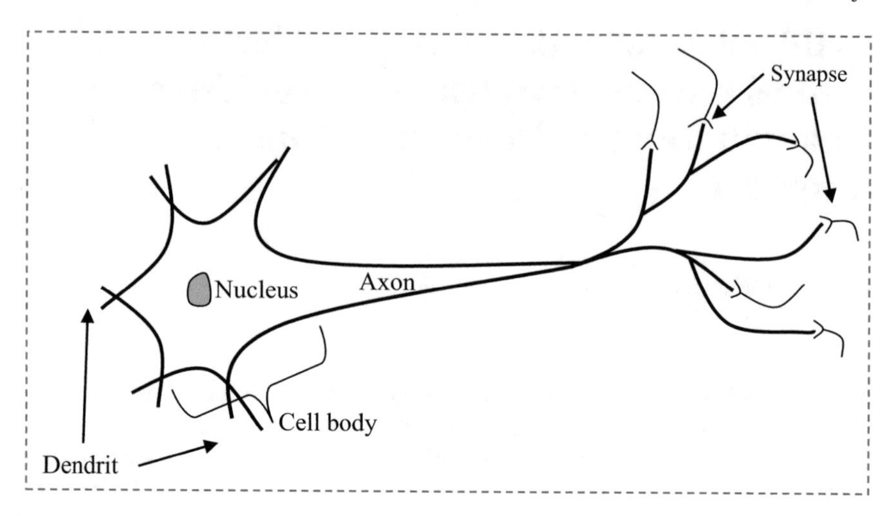

Fig. 1 A biological neuron

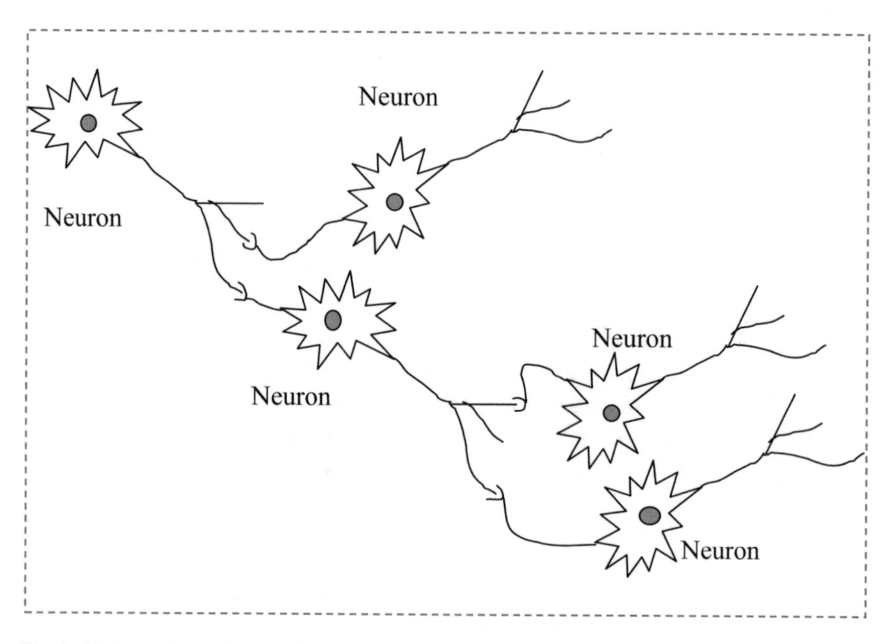

Fig. 2 Biological neural network

A neuron does not respond instantly and linearly but takes time to receive sufficient amount of input to trigger its operation. Therefore, there is a threshold for its response. Generally, inputs (neurotransmitters) generate inhibitory effects, excitatory effects, or modulatory effects depending on the neurotransmitters (a chemical). An excitatory transmitter stimulates to generate electromagnetic signal (called action potential), and inhibitory one prevents this stimulation. Neuromodulators control the populations of the neurons and operate slowly compared to other types of neurotransmitters. The neurotransmitters work through synaptic cleft. It should be remembered that the interconnections have unequal weights in terms of priority to receive inputs.

The working of biological neural network can be summarized as below:

1. The neurons work in the principle of "all- or- nothing".
2. A threshold is required for a neuron to be excited.
3. Synaptic delay is the only delay is of significance
4. Inhibitory neurons prevent the activation of a neuron, and
5. Network structure remains unaltered with time.

Next sections will describe design considerations of artificial neural networks (ANN), photonic or optical neurons, and a new optical neuron QDSOA based on tera hertz optical asymmetric demultiplexer (TOAD).

2 Artificial Neural Networks (ANN)

ANN is nothing but a simplified model of naturally occurring neural network present in animal brain. In the previous section, the basic model of biological neural network is presented. In this section, a basic model of artificial neural network will be presented. More than half a century ANN has developed significantly [2] and finds applications in almost all branches of science and technology. ANN has the capabilities of self learning, complex nonlinear relationship, and intense nonlinear fitting [3, 4] and process huge amount of data very quickly. An artificial neuron has two basic operations linear and nonlinear [5]. Figure 3 shows a basic multi input single output artificial neuron model. The neuron consists of input signals, weight and bias,

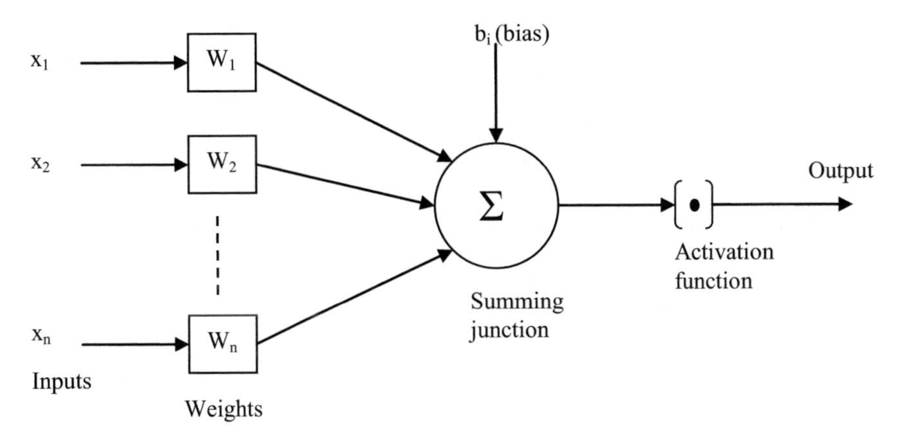

Fig. 3 Artificial neuron model

summing junction, and activation function as basic elements. The weighted sum constitutes the linear operation given by,

$$S_i = b_i + W_{ij} x_i \tag{1}$$

where b_i represents the bias and W_{ij} represents the weight.

The weighted sum of Inputs from other neurons (x_1, x_2, x_3, etc.) is passed through the activation function (having a nonlinear response) to generate output. The most commonly used activation function is sigmoid function:

$$\emptyset(x) = \frac{1}{1 + e^{-(ax-b)}} \tag{2}$$

where a, and b are the parameters. Figure 4 gives plots of Eq. (2) for two different values of $a = 5$, 10 and $b = 0$. The differentiability of the sigmoid function is an important characteristic suitable for artificial neurons [6].

There are different types of neural network architecture [6]:

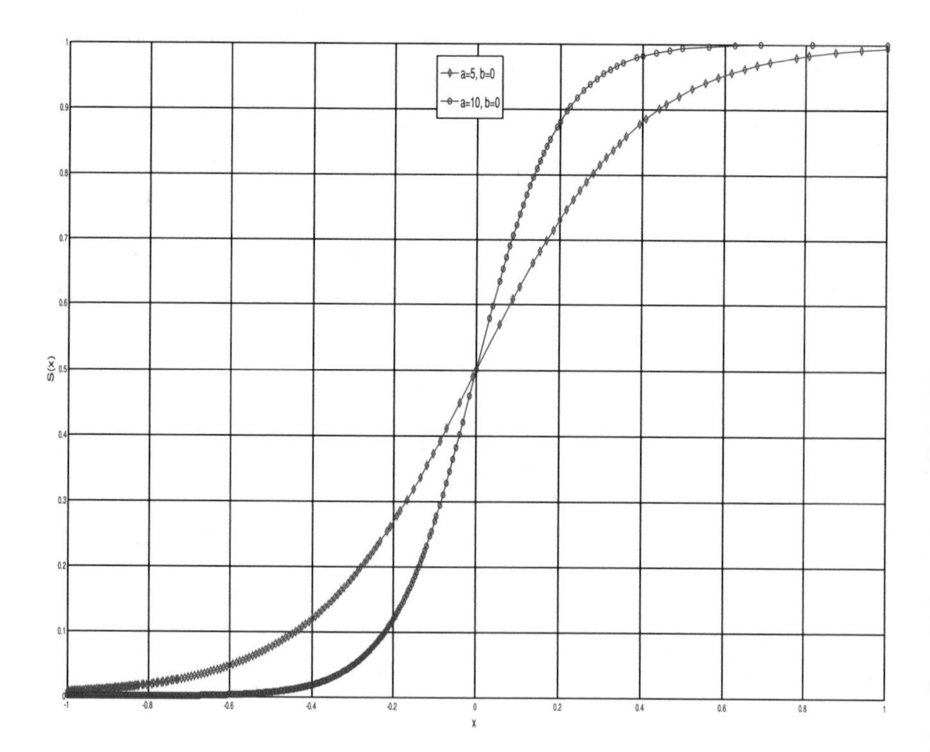

Fig. 4 Sigmoid functions, $S(x)$

(i) Single-layered feed forward: Input layers of neurons directly project onto the output layer.
(ii) Multi-layered feed forward network: There are hidden layers in between input and output layers.
(iii) Recurrent network: In this category at least one feedback loop must be present.

3 Optical Neural Network

Present day high speed communication systems require optical technologies for overcoming electronic bottlenecks. This is possible because photons are better information carriers than electrons [7, 8]. Photonic systems are not only low power consumption but photons do not interfere with each other during transmission and processing. Moreover, photonic signals have several controllable properties like intensity, phase, polarization, and frequency suitable for all optical processing. Therefore, replacing electrical signals with optical signals is a promising technology for modern day communication systems [5]. Several logic gates and optical processors use different kinds of optical nonlinearities based optical switches [9–15]. Photonic processing will enhance the computational speed and power efficiency of optical neural networks (ONN) [16]. Moreover, many optical nonlinear properties may be used to implement optical neuron activation functions. Optical neural networks using photonic technology have become important due to its scalability, compatibility, and stability [17]. Several optical neurons using different techniques are also implemented or proposed in last decade [18–27] also proves the popularity of optical technologies in this field of study. Different kinds of activation functions with sinusoidal, Heaviside step, or sigmoid responses are used to implement the neurons. For powerful recurrent neural network and long short term memory architectures, sigmoid activation function is helpful [20]. In [20], the neuron utilizes semiconductor optical amplifier (SOA) in Mach-Zehnder interferometer (MZI) configuration. The work [23] deals with neurons using quantum dot mode locked lasers have the capability of emulating both inhibition and excitation operations. Nonlinear polarization rotation (NPR) in SOA based neurons with sigmoid activation function is designed in [28]. Optical fiber based neuron is theoretically demonstrated in [29] based on reservoir computing with optical amplifier as activation element. Therefore, optical activation using this nonlinearity is an essential part of the neuron and can be achieved using different techniques like Kerr effect, saturable absorption, nonlinear thermo optic effect, etc. [2]. Reconfigurable activation function can be designed to program the optimized performance of different neuromorphic tasks and such neuron is implemented in [30]. In [31], SOA based cross gain modulation (XGM) is utilized for neuron activation function. The present chapter deals with how TOAD is used to design neuron with sigmoid activation function with quantum dot SOA.

4 Quantum Dot Semiconductor Optical Amplifier (QDSOA)

SOA is a versatile medium with controllable nonlinear properties and finds applications in various kinds of all optical logic processors [32–38]. In QDSOA the active region is deposited with quantum dots of size 10–20 nm [37]. QDSOA is an important device for all optical switching, and signal processing and has many advantages over conventional SOA. These advantages include fast gain recovery process, low current density, higher gain bandwidth, higher input power dynamic range (IPDR), higher burst mode tolerance, lower noise effect, and compactness in size [39, 40].

Figure 5 shows a diagram of QDSOA. The active layer consists of quantum dots, is sandwiched between P type and N type layers. A certain biasing current gives rise to an unsaturated gain when no optical is signal present. Applications of input light cause gain to be reduced due to carrier dynamics. A comparatively high intense signal (pump) can modulate gain and phase of a comparatively low intense signal (probe) by the combination of the processes of cross gain modulation (XGM) with cross phase modulation (XPM) [36]. These processes can be modeled by the rate equation model used in our previous works [35–38]. The dynamics are given by the rate equations:

$$\frac{\partial f}{\partial t} = \frac{(1-f)h}{\tau_{21}} - \frac{f(1-h)}{\tau_{12}} + \frac{f^2}{\tau_{1R}} - \frac{L_w g_{max}(2f-1)P}{N_Q A_{eff} h\nu} \tag{3}$$

$$\frac{\partial h}{\partial t} = -\frac{h}{\tau_{2w}} - \frac{N(1-h)L_w}{\tau_{w2} N_Q} + \frac{(1-f)h}{\tau_{21}} - \frac{f(1-h)}{\tau_{12}} \tag{4}$$

$$\frac{\partial N}{\partial t} = \frac{J}{eL_w} - \frac{N(1-h)}{\tau_{w2}} + \frac{N_Q h}{\tau_{2w} L_w} - \frac{N}{\tau_{wR}} \tag{5}$$

$$\frac{\partial P}{\partial z} = \frac{[g_{max}(2f-1) - \alpha_{int}]P}{A_{eff} h\nu} \tag{6}$$

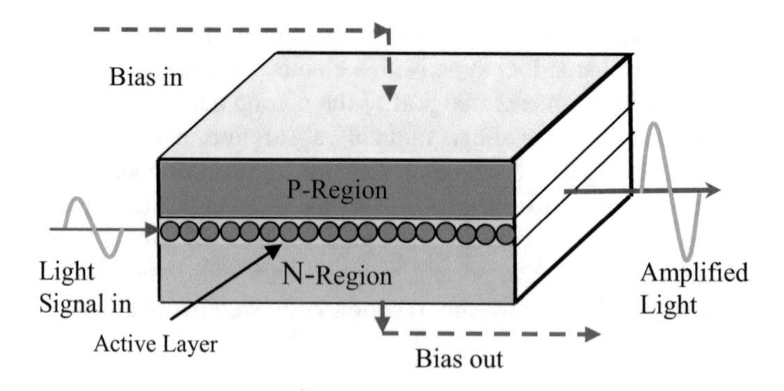

Fig. 5 Structure of QDSOA

Table 1 Parameters used for simulations [35–38]

Parameters
0.2 ns, the Spontaneous lifetime in the WL(τ_{wR})
3 ps, Relaxation time of electron (WL to ES(τ_{w2}))
0.16 ps, Relaxation time of electron (ES to GS(τ_{21}))
8.3×10^7 m/s, Group velocity (V_g)
1 ns, Escape time of electron (ES to WL(τ_{2w}))
1.2 ps, Escape time of electron (GS to ES(τ_{12}))
0.4 ns, Radiative lifetime (spontaneous) in Quantum Dot (τ_{1R})
g_{max}, Material gain coefficient = 14 cm^{-1},
Internal loss,$\alpha_{int} = 2$ cm^{-1},
$\alpha(l) = 4$, line width enhancement factor,
Injection current density, $J = 1$ kA/cm^2
$L_w = 250$ nm, active layer effective thickness
$N_Q = 5.0 \times 10^{10}$, Transparency current density
$A_{eff} = 0.75$ μm^2, Effective area

$$\frac{\partial \varphi}{\partial z} = -0.5\alpha(l) g_{\max}(2f - 1) \tag{7}$$

where N, electron density in the wetting layer, f, h are electron the occupation probabilities of ground state, excited state respectively. Table 1 shows the corresponding parameters for numerical simulations.

5 Tera Hertz Optical Asymmetric Demultiplexer

TOAD is an SOA based high speed optical switch [41] finds applications in different aspects of all optical signal processing (AOSP) [37, 42–46]. QDOSA is utilized in the designing of TOAD in the works [37, 46] enhances the performance because of superior qualities of QDSOA over SOA. A QDSOA based TOAD is shown in Fig. 6. In the TOAD shown in Fig. 6, QDSOA is asymmetrically placed in the fiber-loop, with an input control signal. The probe or data signal is introduced through a circulator and coupler into the loop. This data signal is broken into two counter propagating parts, traveling the loop in opposite direction and passing through the QDSOA at different times. They experience unequal gain and phase variations and interfere at the coupler. Filters (F, F) are centered at the data or probe wavelength. The transmitted port (P_T) gives constructive interference output and reflected port (P_R) gives destructive interference output. The gain of the QDSOA is calculated by solving the coupled differential Eqs. (3) to (7) by numerical technique.

Output of two ports are given by [46],

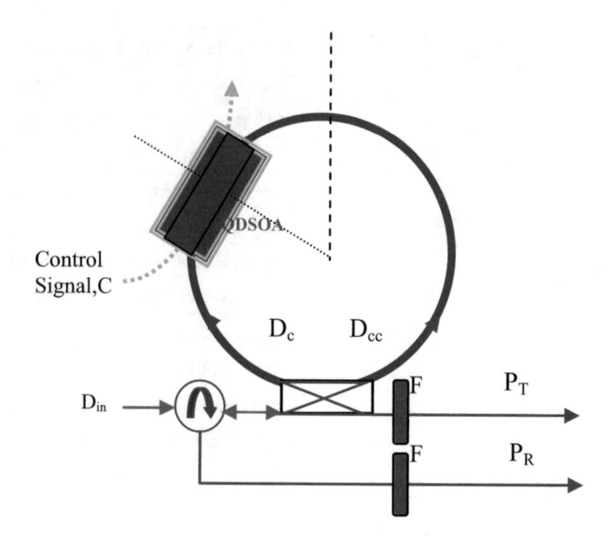

Fig. 6 TOAD using QDSOA

$$P_{T,R} = 0.25 P_{in}[G_0 + G(t) \pm 2\sqrt{G_0 G(t)} \cos(\Delta\theta), \tag{8}$$

where P_{in} is the power of the data signal D_{in}, G(t) and G_0 are dynamic and unsaturated gain of QDSOA. Two counter propagating signals have phase differences [37].

$$\Delta\theta = -0.5\alpha(l) \ln(G(t)/G_0), \tag{9}$$

α(l) is line width enhancement factor. Amplified spontaneous emission (ASE) noise is given by

$$P_{ase} = \frac{N_{sp} h c^2 (G_0 - 1)\Delta\lambda}{\lambda^3} \tag{10}$$

where h is Planck's constant, c is speed of optical signal, λ is wavelength, N_{sp} is ASE noise factor and $\Delta\lambda$ is optical bandwidth of (F, F). When control signal is not present, QDSOA gain is unsaturated, i.e., high. The D_{c} and D_{cc} experiences same gain and phase difference resulting high output at transmitted port and zero output at reflected port. A considerable power at control is sufficient to make a π phase change between oppositely propagating signals reduces the transmitted port output and enhances the reflected port output.

6 Optical Neuron Using TOAD

The neuron is designed using the reflected port of the TOAD shown in Fig. 6. Figure 7 shows the neuron with logistic or sigmoid activation function. Inputs to the TOAD are differently weighted optical signals (VOAs are used) at wavelength other than

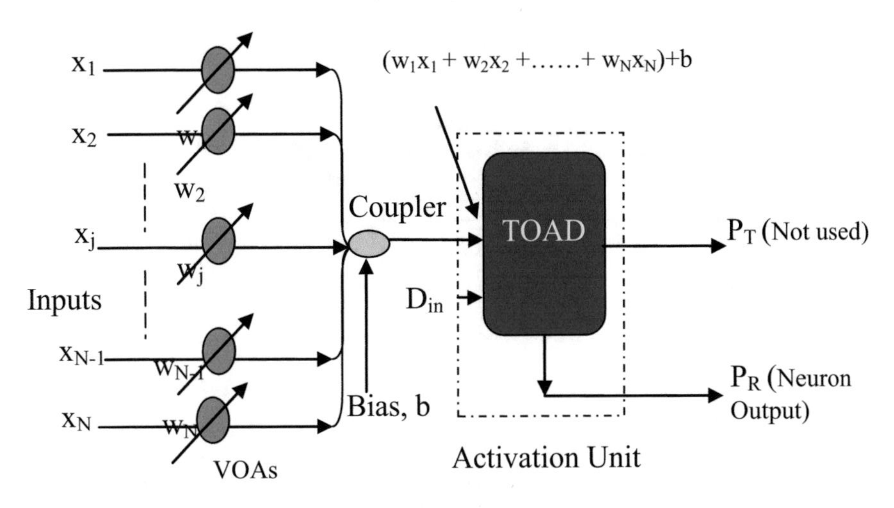

Fig. 7 Design of the TOAD based neuron using QDSOA

probe signal wavelength acts as a control signal. Another constant input signal called bias is added with all the weighted sum of the input signals x_1 to x_N.

Total input power is inserted as a control to the TOAD. As mentioned earlier, a zero control input causes no signal to be present in the reflected port. With application of control, QDSOA starts to decrease its gain, and hence nonzero phase difference between the two counter propagating signals grows up. Since QDSOA is placed asymmetrically, these two signals pass the QDSOA at different times. The clockwise data signal D_c experiences reduced gain. However, the counter-clockwise component D_{cc} experiences a higher gain as it reaches the QDSOA at a later time when the gain of the QDSOA is recovered. These two differently amplified signals acquire a phase difference and interfere at the respective ports. With nonzero control signal, since there is phase difference, the output in the reflected port becomes nonzero. If now the total input is varied stepwise, output of the neuron changes following a transfer function shown in Fig. 8. The curve is fitted with sigmoid function is given by Mourgias-Alexandris et al. [20].

$$y = A_1 + \frac{(A_2 - A_1)}{\left(1 + e^{[(x - x_0)/dx)]}\right)} \tag{11}$$

Simulation results show the values of parameters $A_1 = -0.00539, A_2 = 1.02201$, $x_0 = 0.56832$, and $\mathrm{d}x = 0.11527$. This confirms the fact that the transfer function is almost sigmoid in nature. The output changes from 10 to 90% for input variation from 0.3 to 0.8 mW describe sharp response of the TOAD to the input activation.

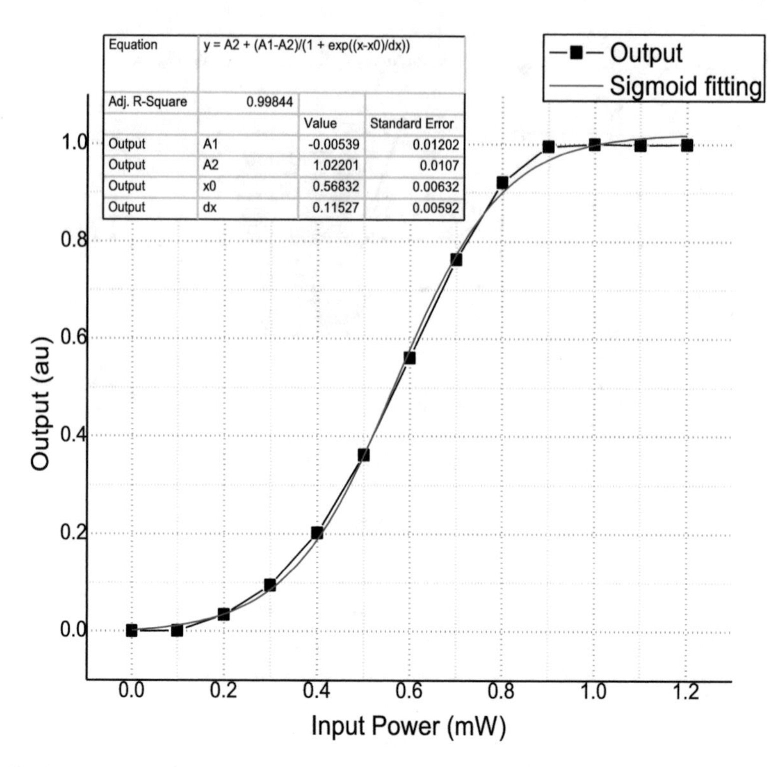

Fig. 8 Transfer function of TOAD based artificial neuron

Figure 9 shows the dynamics response of the TOAD based activation function for total input control signal of 1 dBm (Fig. 9a) and 7 dBm (Fig. 9b). It is interesting to note that for a considerable value of input control power the high and low outputs show significant extinction between them signifies proper threshold operation of the neuron activation function.

7 Conclusions

ANN is briefly introduced with a little survey of optical neural networks. In this chapter for the first time, a neuron with logistic sigmoid activation function is designed using QDSOA based TOAD. The transfer function of the TOAD is shown to be well fitted with sigmoid function proves efficient operation. Time dynamics show efficient performance of the QDSOA based TOAD as activation unit. This elementary neuron can be utilized to design complex neural network.

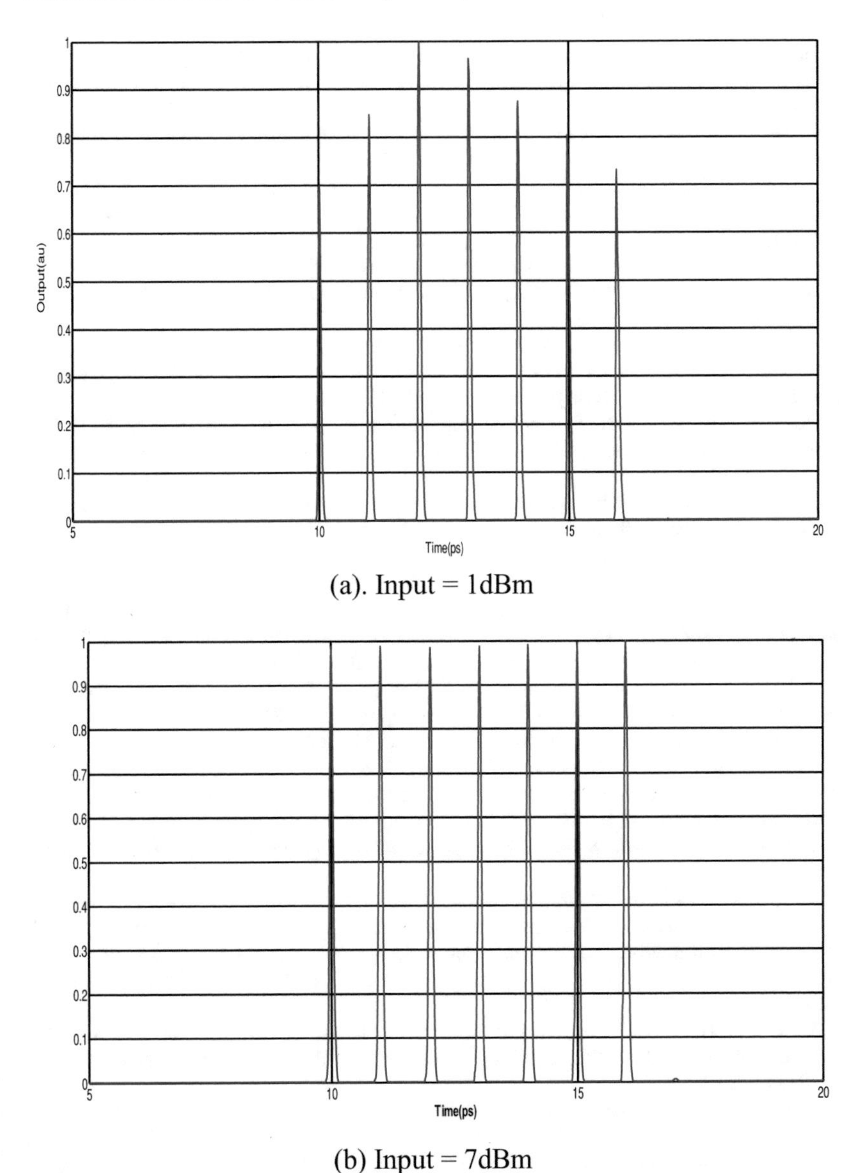

(a). Input = 1dBm

(b) Input = 7dBm

Fig. 9 Dynamic response of QDSOA based TOAD to input signal

References

1. W.S. McCulloch, W. Pitts, A logical calculus of the ideas immanent in nervous activit'. Bull. Math. Biophys. **5**(4), 115–133 (1943)
2. X. Sui, Q. Wu, J. Liu, Q. Chen, G. Gu, A review of optical neural networks. IEEE Access **8**, 70773–70783 (2020). https://doi.org/10.1109/ACCESS.2020.2987333

3. L. Jin-Yue, Z. Bao-Ling, Research on the non-linear function fitting of RBF neural network, in Proceeding of International Conference on Computational and Information Sciences, (June, 2013), pp. 842–845
4. S.-C. Huang, B.-H. Do, Radial basis function based neural network for motion detection in dynamic scenes. IEEE Trans. Cybern. **44**(1), 114–125 (2014)
5. J. Liu, Q. Wu, X. Sui, et al. Research progress in optical neural networks: theory, applications and developments. PhotoniX 2, 5 (2021).https://doi.org/10.1186/s43074-021-00026-0
6. S.O. Haykin, Neural networks and learning machines, Prentice-Hall, (2009)
7. K. Mukherjee, All optical read only memory with frequency encoded addressing technique. Optik—Int. J. Light Electron Opt. **122**(16), 1437–1440 (2011). https://doi.org/10.1016/j.ijleo.2010.09.024
8. K. Mukherjee, A method of implementation of frequency encoded all optical logic gates based on non-linear total reflectional switch at the interface. Optik—Int. J. Light Electron Opt. **122**(14), 1284–1288 (2011). https://doi.org/10.1016/j.ijleo.2010.08.017
9. R. Rigi, H. Sharifi, K. Navi, Configurable all-optical photonic crystal XOR/AND and XNOR/NAND logic gates. Opt. Quant. Electron. **52**, 339 (2020). https://doi.org/10.1007/s11082-020-02454-x
10. E.G. Anagha, R.K. Jeyachitra, An investigation on the cascaded operation of photonic crystal based all optical logic gates and verification of De Morgan's law. Opt. Quant. Electron. **52**, 293 (2020). https://doi.org/10.1007/s11082-020-02384-8
11. P. Kumar Nahata, A. Ahmed, S. Yadav, N. Nair, S. Kaur, All optical full-adder and full-subtractor using semiconductor optical amplifiers and all-optical logic gates, in 2020 7th International Conference on Signal Processing and Integrated Networks (SPIN), (2020), pp. 1044–1049. https://doi.org/10.1109/SPIN48934.2020.9071009
12. D.G.S. Rao, S. Swarnakar, V. Palacharla et al., Design of all—optical AND, OR, and XOR logic gates using photonic crystals for switching applications. Photon. Netw. Commun. **41**, 109–118 (2021). https://doi.org/10.1007/s11107-020-00916-6
13. S.S. Mainka, R. Zafar, M.H. Mahdieh, G. Singh, M. Salim, High Contrast Ratio Based All-Optical OR and NOR Plasmonic Logic Gate Operating at E Band. in *Optical and Wireless Technologies. Lecture Notes in Electrical Engineering*, eds. by V. Janyani, G. Singh, M. Tiwari, A. d'Alessandro, vol 546 (Springer, Singapore, 2020). https://doi.org/10.1007/978-981-13-6159-3_35
14. P. Sami, C. Shen, M.H. Sani, Ultra-fast all optical half-adder realized by combining AND/XOR logical gates using a nonlinear nanoring resonator, Appl. Opt. **59**, 6459–6465 (2020)
15. F. Parandin, R. Kamarian, M. Jomour, A novel design of all optical half-subtractor using a square lattice photonic crystals. Opt. Quant. Electron. **53**, 114 (2021). https://doi.org/10.1007/s11082-021-02772-8
16. Y. Shen, N.C. Harris, S. Skirlo, M. Prabhu, T. Baehr-Jones, M. Hochberg, X. Sun, S. Zhao, H. Larochelle, D. Englund, M. Soljaciˇ c, Deep learning with coherent nanophotonic circuits, Nat. Photonics, **11**, (2017)
17. H. Zhang, M. Gu, X.D. Jiang et al., An optical neural chip for implementing complex-valued neural network. Nat. Commun. **12**, 457 (2021). https://doi.org/10.1038/s41467-020-20719-7
18. G. Mourgias-Alexandris, G. Dabos, N. Passalis, A. Tefas, A. Totovic, and N. Pleros, All-optical recurrent neural network with sigmoid activation function, in *Optical Fiber Communication Conference* (OFC) 2020, OSA Technical Digest (Optical Society of America, 2020), paper W3A.5
19. T.S. Rasmussen, Y. Yu, J. Mork, All-optical non-linear activation function for neuromorphic photonic computing using semiconductor Fano lasers, Opt. Lett. **45**, 3844–3847 (2020)
20. G. Mourgias-Alexandris, A. Tsakyridis, N. Passalis, A. Tefas, K. Vyrsokinos, N. Pleros, An all-optical neuron with sigmoid activation function. Opt. Express **27**, 9620–9630 (2019)
21. H.T. Peng, M.A. Nahmias, T.F. De Lima, A.N. Tait, B.J. Shastri, P.R. Prucnal, Neuromorphic photonic integrated circuits. IEEE J. Sel. Top. Quant. Electron. **24**, 1–16 (2018)
22. I. Chakraborty, G. Saha, A. Sengupta, K. Roy, Toward fast neural computing using all-photonic phase change spiking neurons. Sci. Rep. **8**, 1–10 (2018)

23. C. Mesaritakis, A. Kapsalis, A. Bogris, D. Syvridis, Artificial neuron based on integrated semiconductor quantum dot mode-locked lasers. Sci. Rep. **6**, 1–10 (2016)
24. M.A. Nahmias, A.N. Tait, L. Tolias, M.P. Chang, T. Ferreira De Lima, B.J. Shastri, P.R. Prucnal, An integrated analog O/E/O link for multi-channel laser neurons, Appl. Phys. Lett. **108**, (2016)
25. M.A. Nahmias, B.J. Shastri, A.N. Tait, P.R. Prucnal, A leaky integrate-and-fire laser neuron for ultrafast cognitive computing, IEEE J. Sel. Top. Quant. Electron. **19**, 1–12 (2013)
26. K.S. Kravtsov, M.P. Fok, P.R. Prucnal, D. Rosenbluth, Ultrafast all-optical implementation of a leaky integrate-and-fire neuron. Opt. Express **19**, 2133 (2011)
27. D. Rosenbluth, K. Kravtsov, M.P. Fok, P.R. Prucnal, A high performance photonic pulse processing device. Opt. Express **17**, 22767 (2009)
28. Q. Li, Z. Wang, H. Wang et al., Artificial neuron based on nonlinear polarization rotation in a semiconductor optical amplifier. Opt. Commun. (2018). https://doi.org/10.1016/j.optcom.2018.11.074
29. T.-Y. Cheng, D.-Y. Chou, C.-C. Liu, Y.-J. Chang, C.-C. Chen, Optical neural networks based on optical fiber-communication system. Neurocomputing (2019). https://doi.org/10.1016/j.neucom.2019.07.0
30. A. Jha, C. Huang, P.R. Prucnal, Reconfigurable all-optical nonlinear activation functions for neuromorphic photonics. Opt. Lett. **45**, 4819–4822 (2020)
31. M. T. Hill, E. E. E. Frietman, H. de Waardt, Giok-djan Khoe, H.J.S. Dorren, All fiber-optic neural network using coupled SOA based ring lasers, IEEE Trans. Neural Networks, **13**(6), 1504–1513, (Nov, 2002). https://doi.org/10.1109/TNN.2002.804222
32. M.R. Salehi, S.F. Taherian, Performance analysis of a high-speed all-optical subtractor using a quantum-dot semiconductor optical amplifier-based Mach-Zehnder interferometer. J. Opt. Soc. Korea **18**(1), 65–70 (2014). https://doi.org/10.3807/JOSK.2014.18.1.065
33. T. Ohtsuki, M. Matsuura, Wavelength conversion of 25-Gbit/s PAM-4 signals using a quantum-dot SOA, IEEE Photonics Technol. Lett. **30**(5), 459–462 (2018). https://doi.org/10.1109/LPT.2018.2798645
34. A. Rostami, H.B.A. Nejad, R.M. Qartavol, H.R. Saghai, Tb/s optical logic gates based on quantum-dot semiconductor optical amplifiers. IEEE J. Quant. Electron. **46**(3), 354–360 (2010). https://doi.org/10.1109/jqe.2009.2033253
35. K. Mukherjee, Design and analysis of all optical frequency encoded X-OR and X-NOR gate using quantum dot semiconductor optical amplifier Mach-Zehnder interferometer, Opt. Laser Technol. **140**, (2021). https://doi.org/10.1016/j.optlastec.2021.107043
36. K. Mukherjee, All Optical Universal Logic TAND Gate Using a Single Quantum Dot Semiconductor Optical Amplifier at 2 Tbit/s. in (eds) *Emerging Trends in Terahertz Engineering and System Technologies*, eds. by A. Biswas, A. Banerjee, A. Acharyya, H. Inokawa (Springer, Singapore, 2021). https://doi.org/10.1007/978-981-15-9766-4_8
37. K. Mukherjee, S. Dutta, S. Roy et al., All-optical digital to analog converter using Tera Hertz optical asymmetric demultiplexer based on quantum dot semiconductor optical amplifier. Opt. Quant. Electron. **53**, 242 (2021). https://doi.org/10.1007/s11082-021-02900-4
38. K. Mukherjee, A terabit-per-second all-optical four-bit digital-to-analog converter using quantum dot semiconductor optical amplifiers. J. Comput. Electron. (2021). https://doi.org/10.1007/s10825-021-01675-x
39. N.K. Dutta, Q. Wang, Semiconductor optcal amplifiers, 2nd edn. (World Scientific, 2013)
40. T. Vallaitis, R. Bonk, J. Guetlein, D. Hillerkuss, J. Li, R. Brenot, F. Lelarge, G.H. Duan, W. Freude, J. Leuthold, Quantum dot SOA input power dynamic range improvement for differential-phase encoded signals. Opt. Express **18**, 6270–6276 (2010)
41. Sokoloff et al, A Tera herz optical asymmetric demultiplexer, IEEE Photonics Technol. Lett. **5**(7), (1993)
42. K. Maji, K. Mukherjee, Performance analysis of optical logic XOR gate using dual-control Tera Hertz optical asymmetric demultiplexer (DCTOAD), in 2019 *Devices for Integrated Circuit (DevIC)* (Kalyani, India, 23–24 March 2019)
43. K. Maji, K. Mukherjee, M.K. Mandal, Analysis of all optical dual control dual SOA TOAD based 2's complement generator, in *2020 IEEE VLSI Device Circuit and System (VLSI DCS)* (Kolkata, India, 2020), pp. 119–123. https://doi.org/10.1109/VLSIDCS47293.2020.9179909

44. K. Mukherjee, K. Maji, M.K. Mandal, Design and analysis of all—optical dual control dual SOA Tera Hertz asymmetric demultiplexer based half adder. Opt. Quant Electron. **52**, 402 (2020). https://doi.org/10.1007/s11082-020-02522-2
45. K. Maji, K. Mukherjee, A. Raja, Design and performance analysis of all optical 4-bit parity generator and checker using dual-control dual SOA terahertz optical asymmetric demultiplexer (DCDS-TOAD), J. Opt. Commun. (published online ahead of print 2020), 000010151520190098, (2020). https://doi.org/10.1515/joc-2019-0098
46. K. Mukherjee, Tera-Hertz Optical asymmetric demultiplexer (TOAD) using quantum dot semiconductor optical amplifier, EMNSD (15–16 Dec 2020), Central Institute of Technology Kokrajhar

Terahertz E-Healthcare System and Intelligent Spectrum Sensing Based on Deep Learning

Parnika Kansal, M. Gangadharappa, and Ashwni Kumar

Abstract The Terahertz frequency-based communications from 300 GHz upto 10 THz are crucial for future generation wireless systems beyond fifth-generation networks. The pandemics and epidemics occurring because of infectious diseases like the current COVID-19 leave a formidable effect on people and societies. The main aim in such situations should be providing timely treatments to prevent losses due to virus dispersal. But in prevailing healthcare systems, there are many deficits in distributing medical resources, consumption of enormous power, lower data rates, and no proper consideration of the patients' emergency conditions. Also, the growing use of wireless technologies overcrowds the spectrum that the existing devices and systems use. This book chapter proposes a unique Terahertz E-health system operating on intelligent deep learning-based spectrum sensing techniques to alleviate the problems mentioned above. This unique Terahertz E-Health system is designed using a novel Perceptive Hierarchal Networking framework (PHN) that solves spectrum scarcity by sensing free spectrum slots for transmitting the medical data using pervasive sensing mechanisms operating at Terahertz Frequency. The proposed PHN architecture comprises novel intelligent blocks that deploy deep learning-based sensing mechanisms to provide a risk classification of the patients as high, medium, and low based on their respective medical data and then perform intelligent spectrum allocation to patients to transmit their medical data. These sensing schemes utilize efficient deep learning models like Long Short-Term Memory (LSTM), Recurrent Neural Networks, Convolutional Neural Networks, and CNN-LSTM. This book chapter will present a progressive vision of how the traditional "THz gap" will transform into a "Rush of THz" for the next few years.

Keywords Terahertz · E-healthcare · Deep learning · Spectrum sensing · LSTM · CNN · RNN · CNN-LSTM

P. Kansal (✉) · A. Kumar
Department of Electronics & Communication, Indira Gandhi Delhi Technical University, James Church, New Church Rd, Opp. St, Kashmere Gate, New Delhi, Delhi 110006, India

M. Gangadharappa
Department of Electronics & Communication, Netaji Subash University of Technology, East Campus, Krishna Nagar Road Chacha Nahru Bal Chikitsalaya, Geeta Colony, New Delhi, Delhi 110031, India

S. Das et al. (eds.), *Advances in Terahertz Technology and Its Applications*,
https://doi.org/10.1007/978-981-16-5731-3_18

1 Introduction

Infectious diseases and pandemics have a daunting impact on individuals and societies. But virus dispersal can be stopped by catering timely responses and treatments. Healthcare systems have been saving people's lives since the influenza cases in 1918 to the ongoing pandemic of COVID-19. Nowadays, intelligent healthcare systems are a crucial necessity, and wireless communication is the pillar of the developments in smart healthcare systems. To deliver fast healthcare services and provide patients with treatments within hospitals or remotely, all the healthcare systems must have provision to advanced technologies.

In most developed and developing countries, COVID-19 lead to a boom of patients, a lack of medical facilities, and delayed treatments, which collapsed the healthcare framework [1]. The prevailing healthcare systems depend upon fourth-generation networks for smart healthcare applications. But due to insufficient bandwidth and spectrum resources, these networks cannot cater to the growing demand for wireless services in the long run. In contrast to the 4G services, 5G and Beyond Fifth-generation services supply better bandwidth, robustness, credibility, latency, and data rate [1]. From the viewpoint of healthcare systems, 5G and beyond generations will deliver—

(i) internet connectivity to the internet of things based medical systems via massive machine-type communications
(ii) supply excellent quality video calling for remote treatments, telemedicine, augmented reality on an excellent diagnosis via enhanced mobile broadband
(iii) supply ultra-reliable low latency communications for drones and self-governing vehicles to carry out emergency inspection and monitoring.

The crucial motivator for the current pandemic scenario and future should be developing an E-Health system equipped with the most advanced technologies like deep learning and cognitive radio at Terahertz band to improve the quality of life in all aspects [2–4]. In this respect, numerous technologies are anticipated in the literature like tactile internet, self-governing vehicles, visible light communication, aerial networks, and Terahertz [5]. A cognitive radio system can be utilized for intelligent healthcare systems by taking the distinct quality of service requirements into account. The medical equipment in the healthcare systems is considered as preserved users who need a good quality of service for various e-health applications like telemedicine and hospital data system applications.

Telemedicine applications can be regarded as primary users as they have higher priority, while hospital data systems can be called secondary users as they have lower priority. For E-health systems, the cognitive systems should manage the primary users and the preserved users. These preserved users can be passive or active. Examples of passive devices are incubators, anesthesia machines, infusion pumps, etc. Active devices like ECG, wireless monitors, telemetry monitors can transmit information wirelessly. An extensive band is required to achieve all E-health applications and efficient cognitive radio system enablers for the health systems, which cannot be easily found in spectrum bands below 90 GHz. Therefore, in this book chapter E-health system based on cognitive radio system in THz has been proposed. The main advantages of Terahertz communications for E-health applications are—

(1) The high terahertz frequency offers a massive accessible bandwidth and highly promising data rates.
(2) Directional antennas are anticipated for compulsory use to tackle high path loss by providing narrow beamwidths and restricted interference. Hence a high data rate per area can be expected.
(3) The high data rates provide low delays due to effective beam search and alignment techniques.

In prevailing times, medical information and records of exabyte size are being digitized at various organizations like hospitals, clinics, diagnostic centers, labs, and nursing centers. The structure of this medical data is messy and improper, which is not suitable for the current healthcare systems operating on statistical modeling and analysis. Hence, shifting to deep learning-enabled platforms that operate on terahertz communications will connect to many hospital databases. By linking to many health records and databases, the analysis can be carried out for a complex mixture of data types like medical history, vital signs data, ECG records, blood test reports, etc. Pervasive medical devices can estimate patient's vital signs and parameters for cognitive radio applications in smart healthcare systems. E-health care data transmission is a significant application of intelligent cognitive radio networks based on Artificial intelligence and deep learning. Pervasive spectrum sensing at Terahertz frequency provides efficient energy consumption and low computational complexity to transmit health data. The patients receiving medical attention and treatments can be segregated as follows—

(1) Primary users—These users carry out shared data transmission when accessing the spectrum band.
(2) Urgent Primary users—These users belong to the highest priority and require medical attention on an emergency basis. So, they need spectrum urgently to share their medical data for remote treatments or analyze the disease patterns in their bodies.
(3) Secondary users—These individuals can use the spectrum when primary users are silent.

The fifth-generation and beyond wireless generations can deploy the Terahertz band and deep learning techniques to transmit and receive medical data for treating patients by a doctor from a remote location, identifying lung cancer, heart diseases, etc. [6–8]. Hillger et al. [9] discusses a review of the advancements in THz integrated circuits research for medical imaging. The work [10] presents a broadband THz absorber planned and invented by electrohydrodynamic printing technology. Ben Krid et al. [11] proposes a THz bean that is diffraction-free designed using a lens group system. The crucial medical records can be sent via terahertz communication appropriately to a secured cloud location to be aggregated and processed. Research works [12–16] present advancements in applications of THz frequency in various domains of healthcare like spectroscopy of tumor samples, in vivo imaging of human cornea, breast cancer detection with THz imaging and detection of bio samples aqueous in nature using THz sensors. Incorporating deep learning can help scale the Terahertz E-health systems to include billions of patient records and predictions for the large amount of healthcare data generated in today's scenario. This book chapter proposes a unique Terahertz E-Health system operating on intelligent deep learning-based spectrum sensing techniques. This unique Terahertz E-Health system designed using a novel Perceptive Hierarchal Networking framework (PHN) solves spectrum scarcity by sensing free spectrum slots to transmit the medical data using pervasive sensing mechanisms operating at Terahertz frequency. These sensing schemes utilize efficient deep learning models like Long Short-Term Memory (LSTM), Recurrent Neural Networks, Convolutional Neural Networks, and CNN-LSTM.

1.1 Contributions of the Book Chapter

The contribution of this book chapter is manifold. It proposes a unique approach to a smart healthcare system deploying deep learning at THz frequencies. Following are the main contributions of the book chapter—

- A novel THz E-Health System is proposed, an approach to smart healthcare for transmitting patients' medical data for remote treatments and smart diagnosis.
- A unique Pervasive Hierarchal Networking (PHN) framework is proposed, which comprises intelligent blocks to design the THz E-Health system based on Deep learning models.
- Efficient Deep learning-based spectrum sensing mechanisms utilizing models like LSTM, CNN, RNN, and CNN-LSTM are proposed and deployed in the THz E-health system for intelligent spectrum allocation to transmit patients' medical data.
- The patients are classified based on their health history and risks into high, medium, and low using deep learning predictions. Patient's medical data with emergency cases can be transmitted with the highest priority on the free spectrum slots sensed using deep learning mechanisms.

2 Role of THz Frequency in Forthcoming Wireless Generations and E-Health Systems

The THz frequency communications will play a crucial part in the forthcoming wireless generations, namely the fifth generation and beyond. Ultra-high bandwidth systems can be enabled in this frequency band. Even though the defined frequency range for THz communications is 300 GHz–10 THz [17, 18], the researchers have found it expedient to determine the frequency ranges above 100 GHz as the THz-based communications. Data rates of terabits/sec can be achieved due to THz communications' vast, accessible bandwidth [19] without extra spectral efficiency-boosting schemes. THz-based systems can provide higher link directionality [20] due to the shorter wavelengths of this frequency band. The THz band systems are also less vulnerable to free space diffraction and inter-antenna interference. They also have the advantages of being immune to eavesdropping and can be illustrated in much smaller footprints. Moreover, the THz band has many benefits like advanced user concentrations, more excellent reliability, low latency, improved spectrum utilization, boosted energy efficiency, higher positioning accuracy, and augmented adaptability [21] to propagation situations. Table 1 lists the electromagnetic band segregation and their respective advantages and disadvantages.

Table 1 Electromagnetic spectrum band segregation and respective advantages and disadvantages

Name of spectrum band	Appointed frequency band	Frequency range	Advantages (A) and disadvantages (D)
Radio spectrum	Very low frequency	3–30 kHz	**A**: Ability to infiltrate buildings and solid objects to carry out broadcasting, satellite, mobile communication, and radar-long distance communication **D**: This frequency band is very busy, and obtainable bandwidth is only a few MHz
	Low frequency	30–300 kHz	
	Medium frequency	300 Hz–3 MHz	
	High frequency	3–30 MHz	
	Very high frequency	30–300 MHz	
Microwave spectrum	Ultra High frequency	300 MHz–3 GHz	**A**: Applicable to medium distance communications like radar and Wi-Fi **D**: Higher bandwidths are prone to atmospheric absorption
	Cm wave frequency	3–30 GHz	
	Mm wave band	30–300 GHz	
Infrared spectrum	Sub-mm wave band	300 GHz–3 THz	**A**: This spectrum band supports green field communication and is immune to interference **D**: Its usage is restricted to indoor and point-to-point scenarios called short-range communication due to high path loss and noise
	Infrared-C	300 GHz–100 THz	
	Infrared-B	100–215 THz	
	Infrared-A	215–430 THz	
Visible light	–	430–790 THz	
Anticipated THz spectrum band		95 GHz–10 THz	Combination of microwave and infrared spectrums

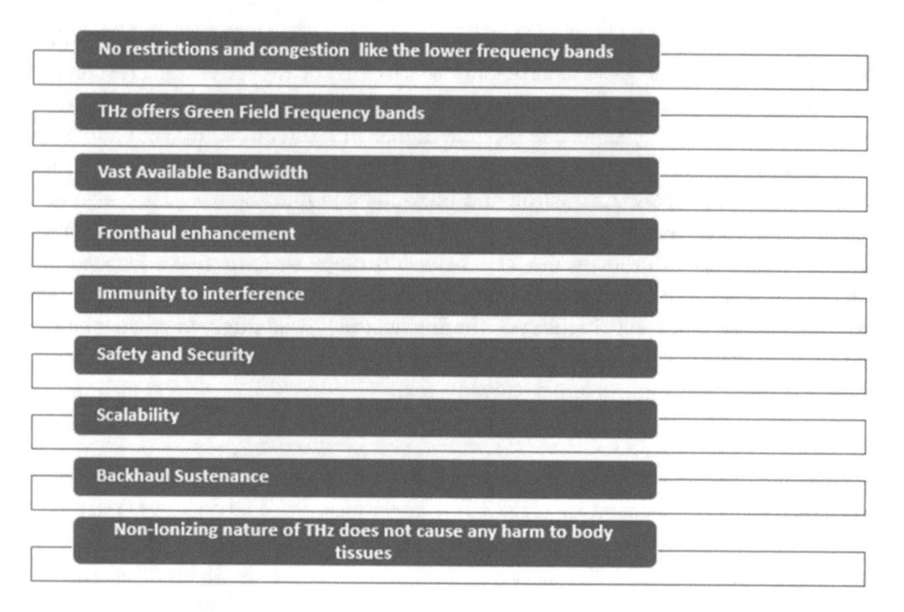

Fig. 1 Unique characteristics of THz spectrum for E-health system and spectrum sensing applications

Figure 1 illustrates the THz spectrum's different unique features, which can be utilized for E-Health systems and Spectrum sensing applications.

2.1 *THz Technology for Medical Applications*

The Terahertz frequency is the end slot of the electromagnetic spectrum, which has not been thoroughly researched and exploited, also called the "Terahertz gap. The frequency range of THz is from 95 GHz to 10 THz. THz has a non-ionizing nature making it alluring for the health industry. Imaging in the THz band provides a non-invasive and non ionizing substitute to X-rays and delivers higher contrast in medical treatments [22, 23]. Dental Imaging, wound and burn examinations, spectroscopy, detection of cancer tumours, blood and breath analysis, and air quality inspections are some of the potential researches [24] being carried out. The See-through characteristic of THz makes it very appealing as it can infiltrate most of the dielectric materials like plastic, paper, wood, and clothing.

The X-ray photons have KeV energies which are drastically higher than the meV energies of THz photons. In contrast to the ionizing nature of the X-rays, the non ionizing property of THz is deficient, with inadequate energy to remove tightly bound electrons. So THz radiation is having the benefit of not causing any harm to the body

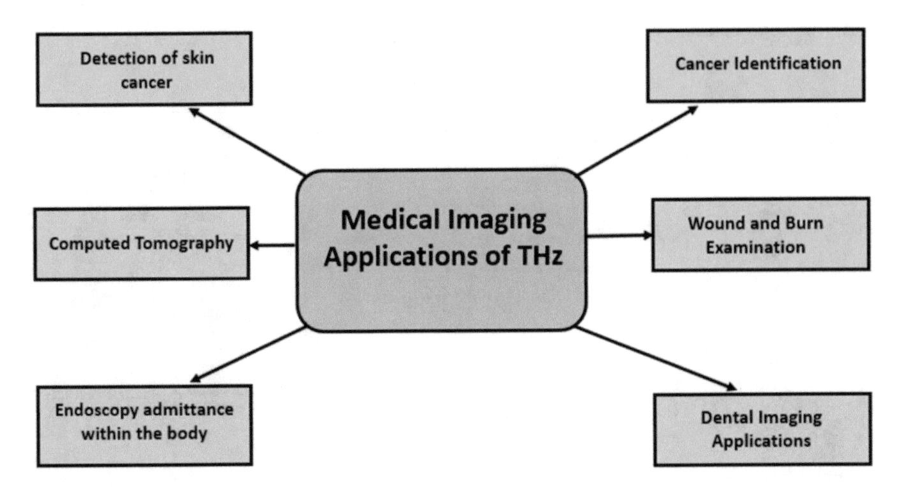

Fig. 2 Medical imaging applications for THz

tissues and DNA and is regarded safe to human bodies. The THz is also an eco-friendly and safer substitute for health applications. So all these properties make THz a crucial contender paving its way for E-health and smart healthcare systems.

Figure 2 depicts the medical imaging applications of THz, which are being researched and implemented with much potential.

3 Terahertz E-Health System—A Novel Approach to Smart Healthcare

A novel architecture for the E-health system is proposed based on Terahertz technology in this section. The proposed novel architecture follows an internet of things-based framework that aims to provide Terahertz-based intelligent engines concerning synchronized medical devices with diverse requirements. The novel architecture introduces sub-processes that intend to gather, collect, analyze and visualize all the data of medical devices through five different stages, as illustrated in Fig. 3. The internet of things enabled devices is primarily revealed and linked to the mechanism through the available Radio Access Network of Beyond 5G networks, followed by gathering their data integrated with the diverse requirements that all these devices possess. Different engines are formed for fulfilling the management needs based upon the data and the requirements. The moment these engines are deployed in the Terahertz E-Health system, the analysis of the collected data takes place, executing intelligent and pervasive deep learning techniques upon them. The analysis using deep learning techniques is followed by analyzing the results in an ultra-reliable manner. The different components of the architecture and their functionality is given as follows—

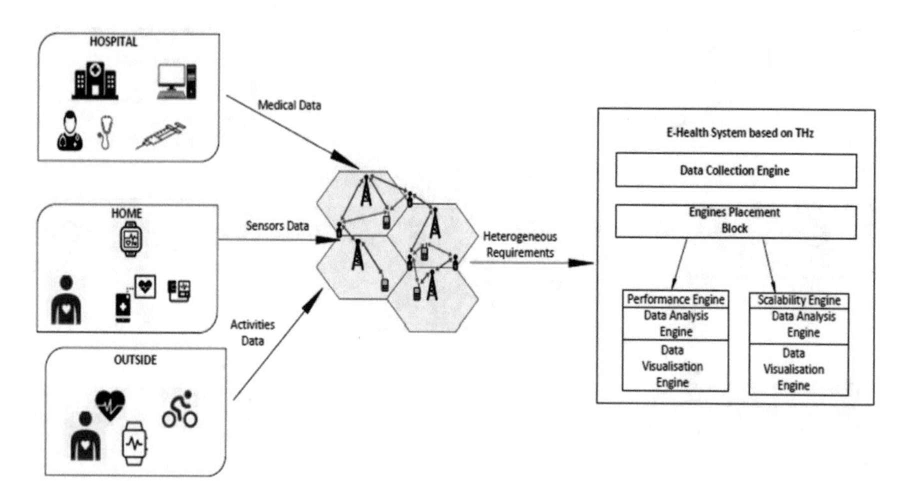

Fig. 3 Terahertz E-health system

(i) **Beyond Fifth-Generation Wireless Network Connectivity and Terahertz Technology**—The E-health system architecture comprises two crucial steps. Firstly, the physical and virtual environments are linked together by permitting data collection at high reliable edge nodes. The data analysis inside the Terahertz platform takes place through suitable integration of intelligent deep learning techniques enabled by deploying an edge node in each medical device linked through a radio access network. The internet of things medical devices are recognized through the established radio access communication network. The radio access network is designed to function in a vast bandwidth of the Terahertz spectrum with various features like channel bandwidth and propagation environment to cater to the diverse and complicated medical data and healthcare services requirements. The novel THz E-health system partly centralizes the functionalities of the radio access network reliant upon the concrete needs, as well as the network features, being able to maneuver vast amounts of data, at high speed with low cost, delivering on-demand resource furnishing, delay conscious storage and increased network capacity wherever and whenever needed. The data management techniques provided (data collection, engine placement, data analysis, and visualization) are created in the form of dedicated engines for each of these services. This provides efficient operation of the E-health system based on Thz, equipping it with flexibility, cost-efficacy, scalability, and being virtualized in distinct E-Health systems running in various hospitals, clinics, labs, etc.

(ii) **Data Collection Engine**—This is the second phase in the THz E-health system. Hence the Data collection Engine is accountable for gathering the requirements by the various linked medical devices and their heterogeneous e-health data. The internet of things enabled devices and smartphones are connected to the internet through the radio access network. The data collection engine has duties of (a) recognizing possible contradictory needs (b) resolving these contradictions by delivering an optimal- aggregated outcome. So the multi-objective optimization methods are considered by resolving the contradictions for diverse requirements of different health devices and parallel medical data. The different sources for the collected medical data are—(a) health monitoring equipment used by the users at their homes (b) clinical and medical devices, Electronic health records, PHRs that the patients utilize, doctors, and hospital professionals for storing the patients' measurements in the hospitals and diagnostic labs (c) health tracking devices and monitors used by the patients for their outside activities.

(iii) **Engines Placement Block**—The Engines placement block plans the end-to-end deployment. The output of the Data collection engine is the starting point for the Engines placement block. The implemented feasibility scrutiny decides about the attainability of the demanded resource requests for the active users, linked medical devices, etc. So, the feasibility analysis inevitably decides about the heterogeneous requirements for deploying the essential engines into the E-health system. The engines placement block guarantees that the delivery of the resources, management, and transposition are viable for synchronized usage in the respective engines in the health system. For attaining the optimal resource placement, the dynamic engine enabling part occurs when the feasibility of the resources is guaranteed to support the defined requirements of the health system. Using the accumulated and feasible requirements coming from the synchronized medical devices, EHRs, PHRs, etc., the primary duty of the Engines placement block is to support the dynamic provisioning of the various engines in the system. So, the engines placement block automatically scans and analyzes the requirements and obtainable infrastructure resources and then decides upon the most suitable ones.

(iv) **Data Analysis Engine**—The fourth stage in the Terahertz E-health system is the Data Analysis Engine. The collected data is analyzed using intelligent deep learning methods like Long Short Term memory, Recurrent Neural networks, Convolutional Neural networks, and CNN-LSTM. The medical data attained from datasets comprising various patients' health conditions, diagnosis, and activity status is given input to the data analysis engine. The engine contrasts these data with comparable data from other patients who had similar health issues. The engine approximates the likelihood of identical living conditions and diagnoses based on the health enhancement or worsening of prior users. The confidence rate regarding user actions is set at 75%, and if it is greater than 75%, only then the data from these activities are stored and analyzed. The periodic unlabeled information from the sensors of medical devices is collected and used for training the deep learning classifiers upon the patients' activities. Each device records the sensor data for a particular time period, like 24 h for each patient. Various features and statistics are calculated for the recorded sensor data, and then these features act as input parameters to the deep learning classifier models. The deep learning models categorize the data received from input into classes like walking, running, etc., to predict probable treatments of the patients. The main categories of medical decision support are (a) proof-based medicine and (b) treatment-based support. The proof-based medicines follow the perceptions extracted from the medical data (mainly diagnosis, treatment, and practice) integrated with a knowledge base with analogous cases and used to find the most apt treatment for each patient and predict and evade possible aggravation difficulty and risks.

(v) **Data Visualization Engine**—The fifth stage comprises a data visualization engine that captures the users' information and uses it suitably. The gathered data is changed incoherently, offering the medical professionals the ability to attain perceptions regarding their patients' disease and health patterns. Valuable data performance and visual analytics mechanisms are utilized like charts, graphs with proper sizes to depict contrasting diagrams, and accurate information labeling to decrease possible misconceptions. It uses bar charts, pie charts, scatterplots, heat maps, and histograms to illustrate concepts and information according to their specific uses. It provides high-quality visual analytics methods crucial to managing complicated datasets, increasing swiftly in capacity and complexity. Medical professionals attain admittance to the patients' location, activity status, and medical condition during the anticipated time period through the data visualization engine and patients' information through charts. The patients' information is gathered through indoor/outdoor location trailing, outdoor activity training, and health numbers based on measurements. The influence of outdoor activities can be observed through the data visualization techniques on the patients' health progress and which specific outside routine can enhance the health condition. Goal achievement charts are also delivered in the data visualization to depict the degree to which the personal goal of a user is attained by integrating time-series data and assessing these data in one chart.

3.1 Proposed Perceptive Hierarchal Networking Architecture and Intelligent Spectrum Sensing for Medical Data Transmission

The existing state-of-the-art healthcare systems face three challenges: (1) ultra-low latency, (2) ultra-high reliability (3) healthcare cognitive intelligence. A novel Perceptive Hierarchal Networking Architecture (PHN) has been proposed based on Beyond 5G communication in the THz band in response to these challenges. For meeting the demands mentioned above, the proposed architecture incorporates information cognitive block and resource cognitive block. The resource cognitive block achieves the required ultra-low latency and ultra-high reliability by using the cognitive radio spectrum sensing function to cognize the network's resources. The information cognitive block achieves the system's required medical intelligence by leveraging the deep learning-based models to analyze the healthcare data and categorize it as high, medium, and low. It ensures reliable and latest patient data to the doctor. Figure 4 represents the proposed PHN architecture. There are the following levels in the Proposed framework-

(i) **Level 1 Cognitive Radio Users (CU)**—At this level, a combination of primary and secondary users occurs. These patients are classified into primary, emergency primary, and secondary users for whom the spectrum would be allotted for transmitting their vital signs data based on their conditions. At this level, there are *N* networks of CR users represented by $n1$, $n2$ … nN.

(ii) **Level 2 Information Cognitive Block and Resource Cognitive Block**—In the information cognitive block, data supply is critical. The cognitive medical application relies on the sustainable provisioning of healthcare data. With this data, the information cognitive block can achieve environmental perception and human cognition with deep learning. Perception and cognition are based on the data flow in the PHN system, while the information cognitive block analyzes the data flow to detect various healthcare requests. Here a multi-user spectrum allocation uses necessary static data about patients and the vital signs risk level information updated by the users in real-time. Vital signs of physiological parameters like Blood pressure, Heart Rate, and Body Temperature are collected through the wearable sensors on the patient's body. These critical signs serve as features for training the deep learning models. The resource cognitive block can achieve resource optimization through perception and learning of network contexts (such as network type, data flow, communication quality, and other dynamic environmental parameters) and user information. Utilizing the resource cognitive block, PHN can achieve green communications and energy efficiency, critical for its infrastructure to fulfil healthcare applications' requirements. Furthermore, the intelligence generated by cognizing hardware systems and various available resources in computing, communications, and networking, can achieve high reliability, high flexibility, ultra-low latency, and scalability. The patients with high-risk levels are allotted spectrum first, then medium and low priority patients.

(iii) **Level 3 Cognitive Gateways**—These gateways employ cloud processing to attain rewards, reduced computing costs, combining resources and flexibility. Deep learning is used for consolidated data processing. The information gathered by the cognitive gateways is transmitted to the cloud server, where the data is stored centrally for medical data transmission.

(iv) **Level 4 Cloud Platform**—The medical data is transmitted after the spectrum is allocated to the patients classified and categorized using the deep learning model. This data is stored on the cloud platform and swiftly transferred to the doctors and other required destinations to diagnose patients through THz communication.

In smart health data transmission, several sensors can be connected for patient monitoring, and their data can be stored in a cloud network for future use. With the rising number of diseases, environmental issues, and the number of growing

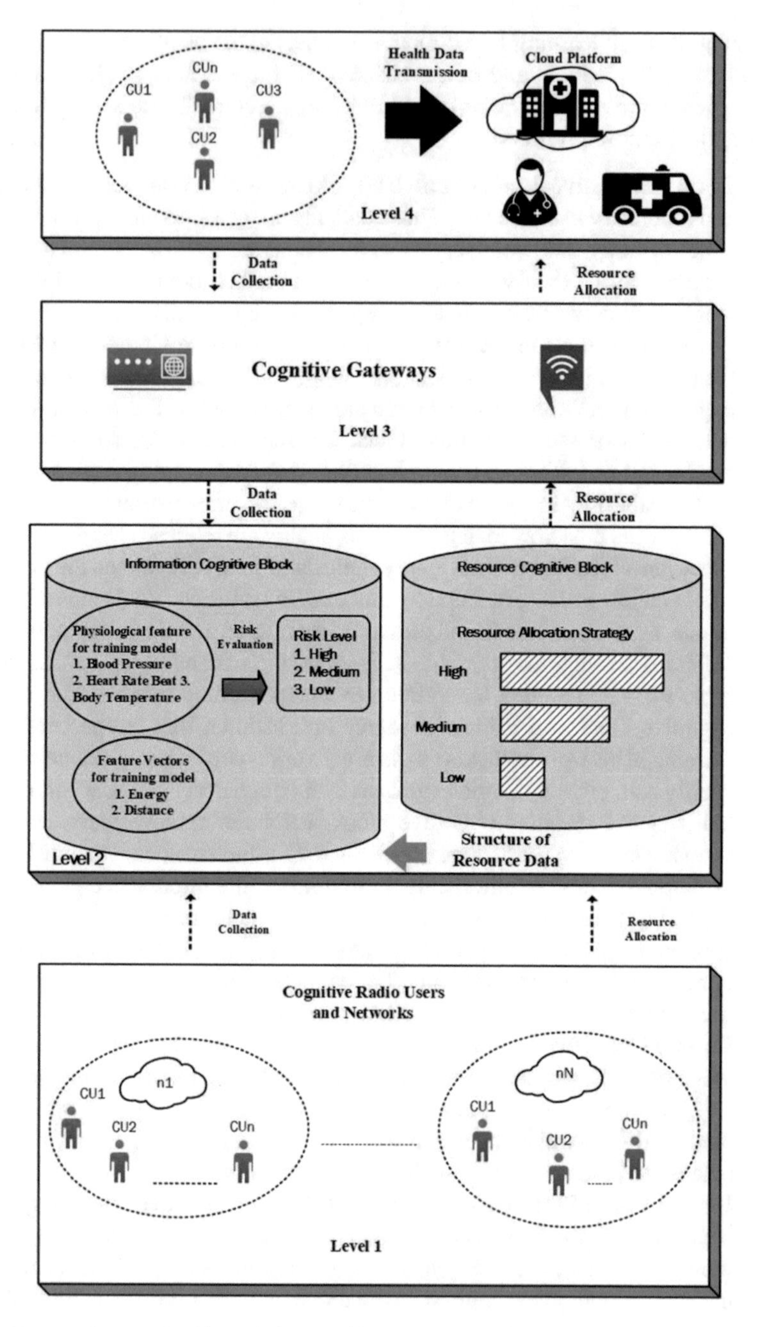

Fig. 4 Proposed perceptive hierarchal networking architecture for medical data transmission

Table 2 Blood pressure categorization

Blood pressure category	Systolic mm Hg (upper level)	Diastolic mm Hg (lower level)
High priority	130 and higher	80 and higher
Medium priority	120–129	Less than 80
Low priority	Less than 120	Less than 80

Table 3 Heart rate categorization

Heart rate category	Range (beats per minute bpm)
High priority	Greater than 130 or less than 60
Medium priority	101–130
Low priority	60–100 bpm

Table 4 Body Temperature categorization

Body temperature category	Range (F)
High priority	103 or above
Medium priority	100.4–103
Low priority	97–99

patients, healthcare networks' serviceable area has become very dense, for example, approximately 10^6 connections per km^2. The ultra-densification of connections with very ultra-low latency can be fulfilled by providing a vast bandwidth through THz communication to transmit a tremendous amount of healthcare data. Physiological parameters of vitals are used as feature parameters for training the deep learning models to classify patients into high, medium, and low priority. The deep learning classifier can identify the patient with the most critical condition and detect an available spectrum on the highest priority. According to the American Heart Association (AHA), these vital signs parameter ranges are given in Tables 2, 3, and 4. According to these tables, the patients can be classified as high, medium, or low based on the vital signs data collected from the sensors through the THz network. A patient with a blood pressure of 130 or higher in the upper level and 80 or higher in the lower level, heart rate greater than 130 or less than 60, and body temperature of 103 F or higher have the highest priority for spectrum sensing and allocation for sending their health data. The proposed PHN framework senses the free spectrum using deep learning techniques and allocates it for a critical patient so that its medical data can be transmitted for remote treatment at a remote location or urgent diagnosis by the doctors. The spectrum is sensed, and if the primary users or patients are not active, then the secondary patients can utilize the spectrum.

In addition to these vital signs parameters, many other healthcare features like lab reports, diagnostic tests, ECGs, MRI scans, ultrasound results, blood tests, X Rays, and other Electronic Health records. So, all these variations of the healthcare dataset can be transmitted using the PHN architecture on an available spectrum slot on a THz band.

4 Intelligent Spectrum Sensing Based on Deep Learning Classifiers for THz E-Healthcare System

Secondary users and primary users can coexist and manage the spectrum band intelligently using a promising technology called cognitive Radio. It permits the secondary users to unscrupulously access the licensed band of licensed users when the primary users or authorized users are absent. The crucial roles of cognitive radio wireless spectrum analysis, spectrum management, estimation of channel conditions, power control, etc. Hence spectrum sensing methods leveraging deep learning classifiers [25, 26] will diminish the flaws in classification and recognition of channels as the deep learning classifiers are not dependent on signal features, but relatively, it inevitably learns the features. So, deep learning will enhance the performance metrics of spectrum classification. Deep Learning is composed of numerous transitional layers of non-linear processing for modeling multifaceted representations in data. Deep learning possesses big data investigation that makes it more apt to identify patterns in several applications of economics, computer vision, natural language processing, bioinformatics.

The intelligent machine will be aware of its surroundings and conduct activities to increase its efficacy. The necessary actions of deep learning involve inference, perception, problem resolution, knowledge illustration, and learning. The chief processes in the learning methodology of cognitive radio based on deep learning are illustrated in Fig. 5. The learning methodology can be given as follows: (i) sensing the spectrum bands and detecting the channel specifications like channel quality (ii) observing, perceiving, and analyzing the feedback like Acknowledgment responses (iii) learning (iv) storing the outcomes and observations for apprising the model and attaining better accuracy in imminent decision formulation (v) Determining the issues of spectrum management and adapting the transmission errors consequently. The Deep Learning Classifiers used for Intelligent Spectrum sensing are discussed as follows.

4.1 Recurrent Neural Network-Based Intelligent Spectrum Sensing

The Recurrent neural networks comprise an internal memory which is a simplification of the feedforward neural network. It has a repeating nature, as the name suggests, as it executes the same function for each input data while the outcome of the existing input is reliant on the prior one calculation. When the outcome is generated, it is replicated and transferred back into the recurrent network [27, 28]. The existing input and outcome learned from the prior input are considered for decision formulation. The recurrent neural networks utilize their internal memory to process the input arrangements. All the inputs are associated with each other.

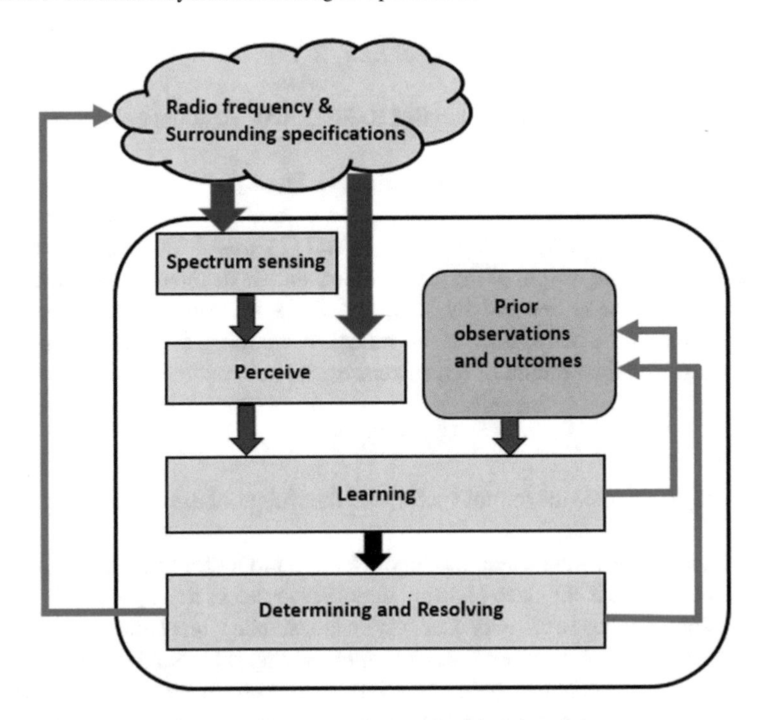

Fig. 5 Learning methodology in cognitive radio based on deep learning

Figure 6 depicts the block diagram of a recurrent neural network. In the first stage, the recurrent neural network takes X_O as the input sequence and gives the output as h_o. So the output of the first stage, which is $h_{o,}$ acts as an input, and the X_1 for the second stage. So this keeps on continuing to train the model. Hence the current state in a recurrent neural network can be formulated as—

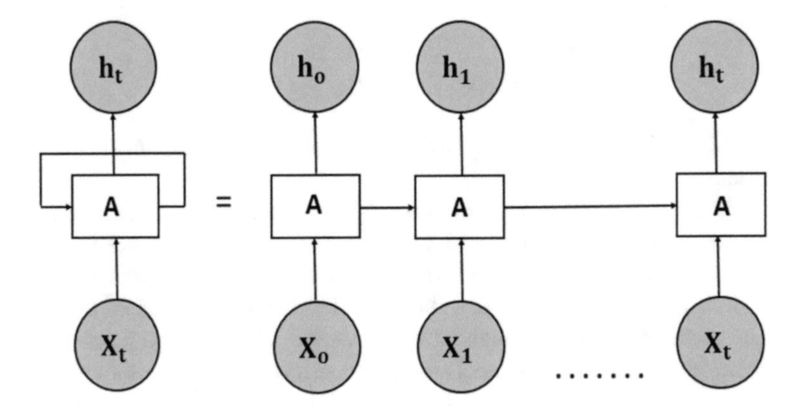

Fig. 6 Block diagram of recurrent neural network

$$h_t = f(h_{t-1}, X_t) \quad (1)$$

Activation function (*A*), when applied to Eq. 1, will yield the current state as—

$$h_t = \tanh(W_{hh}h_{t-1} + W_{xh}X_t) \quad (2)$$

The meanings of the symbols in Eqs. (1) and (2) are—W is weight vector, h is the hidden vector, the weight of the prior hidden state is denoted by W_{hh}, the weight of the existing state is denoted by W_{xh}, and the activation function here is tanh which transforms the activations of the model in the range $[-1, 1]$, implementing non-linearity. The final output of a recurrent neural network can be defined as—

$$Y_t = W_{hy}h_t \quad (3)$$

In Eq. (3) Y_t denotes the output vector, and the weight of the output is represented by W_{hy}.

Based on the above-presented model of RNN, a model for spectrum sensing can be formed using the RNN deep learning classifier to detect free spectrum for the E-Health system. Suppose a primary user system is assumed with N number of channels where $n \in \{1, 2, 3, \ldots N\}$ depicts the nth channel having B Hz bandwidth allocated to each channel. These channels belonging to primary users are mutually independent of each other. So the aim is to detect spectrum holes or free bands for a specific time period where the secondary users can access the spectrum bands. It can be presumed that the primary user interaction follows the hidden Markov model, where the time duration is exponentially distributed. If the primary users act as intermingling objects with one primary user denoted as P_1 and the other primary user as P_2, four interacting states can be between them from 00 to 11. 01 state indicates that the primary user P_2 sends information to $P_{1.}$ State 10 denotes that P_1 transfers information to $P_{2.}$ State 00 and 11 depict that both the intermingling primary users are free and busy, respectively. Hence the primary user categorization can be depicted as a sequence as follows—

$$P_{nt} = \{P_{nt}(1), \ldots P_{nt}(t), \ldots, P_{nt}(T)\} \quad (4)$$

Here $P_{mt}(t)$ denotes the primary user incumbency state at the nth channel in the tth time slot.

The primary user incumbency and interaction with each other are dependent upon time and are therefore time-correlated. Hence, recurrent neural network features can be used to represent the spectrum states and primary user model. The binary sequence of the primary user interactions can illustrate the past information of the primary user state in the frequency channel and acts as the input. So the outcome of the recurrent network acts as the channel state at the succeeding time slot. Incumbencies of various channels simultaneously is fed into the system as a time step, and the past incumbency state information for the prior M time slots is utilized to predict the channel condition at the time slot $M + 1$. The anticipated channel condition at the $M+1$ time slot and prior $M-1$ slot past information is used to indicate the channel condition at the time slot $M+2$ and so on. Therefore the output layer of the spectrum incumbency condition predictor can be formulated using the sigmoid activation function as follows—

$$\text{output} = \frac{1}{1+e^{-x}} \tag{5}$$

As the output function formulated in Eq. (5) involves the sigmoid activation function, the outcome of the whole system lies in the range [0, 1]. If a decision threshold d is set, then the output greater than d indicates that the primary user's anticipated time slot is busy. Otherwise, the anticipated time slot is free and available. Hence, the RNN deep learning model helps the secondary user find the available time slot intelligently based on the modelled incumbencies. The THz E-Health system could then utilise the detected free spectrum slot to transmit the medical data of an urgent patient on a high-priority basis to receive a medical diagnosis and remote treatments.

4.2 *Long Short Term Memory Based Spectrum Sensing Model*

The LSTM [29, 30] is an improved version of the recurrent neural network model. A cognitive system can be considered, including the *M* number of secondary users, one primary user, and a fusion centre. The secondary users extract the local signal features of the received signal. The fusion centre or FC combines the features obtained by the secondary user. A hypotheses model can be formulated for the primary user as follows-

$$Y = \begin{cases} \varphi(\theta(n_1), \ldots \theta(n_M))H_O \\ \varphi(\theta(h_1 x_{PU} + n_1), \ldots \theta(h_M x_{PU} + n_M))H_1 \end{cases} \tag{6}$$

In Eq. (6), hypotheses H_O and H_1 depict the absence and presence of the primary user, the primary signal transferred is represented as x_{PU}, $h_i$denotesthechannel coefficient amongst the primary user and ith secondary user, n_i denotes the noise encountered. The signal received by an ith secondary user can be represented as—

$$y_i = [y_i(1), y_i(2) \ldots y_i(l)] \tag{7}$$

'l' represents the span of the received signal. Therefore the received signal of the secondary user is $y_i = h_i x_{PU} + n_i$ when the primary user is present. If the primary user is not present, then the received signal can be formulated as $y_i = n_i$. The recurrent neural network model is dependent on sequences. But there are many issues with the RNN classifier, such as loss of gradient and its explosion. So the LSTM model is applied to solve the problems occurring because of recurrent neural

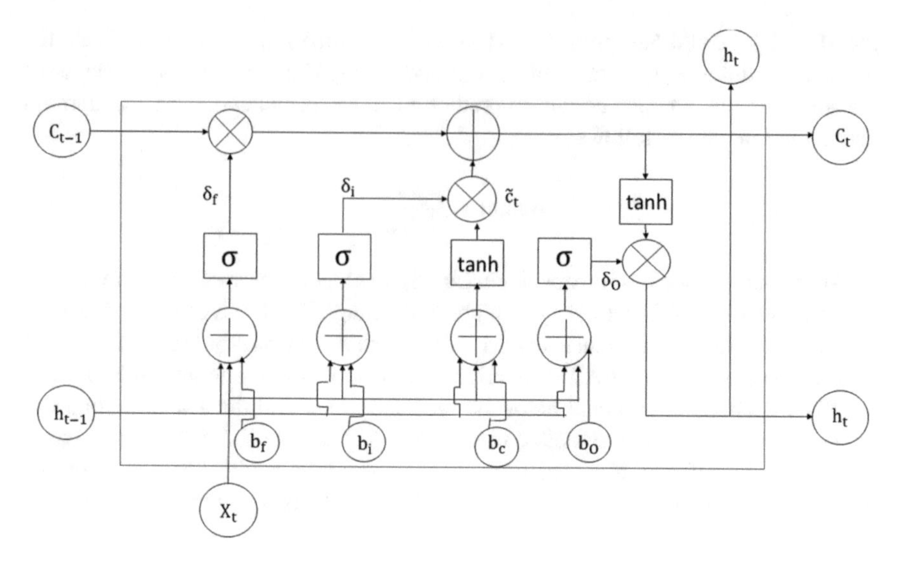

Fig. 7 Architecture of LSTM cell

networks by including a memory unit. The LSTM consists of a hidden layer. Each recurring component has a tanh activation function. The storage cells in the hidden layer ensure that the LSTM is present in the recurrent model in an arrangement of weights and transient activations.

Figure 7 illustrates the Architecture of a Long Short Term Memory cell. The different crucial parts of the LSTM cell are—

i. Input gate—This is the gate that chooses when to appraise the existing cell condition and decides how much new data can transfer into the memory unit. It is represented by δ_i.
ii. Forget gate—This component comes to an outcome about when to discard the existing cell. It is represented by δ_f.
iii. Output gate—This gate is denoted by δ_O It manages the amount of data for output activation of the memory cell and then transfers it into the remaining network.

In the architecture, the input is denoted by X_t, the output othe LSTM cell is denoted by h_t, h_{t-1} is the output of the last unit of LSTM, C_t and C_{t-1} are the current and previous cell states respectively, tanh function is used as the activation function of the long short term memory cell, which is like a squashing function to convert values between -1 and 1, σ is the sigmoid layer and b_c, b_f, b_O, b_i are the bias vectors of contender vector, forget gate, output gate, and input gate.

The tanh activation function can be given as-

$$\tanh = \frac{e^z - e^{-z}}{e^z + e^{-z}} \tag{8}$$

For updating the cell state, $\tilde{c}_t$ vector of contender values is formed-

$$\tilde{c}_t = \tanh[W_c(h_{t-1}, X_t) + b_c] \tag{9}$$

where b_c is the bias. The sigmoid activation can be utilized to calculate the values of the update, forget and output gates as follows—

$$\delta_i = \sigma[W_i(h_{t-1}, X_t) + b_i] \tag{10}$$

$$\delta_f = \sigma[W_f(h_{t-1}, X_t) + b_f] \tag{11}$$

$$\delta_O = \sigma[W_O(h_{t-1}, X_t) + b_O] \tag{12}$$

$$\sigma(z) = \frac{1}{1 + e^{-z}} \tag{13}$$

Here W_i, W_f, W_O are the matrices of weight vectors. Also, an element-wise multiplication is taken between the forget gate δ_f and cell state of the previous timestamp C_{t-1} and also, between the input gate δ_i and $\tilde{c}_t$ [31, 32]. The output gate δ_O and the hyperbolic tangent $\tilde{c}_t$ give the output h_t by using elementwise multiplication given in Eqs. 16 and 17—

$$C_t = \delta_u o \tilde{c}_t + \delta_f o C_{t-1} \tag{14}$$

$$h_t = \delta_O o \tanh(C_t) \tag{15}$$

The Hadamard product is denoted by o, and the element-wise addition is denoted by '+.'The different features of the secondary user are combined in the fusion center using the eigen vector as follows-

$$F_{SU} = [F_{SU1}, F_{SU2}, \ldots F_{SUi}] \tag{16}$$

Back propagation and gradient descent algorithms are used to manage the weight parameters of the connections to combine the features of all the secondary users in the process of training. The final output of the LSTM network is computed by the softmax function that depicts output as a two-dimensional vector$[l, m]^T$. The vector $[0, 1]^T$ indicates the existence of a primary user, and $[1, 0]^T$ denotes the nonappearance of the primary user. A Euclidean distance can be computed between the output and the vector $[0, 1]^T$, which is formulated as below-

$$E = \sqrt{(l - 0)^2 + (b - 1)^2} \tag{17}$$

The detection probability increases for a low false alarm probability, and this is ensured by setting an error threshold E_{th}. Hence the primary user is considered to be present if E is lesser than E_{th}; otherwise, the primary user is absent.

4.3 Convolutional Neural Network-Based Spectrum Sensing Model

Convolutional Neural Network is a categorization of deep neural networks that employs a mathematical algorithm called convolution. The CNN network [33, 34] comprises input, hidden, and output layers. The CNN architecture is formed of hidden layers in the middle as the inputs and outputs of the hidden layers are concealed by the activation function and the ultimate convolution operation. The hidden layers are used to perform convolution using a dot product of the convolution kernel with the input matrix of the layer. The convolution operation produces a feature map as the convolution layer goes along the input matrix. This acts as the input for the next layer. The other layers inside the CNN network are pooling layers, fully connected layers, and the normalization layer.

A spectrum sensing model can be formulated using a binary hypothesis model as follows-

$$H_O : y(t) = w(t) \tag{18}$$

$$H_1 : y(t) = x(t) + w(t) \tag{19}$$

H_O and H_1 are the two hypotheses where it represents the absence and presence of primary users respectively in a T time period, *x(t)* is the primary user signal with cyclostationary property, *y(t)* is the received signal, and *w(t)* denotes the additive white gaussian noise possessing variance σ_n^2. The features are extracted for the primary user signal. Then the training and testing data is used to build the spectrum sensing model using CNN architecture. Here cyclostationary property of the primary user and the static nature of the noise signal extract the features using the autocorrelation function. The signal *y(t)* received by the secondary user can be used to compute its cyclic autocorrelation property. The equation of the cyclic autocorrelation can be depicted as-

$$R_y^{\mu} = \frac{1}{T}\int_0^T R_y(t, \tau)e^{-j2\pi\mu t}\mathrm{d}t \tag{20}$$

In Eq. 20, μ is a cyclic frequency which can be formulated as $\mu = N(1/T)$, N is an integer, and the time period is symbolized by T.

The process of spectrum sensing using the CNN framework is depicted in Fig. 8. The extracted features are pre-processed and then divided into training and testing

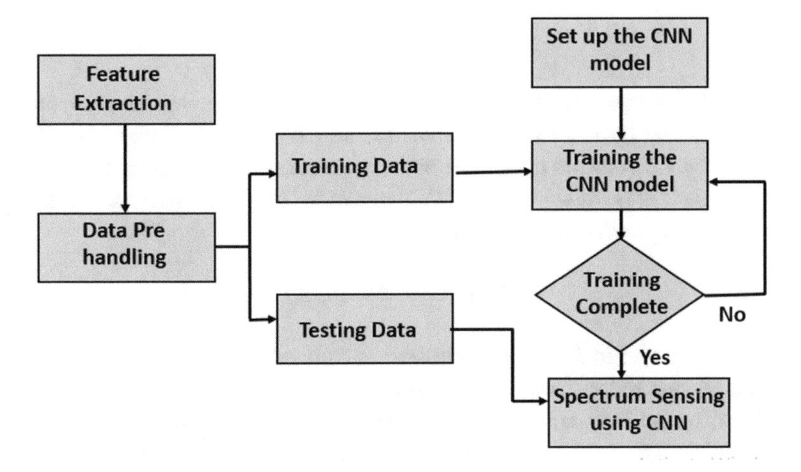

Fig. 8 Flow chart of spectrum sensing using CNN

datasets for establishing the spectrum sensing model. The training data trains the convolutional neural network classifier, and the testing data is used to make the spectrum predictions. The autocorrelation feature can be formulated as-

$$R_y(t,\tau) = E\{y\left(t+\frac{\tau}{2}\right)y^*\left(t-\frac{\tau}{2}\right)\} \tag{21}$$

The autocorrelation function exists in the time domain. So if the Fourier transform of the autocorrelation function is taken, then the spectral correlation function is achieved as follows-

$$S_y^{\mu}(f) = \int_{-\infty}^{\infty} R_y^{\mu}(\tau)e^{-j2\pi\mu\tau}\mathrm{d}\tau \tag{22}$$

The autocorrelation feature used to present the cyclostaionarity nature is standardized as—

$$\gamma_l = \frac{m-\overline{m_l}}{\varepsilon} \tag{23}$$

Here $\overline{m_l} = \sum_{l=1}^{M} m_l$ and $\varepsilon = \sqrt{\frac{\sum_{l=1}^{M} m-\overline{m_l}}{N-1}}$ are the mean and standard deviation of the cyclostationary features extracted by calculating autocorrelation and M number of samples. Therefore, the spectral correlation function is formulated as $S_y^{\mu} = (\gamma_1, \gamma_2, \ldots \gamma_M)^T$. The CNN architecture consists of an input layer formed by the extracted features from primary and secondary users, the convolutional layer, the sampling, and the fully connected layers. The input layer is formed by the cyclostationary features formed and extracted using the autocorrelation function. Then after this, the convolution operation takes place according to the following equation—

$$X^l = f(w^l \otimes X^{l-1} + b^l) \tag{24}$$

In Eq. 24, the layer of neurons is symbolized by X^l, the weight of a particular layer l is denoted by w^l, the bias vector of the l layer is symbolized by b^l, $\otimes$ is the symbol for convolution operation, and f is the activation function. Here the Relu activation function is utilized by the CNN model for spectrum sensing. For an input X, the Relu function can be formulated as follows—

$$\text{Relu}(X) = \max(0, X) \tag{25}$$

The pooling layer in the CNN framework is utilized to eliminate redundant information and decrease the parameter training. After this, finally, in the fully connected layer, the softmax function acts as activation to output the probability value according to the hypotheses H_O and H_1. Out of the probability values obtained from the softmax function, the maximum probability values determine which category the data belongs to according to the binary hypotheses model to estimate the presence of the primary user. If the primary user is present, then channel slots of the spectrum are deemed busy; otherwise, they are accessible. The softmax function for the spectrum sensing can be expressed as follows-

$$\text{softmax}(\varphi) = \frac{1}{\sum_{m=1}^{M} \exp(\varphi_m)} \begin{pmatrix} \exp(\varphi_1) \\ \exp(\varphi_2) \\ . \\ . \\ . \\ exp(\varphi_M) \end{pmatrix} \tag{26}$$

The training and testing data sets obtained are used to train the CNN model to perform spectrum sensing of slots and categorize the E-Health system patients based on their risk classification as high, low, or medium. So the testing data set is input into the CNN model to obtain the spectrum sensing predictions. These predictions are made on the highest priority for a critical patient requiring urgent medical care. The CNN-based spectrum sensing model will identify and classify the spectrum as busy or free for such a patient. The free spectrum slots sensed by the technique are then used to transmit the patient's medical data requiring remote treatments or some analysis of the disease. So, the output of the spectrum sensing is decided according to the hypotheses mentioned above using the softmax function.

4.4 CNN-LSTM Classifier Model-Based Spectrum Sensing

The sequential reliance in a time series data is taken into account by the recurrent neural networks, and then the model is trained using this data. Long short-term memory is a kind of RNN model which considers the dependencies of the time-based information sets. A conventional neural network ignores these dependencies in the data and treats all the data as independent, inaccurate, and produces deceptive outcomes. The Convolutional neural networks use a mathematical operation called convolution to find out the relationships amongst two functions, say f and g, where the convolution integral describes how the shape of one function is altered by the other. The CNN models are conventionally applied in image classification and do not consider sequential dependencies. The convolutional neural networks have a unique feature of dilated convolutions in which they employ filters to calculate the dilations between each unit. The space dimensions between every cell permit the neural networks to predict the time series data better. Hence, LSTM and CNN models are integrated for predicting the time series sequence. The LSTM units consider the sequential dependencies in the time series data, and the convolutional neural network also analyses this procedure by using dilated convolution steps.

Figure 9 illustrates the CNN-LSTM model for spectrum sensing. The block diagram is characterized by implementing CNN units in the front, followed by the LSTM units and a dense layer on the output. The CNN model is used to extract the features for spectrum sensing, and the LSTM units are then executed to analyze the collected features across time steps. For implementing the spectrum sensing based on the CNN-LSTM [35, 36] model, the primary user movement pattern is first studied and collected in different spectrum sensing time periods. The information collected from the prevailing time period and the information gathered from the prior sensing time periods are converted into matrices. Then the CNN model [36] is used to extract

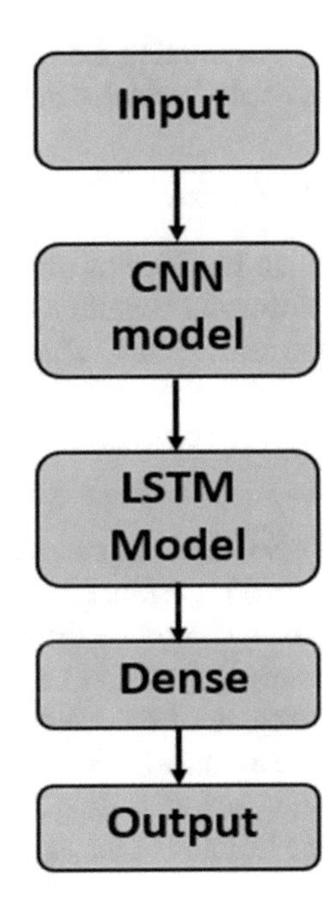

Fig. 9 Block diagram of CNN-LSTM model

the features from these matrices, and the extracted features then act as input to the LSTM units for further analyzing the primary user movement patterns.

A particular spectrum sensing time period can be referred to as a tth sensing frame. M signal samples can be considered to be collected for a specific sensing frame based on the primary user movement analysis. During this sensing frame, the secondary user can decide to remain inactive or transmit for the rest of the sensing period. Hence in a cognitive radio network based on the CNN-LSTM model, the primary user can be considered to bear one antenna, and the secondary user can be considered to have A number of antennas. Therefore, a binary hypothesis model can be framed for the spectrum sensing following the CNN-LSTM architecture as follows—

$$H_O : \{y_t(n)\}_{n=1}^{M} = \{n_t(n)\}_{n=1}^{M} \tag{27}$$

$$H_1 : \{y_t(n)\}_{n=1}^{M} = \{h_t s_t(n) + n_t(n)\}_{n=1}^{M} \tag{28}$$

H_O and H_1 denote the absence and presence of the primary user, respectively. h_t represents the channel coefficient between the primary user and the secondary user. $s_t(n)$ and $n_t(n)$ are the signal transferred by the primary user and the noise received in an nth sample in the tth sensing period, respectively. The complete received signal can be considered to be a stream of the different gathered signals of different sensing samples as follows—

$$Y_t = [y_t(1), y_t(2) \ldots y_t(M)] \tag{29}$$

For training the CNN-LSTM spectrum sensing model, a labeled training dataset is needed, which can be formulated as-

$$T = \{(Y_1, l_1) \ldots (Y_t, l_t) \ldots (Y_S, l_S)\} \tag{30}$$

In Eq. 30, S is the training dataset's size, which defines the actual primary user movement condition corresponding to Y_t. The size of Y_t is $A \times M$ which comprises the sensing data. This data set is then preprocessed to form a matrix as follows—

$$M_t = \frac{1}{M} Y_t Y_t^H \tag{31}$$

So the preprocessed training dataset can be illustrated as $\{(M_1, l_1), \ldots (M_t, l_t) \ldots (M_S, l_S)\}$. Firstly, the training matrix dataset is input into the CNN unit, which comprises the convolution layers. The outcome conforming to each sensing frame is vectorized and then input into the respective LSTM units. The vectorized outcome of the CNN cells comprises the features of each sensing frame, and the LSTM units further analyze them to extract the time dynamic features hidden in the sequence of sensing frames. Finally, the last LSTM unit's output, which comprises the features of the entire input sensing sequence, is input into a dense layer to adapt the output dimensions according to the different data classes. The output of the CNN-LSTM is normalized by a softmax function and formulated as follows—

$$[\mathrm{d}_{\alpha|\mathrm{H_O}}(\mathrm{M_t}), \mathrm{d}_{\alpha|\mathrm{H_1}}(\mathrm{M_t})] \quad (32)$$

Here $d_{\alpha|H_O}(M_t)+d_{\alpha|H_1}(M_t)=1$, M_t corresponds to the state H_O and H_1 determines the presence or absence of the primary user, and α represents the parameters of the CNN-LSTM model. Therefore, after an effective training of the CNN-LSTM model, as a new incoming sequence arrives, its fundamental primary user state can be found by comparing the CNN-LSTM output value $d_{\alpha|H_O}(M_t)$and$d_{\alpha|H_1}(M_t)$.

So, the CNN-LSTM based spectrum sensing scheme helps determine whether the spectrum is free or not based on the primary user movement status. So, the movement status of the primary user can be evaluated, and then in the E-health system, the available spectrum can be utilized to transmit the medical data of a particular patient decided according to their priority. The priority of the patients can be selected based on the severity of their vital signs data and other diagnostic data, which is to be sent on the detected free spectrum slot using the CNN-LSTM model.

4.5 Advantages and Disadvantages of RNN, CNN, LSTM, and CNN-LSTM Models

The different deep learning classifiers, namely recurrent neural networks, convolutional neural networks, long short term memory, and CNN-LSTM, have been contrasted based on their advantages and disadvantages in Tables 5 and 6.

5 Conclusion

This book chapter presents a novel THz E-Health system designed using Deep Learning-based Spectrum sensing mechanisms. The THz band between 300 GHz to 10 THz is crucial for the fifth-generation wireless systems and beyond. The pandemics like the current COVID-19 require timely treatments to prevent losses due to virus dispersal. The different intelligent components engines of the proposed THz E-Health system have been accounted for based on their functionalities. A novel Pervasive Hierarchal Networking (PHN) framework has also been proposed, which is used to transmit patients' medical data according to their health priority. The priority and severity of their health can be evaluated based on their vital signs like blood pressure, body temperature, heart rate ranges, and electronic health records comprising diagnosis reports, clinical lab results, and doctor analysis. The patient with the highest severity is then allotted a free spectrum slot senses by the deep learning sensing mechanisms based on efficient models like LSTM, RNN, CNN,

Table 5 Advantages of different deep learning classifiers

Recurrent neural networks	Long short term memory	Convolutional neural network	CNN-LSTM
1. The RNN model can have any span of input that resolves the neural networks' static window issue 2. Optimization of the weight matrix is done in the RNN model, and the same size matrix is applied at each of the inputs, which was an issue for the neural networks 3. The dimensions of the RNN model remain the same for any input size	1. LSTM model can safeguard the data from the past steps and resolve the vanishing gradient issue of RNN models 2. Very long time gaps can be reduced by the steady error backpropagation within the memory units of the LSTM model 3. The LSTM classifier can manage time lags by handling noise, distributed representations, and continuous values 4. There is no requirement for parameter alteration as it operates well on a broad range of constraints like learning rate, input, and output gate bias	1. The CNN model has good classification performance and does not require parameter alteration to enhance performance 2. Parallelization concept is possible in CNN models and contributes to the efficacy of the model's performance 3. It has very high accuracy and inevitably senses the features requiring no human regulation 4. The CNN model has a weight-sharing feature	1. The advantages of the CNN model and the LSTM model are integrated to form this effective model amalgamation. The CNN model can effectually extract the features from the information, and the LSTM improves the accuracy of the performance by identifying the interrelationship of data in the time-series data. It also inevitably senses the best approach for pertinent data 2. The CNN-LSTM model is efficient in extracting spatial and sequential structures of the input information generated by the spectrum sensing data and learn the movement of the primary user to sense the free spectrum 3. CNN-LSTM is different from the other deep learning techniques as it is not dependent upon any signal–noise model presumptions, which helps it learn the primary users movements and can be trained efficiently to extract the features for spectrum sensing

Table 6 Disadvantages of different deep learning classifiers

Recurrent neural networks	Long short-term memory (LSTM)	Convolutional neural network (CNN)	CNN-LSTM
1. The calculation is very slow in every hidden state and cannot be calculated until the computation of the prior hidden state is accomplished 2. The RNN models cannot be implemented in parallel form, and training of the RNN units takes more resources and time 3. RNN has the vanishing gradient issue in which the prior hidden states have a minor influence on the ultimate hidden state. As the data from the past hidden states is carried forward to newer hidden states, then the data will become reduced at each additional hidden state. Therefore this is called the vanishing gradient as the effect of the prior hidden states keeps reducing until it becomes inconsequential	1. LSTM model does not resolve the parallelization issue of the RNN model as the calculation of each hidden state, and cell state has to be done before the next hidden and cell state arrives 2. LSTM model requires a lengthier time to train and also needs more storage memories 3. The LSTM units are necessarily required to move in one direction, usually left to right. However, during machine translation tasks, the LSTM units start going in both directions, which affects the performance	1. The CNN model needs padding before the first word and the last word 2. The CNN model does not encode the location and alignment of the object 3. It lacks the skill to be spatially unalterable to the input information 4. The CNN model needs a large size of training data set	1. The CNN-LSTM model has difficulty in detecting small-sized objects due to a lack of discriminativeness 2. The CNN-LSTM can only predict a label and not a segmentation box

and CNN-LSTM. The advantages and disadvantages of all the models have been presented, along with the unique characteristics and applications of the THz for Smart Healthcare. So, this book chapter can be envisioned as a beginning stage for the discussions and proposal of THz technology applications in E-Healthcare systems and beyond fifth generation systems.

References

1. V. Chamola, V. Hassija, V. Gupta, M. Guizani, A comprehensive review of the COVID-19 pandemic and the role of IoT, drones, AI, blockchain, and 5G in managing its impact. IEEE Access **8**, 90225–90265 (2020)
2. B. Li, Z. Fei, Y. Zhang, UAV communications for 5G and beyond: recent advances and future trends. IEEE Int. Things J. **6**, 2241–2263 (2019)
3. H.-C. Da Costa Daniel Benevides, Grand challenges in wireless communications. Front. Commun. Net. **1** (2020)
4. S. Dang, O. Amin, B. Shilhada, M.S. Alouini, What should 6G be? Nature Elec. **3**, 20–29 (2020)
5. M.A. Rahman, M.S. Hossain, N.A. Alrajeh, N. Guizani, B5G and explainable deep learning assisted healthcare vertical at the edge: COVID-I9 perspective. IEEE Net. **34**, 98–105 (2020)
6. I.F. Akyildiz, A. Kak, S. Nie, 6G and beyond: the future of wireless communications systems. IEEE Access **8**, 133995–134030 (2020)
7. K. Shafique, B.A. Khawaja, F. Sabir, S. Qazi, M. Mustaqim, Internet of things (IoT) for next-generation smart systems: a review of current challenges, future trends and prospects for emerging 5G-IoT scenarios. IEEE Access **8**, 23022–23040 (2020)
8. S. Lin, K. Chen, Statistical QoS control of network coded multipath routing in large cognitive machine-to-machine networks. IEEE Int. Things J. **3**, 619–627 (2016)
9. E. Selem, M. Fatehy, S.M.A. El-Kader, E-health applications over 5G networks: challenges and state of the art, in *2019 6th International Conference on Advanced Control Circuits and Systems (ACCS) & 2019 5th International Conference on New Paradigms in Electronics & information Technology (PEIT)* (2019), pp. 111–118
10. P. Hillger, J. Grzyb, R. Jain, U.R. Pfeiffer, Terahertz imaging and sensing applications with silicon-based technologies. IEEE Trans. Tera. Sci. Tech. **9**, 1–19 (2019)
11. Y. Wu, Y. Deng, J. Wang, Z. Zong, X. Chen, W. Gu, THz broadband absorber fabricated by EHD printing technology with high error tolerance. IEEE Trans. Tera. Sci. Tech. **9**, 637–642 (2019)
12. Z. Zhang, H. Zhang, K. Wang, Diffraction-free THz sheet and its application on THz imaging system. IEEE Trans. Tera. Sci. Tech. **9**, 471–475 (2019)
13. S. Canovas-Carrasco, R. Asorey-Cacheda, A.J. Garcia-Sanchez, J. Garcia-Haro, K. Wojcik, P. Kulakowski, Understanding the applicability of terahertz flow-guided nano-networks for medical applications. IEEE Access **8**, 214224–214239 (2020)
14. H. Liu et al., Dimensionality reduction for identification of hepatic tumor samples based on terahertz time-domain spectroscopy. IEEE Trans. Tera. Sci. Tech. **8**, 271–277 (2018)
15. S. Sung et al., THz imaging system for in vivo human cornea. IEEE Trans. Tera. Sci. Tech. **8**, 27–37 (2018)
16. T. Chavez, N. Vohra, J. Wu, K. Bailey, M. El-Shenawee, Breast cancer detection with low-dimensional ordered orthogonal projection in terahertz imaging. IEEE Trans. Tera. Sci. Tech. **10**, 176–189 (2020)
17. N. Pandit, R.K. Jaiswal, N.P. Pathak, Towards development of a non-intrusive and label-free THz sensor for rapid detection of aqueous bio-samples using microfluidic approach. IEEE Trans. Biomed. Circ. Syst. **15**, 91–101 (2021)
18. W. Gao, Y. Chen, C. Han, Z. Chen, Distance-adaptive absorption peak modulation (DA-APM) for terahertz covert communications. IEEE Trans. Wirl. Commun. **20** (2021)
19. N. Khalid, O.B. Akan, Experimental throughput analysis of low-THz MIMO communication channel in 5G wireless networks. IEEE Wirl. Commun. Lett. **5**, 616–619 (2016)
20. H. Elayan, O. Amin, B. Shihada, R.M. Shubair, M.S. Alouini, Terahertz band: the last piece of RF spectrum puzzle for communication systems. IEEE Opt. J. Commun. Soc. **1**, 1–32 (2020)
21. M.T. Barros, R. Mullins, S. Balasubramaniam, Integrated terahertz communication with reflectors for 5G small-cell networks. IEEE Trans. Veh. Tech. **66**, 5647–5657 (2017)
22. H. Elayan, C. Stefanini, R.M. Shubair, J.M. Jornet, End-to-end noise model for intra-body terahertz nanoscale communication. IEEE Trans. NanoBiosci. **17**, 464–473 (2018)

23. S. Sung et al., Optical system design for noncontact, normal incidence, THz imaging of in vivo human cornea. IEEE Trans. Tera. Sci. Tech. **8**, 1–12 (2018)
24. Z.D. Taylor et al., THz and mm-wave sensing of corneal tissue water content: electromagnetic modeling and analysis. IEEE Trans. Tera. Sci. Tech. **5**, 170–183 (2015)
25. H. Chen, Z. Wang, L. Zhang, Collaborative spectrum sensing for illegal drone detection: a deep learning-based image classification perspective. China Commun. **17**, 81–92 (2020)
26. R. Sarikhani, F. Keynia, Cooperative spectrum sensing meets machine learning: deep reinforcement learning approach. IEEE Commun. Lett. **24**, 1459–1462 (2020)
27. Y. Chu, J. Fei, S. Hou, Adaptive global sliding-mode control for dynamic systems using double hidden layer recurrent neural network structure. IEEE Trans. Neural Netw. Learn. Syst. **31**, 1297–1309 (2020)
28. H.M. Lynn, S.B. Pan, P. Kim, A deep bidirectional GRU network model for biometric electrocardiogram classification based on recurrent neural networks. IEEE Access **7**, 145395–145405 (2019)
29. P. Liu, Z. Zeng, W. Wang, Global synchronization of coupled fractional-order recurrent neural networks. IEEE Trans. Neural Netw. Learn. Syst. **30**, 2358–2368 (2019)
30. Z. Jing, J. Liu, M.S. Ibrahim, J. Fan, X. Fan, G. Zhang, Lifetime prediction of ultraviolet light-emitting diodes using a long short-term memory recurrent neural network. IEEE Elec. Dev. Lett. **41**, 1817–1820 (2020)
31. Y. Cheng, J. Wu, H. Zhu, S.W. Or, X. Shao, Remaining useful life prognosis based on ensemble long short-term memory neural network. IEEE Trans. Inst. Meas. **70**, 1–12 (2021)
32. R. Xin, J. Zhang, Y. Shao, Complex network classification with convolutional neural network. Tsin. Sci. Tech. **25**, 447–457 (2020)
33. G. Shomron, U. Weiser, Spatial correlation and value prediction in convolutional neural networks. IEEE Comp. Arch. Lett. **18**, 10–13 (2019)
34. T. Li, M. Hua, X. Wu, A hybrid CNN-LSTM model for forecasting particulate matter (PM2.5). IEEE Access. **8**, 26933–26940 (2020)
35. L. Ma, S. Tian, A hybrid CNN-LSTM model for aircraft 4D trajectory prediction. IEEE Access **8**, 134668–134680 (2020)
36. C. Ding, D. Tao, Trunk-branch ensemble convolutional neural networks for video-based face recognition. IEEE Trans. Patt. Anal. Mach. Int. **40**, 1002–1014 (2018)

Overview of THz Antenna Design Methodologies

K. Anusha, D. Mohana Geetha, and A. Amsaveni

Abstract The potential growth in the domain of wireless communication is boundless. Ultra-high speed with low-latency communications will be the foundation for the next-generation wireless systems. Henceforth, the futuristic massively interconnected wireless system immensely relies on terahertz (THz) spectrum devices. The THz band of frequency ranging between 0.1 and 10 THz will be primarily employed in 6G communication for enabling smart interconnections and high data rates. Due to the inherent characteristics of the THz spectrum, the modularity of the antenna is a miniaturized structure. Atmospheric attenuation and free space path loss of the signals in this spectrum are high due to its wave properties. There are many challenges in developing devices operating in the THz range. To overcome these downsides, the antenna designed for this spectrum must possess high gain and directivity. The conventional method of performance enhancement of antenna incorporate variation in material composition, structure, inclusion of defected ground, Photonic Band Gap (PBG) structure or meta-material superstrate. Only few of these methods will be applicable for THz antenna. The availability of antenna capable of providing ultra-wide band (UWB) capability in THz band will be a major bottleneck in the implementation of 6G wireless system. In recent years, research activities of THz antenna design methodologies have gained momentum due to high potential of the antenna in applications like ultra-fast short-range communication, medical imaging, and remote sensing. This chapter highlights the recent trends and the key breakthroughs in the antenna design in THz spectrum.

Keywords THz antenna · Meta-material

K. Anusha (✉) · A. Amsaveni
Kumaraguru College of Technology, Coimbatore, India
e-mail: anusha.k.ece@kct.ac.in

D. Mohana Geetha
Sri Krishna College of Engineering and Technology, Coimbatore, India

S. Das et al. (eds.), *Advances in Terahertz Technology and Its Applications*,
https://doi.org/10.1007/978-981-16-5731-3_19

1 Introduction

The potential growth of wireless communication is boundless. The next generation of 6G communication will necessitate the need of seamless, ultra-high-speed connectivity between the various subsystems. The frequency ranging between 0.1 and 10 THz is collectively identified as THz band [1]. Figure 1 depicts a simplistic representation of electromagnetic spectrum. The sub-THz region, far infrared, and near millimeter wave comprise of the frequencies from 0.1 THz to 0.3 THz [2]. The potential of this band was not explored till recent years. The band has been reported as THz gap due to the unavailability of materials and generators in this frequency spectrum [3]. The recent advancements in the field semiconductor technology have augmented the availability of the terahertz devices. This has steered a huge leap in the research activities in the THz band.

Present wireless systems are constrained with limited resource and capacity. THz band communication aids to overcome these short coming. THz spectrum will be primarily deployed in wireless communication for enabling a smart interconnection and realizing high data rates [4]. The futuristic massively interconnected wireless system will immensely rely on terahertz (THz) spectrum devices.

The non-ionizing radiation characteristics of the THz band make it a safer alternative for applications in which human tissues are involved [5]. In recent days, many applications in biomedical imaging are implemented in the THz band. This spectrum is also utilized in military applications and space explorations. The difficulties in fabrication and testing facilities in the THz band have been a major concern for development of devices in this spectrum. With the advent of innovative fabrication methods, the THz band will be able to support wide range of applications.

The significant contributions in this work are:

- Application of THz band.
- Investigation of the various structures and material used in THz antenna.
- Outline the challenges and research scope in THz antenna.

The chapter is structured as follows: Outline of features and applications of THz antenna is examined in Sects. 2 and 3, Sect. 4 discusses the existing material and

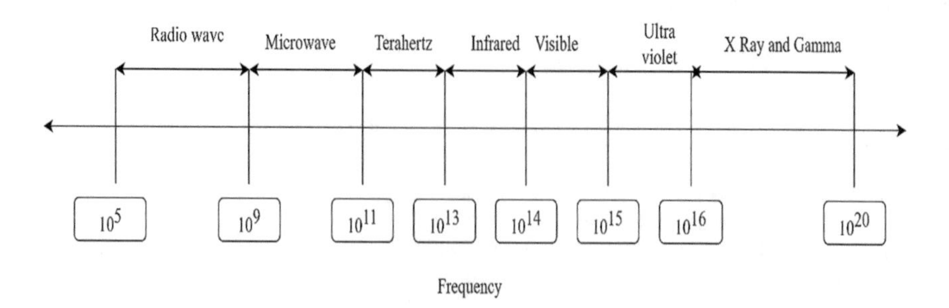

Fig. 1 Electromagnetic spectrum representation

structures deployed in THz antenna, Sect. 5 summarizes the challenges and research scope in THz antenna, and Section 6 focuses on the conclusion.

2 Features of THz Spectrum

THz spectrum is the sub-mm waves engulfed between microwaves and infrared light. The lower frequency spectrum is extensively used in wireless communication and the higher frequencies utilized in the optical domain [6–9]. The features of the THz bad are summarized below:

- *Penetration*: Due to the inherent properties at sub-mm wavelength, the amount of scattering observed in this spectrum scattering is very minimum. As a result, the depth of penetration is high. The waves exhibit an ability to penetrate through amorphous substances.
- *Spectroscopy*: The molecular vibrational modes of many materials are in this spectrum. Hence, the waves can be utilized to detect the nature of the material.
- *Resolution*: In general, the resolution of the image is inversely proportional to the wavelength of operation. Shorter the wavelength, better is the resolution. THz band provides better resolution than the microwave imaging systems.
- *Non-ionizing radiation*: The photon energy level associated with the spectrum is low leading to a non-ionizing property of radiation. When biological tissues are exposed to THz band, the hazardous effects are less in contrast to X-Ray imaging systems
- *Path loss*: The path loss associated with a wave is calculated using Friis transmission equation. Owing to the shorter wavelength, the THz band has a higher atmospheric and path loss. The path loss associated in the THz band varies with transmission distances.

The unique features of THz spectrum extend its usage in many real-time applications. The limiting factor in unleashing the potential in this band is the high path loss that can be overcome with proper design techniques [3]. The THz spectrum property of creating a natural resonance unique to the specific molecule is utilized for molecular detection. The non-ionizing property of the waves is used in many biomedical imaging systems.

3 Applications of THz Antenna

The wave characteristics of the THz band facilitate application in numerous domains, from imaging, communication, quality management, and security. The prominent application in the THz spectrum is illustrated in Fig. 2. Considering the prominence of THz antenna, various applications of THz spectrum associated to optoelectronic systems are discussed in detail in this section.

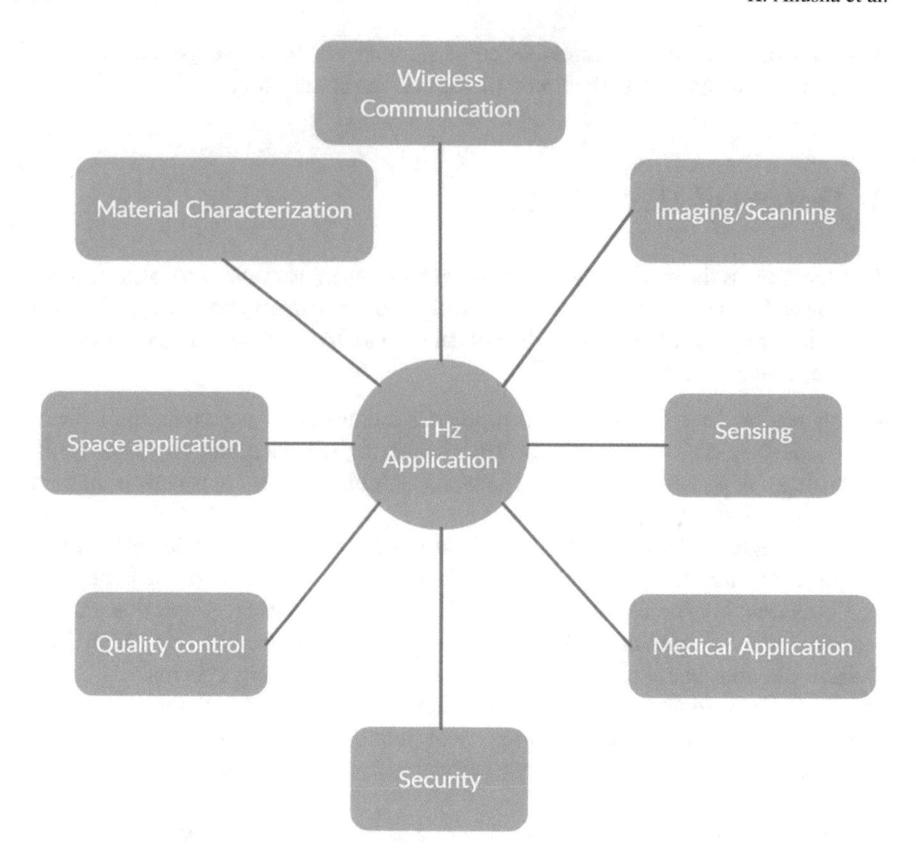

Fig. 2 THz application

3.1 Medical Application

The non-ionizing radiation characteristic of THz band makes it a significant player in medical imaging domain. The THz imaging system can be utilized for cancer cell detection, tissue identification, and for many diagnostic purposes. The better resolution and detection are achieved by employing nanotechnology in THz system. THz waves can penetrate up to a few 100 μm in human tissues and are utilized for medical diagnosis in detection of skin, breast, and mouth cancer [9, 10]. In dentistry, it is used for imaging for detection of dental erosion [11].

Conventional cancer detection causes enormous radiation hazards to the patients. An imaging system based on THz spectrum incorporating nanomaterial-based microstrip antenna provides better resolution at lower cost. A miniaturized antenna with low power levels of 3.58 mV and specific absorption rate (SAR) value of 3.8 W/Kg has been demonstrated that outperforms than the exiting imaging methods [10]. Breast cancer detection using THz imaging has been demonstrated in a simple

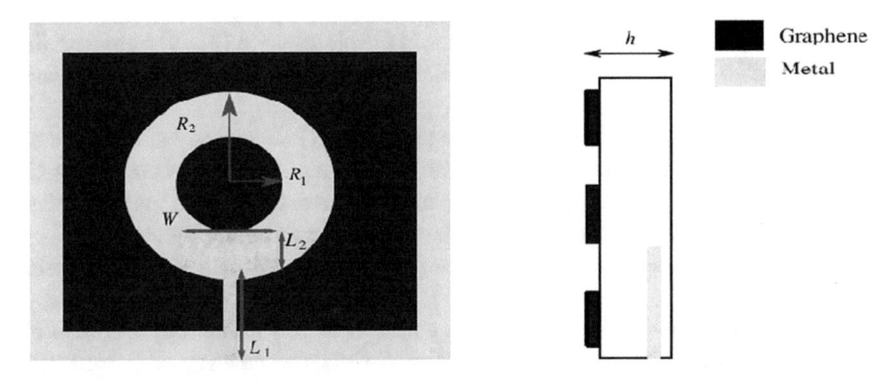

Fig. 3 Graphene-based reconfigurable patch antenna THz spectrum ([13], reproduced courtesy of The Electromagnetics Academy)

rectangular patch [10] and also by incorporating a liner array in a rectangular patch microstrip antenna [12].

The tunability of the imaging system can be improved by utilizing a reconfigurable antenna as shown in Fig. 3 [13].

In medical domain for certain disease diagnosis, clinical procedure of DNA and protein signature analysis are done. Recently, pulsed THz spectroscopy has been used for these clinical diagnosis procedures [14]. In clinical diagnosis, spectroscopy of the bacteria plays a vital role. Due to the penetration ability of the THz spectrum, the radiation can penetrate, and a high-resolution image can be obtained. Split ring resonator (SRR)-loaded meta-material structure implemented for *E-Coli* bacteria detection that provided high sensitivity is used for clinical diagnosis [15]. Pharma industry is an allied field of medical science; there is a need for quality assurance of the materials, chemical composition of the drugs. These quality control measures in the pharmaceutical industry can be implemented by deploying time domain spectroscopy in the THz frequency spectrum [16].

3.2 Wireless Communication

The requirement of high data rate communication in wireless communication domain has increased manifold in recent years. In a wireless environment, we can provide high data rates for short distance communication. The long haul applications necessitate higher power levels and bandwidth at the modulating frequency. The bandwidth of a wireless system is nominally 10% of the operating frequency. THz spectrum can provide an enhanced bandwidth and higher data rates than the existing standards [17].

The natural phenomenon of attenuation and molecular absorption losses that occur in the THz spectrum is a major constrain on the range of communication. Due to the size similitudes of water molecules and oxygen to the wavelength of THz band, results in absorption and scattering of the particles in the atmosphere [3]. To transcend the atmospheric loss, wireless communication antenna must be a high-gain directional antenna [18].

The next generation of wireless communication 6G technology stipulates data rates of 1 Tbps with a latency of 1 ms. Conventional antenna structures such as bowtie [7] and Yagi uda [19] are reengineered as nanoantennas and deployed for application requiring high directional connectivity. A bowtie nanoarray configuration as in Fig. 4 [7] is fabricated using electron beam lithography technique, an metal–insulator–metal (MIM) diode.

Ultra-massive multiple input multiple output (MIMO) technology [20] can be incorporated to provide the required capacity. A nanoantenna array fabricated with graphene material is widely used in many applications and also provides tunability characteristics [20]. In recent times, multilayered antenna loaded with meta-material on different substrate are used to render seamless connectivity [21]. An array antenna utilized for wireless communication in THz spectrum is shown in Fig. 5 [22].

The wireless scenario is certain instances involves beam steering which can be implemented using reconfigurable antenna. The reconfigurable antenna based on meta-surfaces offer better performance in the THz band. To summarize for high data rate communication plasmon-based antenna, massive MIMO, multilayered graphene/Teflon-coated antenna, SRR integrated meta-material antenna, tapered array structure, PBG, defective ground structure (DGS), electronic band gap (EBG) incorporated planar antenna have been used utilized. In recent years, meta-superstrate and substrate variants are also examined for better gain and minimization of radiators in the THz frequency spectrum.

Fig. 4 Bowtie nanoarray antenna structure

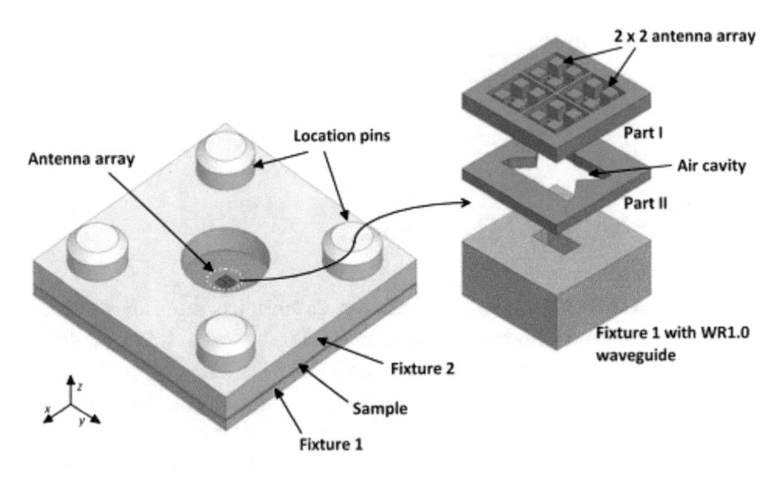

Fig. 5 Antenna array for wireless communication (reproduced courtesy Luk et al. [22])

3.3 Security

The THz frequency region of the electromagnetic spectrum (i.e., below 1 THz) is employed in security applications primarily for identifying hidden hazardous objects. The barrier materials that conceal deadly things carried by potential terrorists are typically clothing that are semi-transparent to THz radiation. It is also necessary to analyze the contents of various sorts of shipments delivered through any modes of freight transit.

The non-hazardous scanning system can be used to detect real-time concealed threat detection. Military applications need high security fool proof communication systems. The several applications in the military have diverse requirements in terms of security and coverage area [23]. At higher frequencies, narrow beam width can be obtained using high-gain directional antenna providing a jam resistant network.

THz imaging can be provided greater security with 2D and 3D imaging schemes. It can be used to detect tiny layers of powder and individual sheets inside an envelope [24]. THz spectroscopy allows the detection of chemical signals even when they are enclosed within a packet or disguised in clothes.

3.4 Space Application

The THz signals in the space are atmosphere-free environment, and the atmospheric attenuation can be neglected. Research in THz technology for space application has gained momentum with the advent of electronic devices operating in this band. Research in the development of sensors and instruments in astronomy in the THz band has been carried out in NASA for past 2 decades [25]. High-gain antennas can provide higher data rates for the uplink and downlink. A massive MIMO antenna in space application can create high-gain narrow beam between the transmitter and receiver.

4 Salient Features of THz Antenna

Depending on the coverage range and terrain conditions, the THz-enabled applications can be classified into nano and macroscale networks. Most of the medical applications are implemented using nanosensors and the data communicated over a wireless network. Macroscale network is crucial in wireless communication, radars, and military security applications [17, 18, 23, 24]. By nature, the resolution of THz imaging is higher and provides better quality images. The antenna design basically depends on the application, coverage area, power requirement, and the layout.

The fundamental necessity for applications in the THz spectrum is ultra-wideband and multiband antenna [3]. UWB antenna requires that the fractional bandwidth of antenna to be more than 50%. The directionality property of the antenna depends on the specific application. Integration of multiband antenna in certain applications enhances the performance and functionality of the system [26]. For next generation, wireless application UWB with circular polarization is the most feasible antenna. In general, the antenna must be compact size and low cost. To counterpart the high propagation losses in THz spectrum, the antenna needs to be a designed with higher gain for long haul applications [20].

To achieve high performance of THz band, antenna requires a complete reevaluation of the conventional antenna designing practices. There is significant increase in ohmic losses with an increase in the frequency range. At higher frequency, the conventional perfect electrical conductors-based approach becomes redundant. At nanoscale dimensions, the existence of surface plasmon polaritons waves is dominant. As a result, the THz spectrum antenna design necessities an understanding of the dimensional nanoscale characteristics, and the design concepts must be revisited for better designed structures [19].

The performance of the application relies on the efficacy of the antenna parameters. The conventional antenna design in the RF region is still regarded as prospective research area. THz antenna is less explored territory, and there are lot of challenges and research scope in this domain. Innovation in antenna design with comprehensive knowledge of the underlaying physical processes in the THz spectrum will pave for better utilization of the resources. A detailed discussion on the materials and structures deployed in THz antenna is presented in this section.

4.1 *THz Antenna Materials*

The lack of material resources capable of operating in this spectrum has stifled the growth of THz communication for several years. The availability of sources, detectors, and measuring devices is critical to establishing communication. The dielectric strength of the substrate determines the performance of the antenna [27]. In general, the dielectric constant of the material is determined from Drude model [7] and is stated as in Eq. 1:

$$\varepsilon(\omega) = \frac{\epsilon_{\infty} - {\omega_p}^2}{\omega^2 - j\omega_{\tau}} \quad (1)$$

ϵ_{∞}—Contribution of the bound electrons to the relative dielectric constant, ω_p—Plasma frequency,ω_{τ}—Damping frequency.

The lower conductivity of the metals at higher frequencies increases the field penetration resulting in degradation of the radiation efficiency [28]. Also, THz band suffers large path loss and selection of material with minimum propagation loss is a key factor in the design of antenna. Copper is by far the most commonly used material in antenna fabrication. Copper cannot be used in the THz band due to ohmic losses and wave properties. This necessitates the investigation of materials suitable for use in the THz band. This section examines the various elements that can be used in THz antennas.

4.1.1 Gold (Au)

Gold is chemically inert metal with very high conductivity, lower skin depth and exhibits plasmonic characteristics in THz frequency band. THz sensors developed using gold layer deposits have been reported in many literatures [29]. The use of gold also aids in the creation of corrosion-resistant and chemically stable antennas. The fabrication can be done using electron beam lithography [30]. A double spiral antenna structure fabricated using electron beam lithography is shown in Fig. 6 [30].

An array of gold planar inverted cone antenna (PICA) demonstrated has a bandwidth of 37.9%, is fabricated with metal deposition techniques, and has been demonstrated in the frequency range of 0.75–1.10 THz [31]. The performance indicators of the radiator like the radiation efficiency and gain are also in the satisfactory range.

Ultra-wideband capability is also demonstrated in the THz frequency band using gold with deposition of other elements. Additional layer of perovskite material laid over the gold has a higher radiation characteristic [32]. The hybrid antenna with perovskite antenna array operated in the band of 0.9–1 THz with a gain of 11.4 dBi. Appropriate preparation process also aids to achieve better electrical properties of the fabricated antenna. These materials have the characteristics that make them a viable option for next-generation wireless systems that need to be miniaturized, wearable, and reconfigurable.

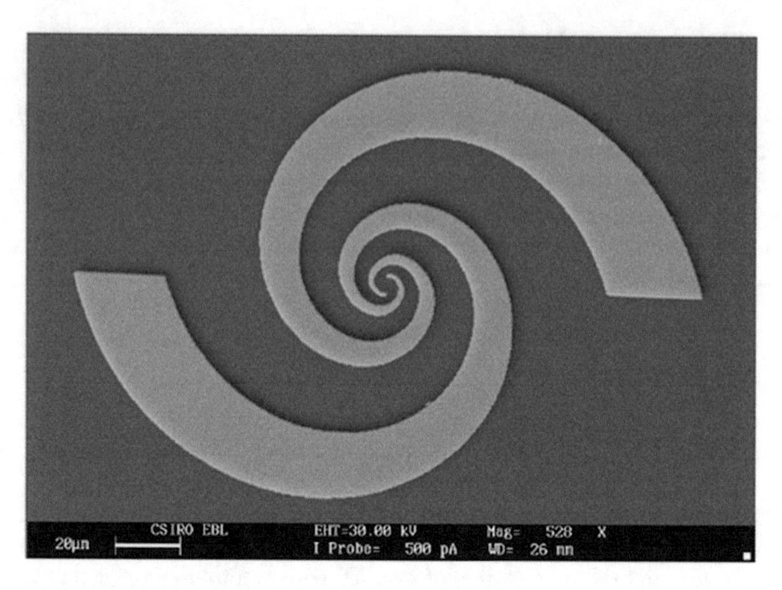

Fig. 6 Double spiral antenna fabricated by electron beam lithography ([30], reproduced courtesy of The Electromagnetics Academy)

4.1.2 Graphene

Graphene's molecular structure is in the shape of a honeycomb hexagonal lattice. High electron mobility, better thermal and electrical conductivity, imperviousness, and a large surface area are unique properties associated with graphene [33]. These exclusive structures of graphene make it an ideal solution for THz antenna fabrication. The complex surface conductivity of graphene can be calculated using Kubo formula, and using chemical doping, it can be controlled in the fabrication process. The total surface conductivity is the sum of inter and intra-surface conductivity represented as in Eq. 2

$$\sigma_{\text{Total}} = \sigma_{\text{Inter}} + \sigma_{\text{Intra}} \tag{2}$$

The expression for surface conductivity in the terahertz band in terms of inter-band is calculated in Eq. 3

$$\sigma_{\text{inter}}(\omega) = \frac{e^2}{4h}\left(H\left(\frac{\omega}{2}\right) + i\frac{4\omega}{\pi}\int_0^{\infty}\frac{H(\varepsilon) - H\left(\frac{\omega}{2}\right)}{\omega^2 - 4\varepsilon^2}\partial\varepsilon\right) \tag{3}$$

$H(\varepsilon)$ is stated as Eq. 4

$$H(\varepsilon) = \frac{\mathrm{Sin}h\left(\frac{h_{\varepsilon}}{K_B T}\right)}{\mathrm{Cos}h\left(\frac{\mu_c}{K_B T}\right) + \mathrm{Cos}h\left(\frac{h_{\varepsilon}}{K_B T}\right)} \quad (4)$$

The intra-band term conductivity is modeled as in Eq. 5

$$\sigma_{\mathrm{Intra}}(\omega) = \frac{2e^2 K_B T_i}{\pi h^2(\omega + i\tau^{-1})} \mathrm{In}\left(2\mathrm{Cos}h\left(\frac{\mu_c}{2K_{\mathrm{B}} T}\right)\right) \quad (5)$$

where e represents electron charge constant, T the temperature, ω the angular frequency, K_{B} the Boltzmann, h the reduced Planck's constant, μ_c the chemical potential, and τ the relaxation time. In the THz spectrum, inter-band conductivity is negligible, and the surface conductivity can be represented as in Eq. 6

$$\sigma(\omega) = \frac{2e^2 K_B T_i}{\pi h^2(\omega + i\tau^{-1})} \mathrm{In}\left(2\mathrm{Cos}h\left(\frac{\mu_c}{2K_B T}\right)\right) \quad (6)$$

For higher-order frequencies, it can be approximated as in Eq. 7

$$\sigma_0 = \frac{\pi e^2}{2h} \quad (7)$$

For THz antenna, simulation based on graphene is considered as material with thickness Δ, and the plasma frequency is characterized as in Eq. 8

$$\omega_P = \sqrt{\frac{2e^2 K_B T}{\Delta \pi \varepsilon_0 h^2} \mathrm{In}\left(2\mathrm{Cos}h\left(\frac{\mu_c}{2K_B T}\right)\right)} \quad (8)$$

Due to the effect of the imaginary portion of graphene conductivity, the transverse electric (TE) and transverse magnetic (TM) mode propagation of surface plasmon polariton (SPP) waves propagation exit in the terahertz band [34]. In terms of the effective index, the SPP dispersion relationship for TM mode can be described as Eq. 9

$$\sqrt{n^2 - n_{\mathrm{eff}}^2} + n^2\sqrt{n^2 - n_{\mathrm{eff}}^2} + \frac{4\pi}{c}\sigma_{\omega}\sqrt{1 - n_{\mathrm{eff}}^2}\sqrt{n^2 - n_{\mathrm{eff}}^2} = 0 \quad (9)$$

where n_{eff} represents the complex refractive index, c is Speed of light, and σ_{ω} is frequency dependent graphene surface conductivity. The SPP wave propagation determines the resonant properties of graphene.

The coupling of incident EM radiation with SPP modes induces resonance, the resonant condition depends on the length of graphene L, and m is the number of resonant modes is defined as per Eq. 10

$$L \approx m\frac{\lambda_{\text{spp}}}{2} \approx m\frac{\pi}{Re\{K_{\text{spp}}\}} \tag{10}$$

where mode wavenumber k_{spp} is determined by from the free space wave number k_o using the relation as in Eq. 11.

$$k_{\text{spp}} = k_o n_{\text{eff}} \tag{11}$$

In the THz spectrum, graphene is employed to implement ultra-wide band antennas [35]. In the initial works, graphene-based THz antennas using different substrates like silicon, silicon dioxide, and quartz FR-4 have been analyzed for THz spectrum utilization [35]. Silicon-based substrate had better radiation performance for graphene-based THz antennas.

The antenna gain increases with the additional superstrate layer. The addition of high dielectric constant superstrate layer increasing the gain at the cost of reduction in bandwidth. The integration of a Teflon superstrate to a graphene circular radiating patch provided environmental protection and improved antenna performance [36].

High-gain and UWB antenna with single element hexagonal-shaped graphene patch antenna implemented with photonic crystal and dielectric grating is reported to have a bandwidth of 9.552 THz [26]. The variation in the shape, periodicity, and the dimensions of photonic crystals do not have a major impact in the antenna performance.

Modern era reconfigurability has become a needed feature in antenna for many applications [37]. Reconfigurable characteristic is possible by using graphene load at THz band. The antenna gain also can be improved in the reconfigurable graphene-loaded patch antenna. Increase in the graphene chemical potential causes a shift in the antenna resonant frequency but there is drastic increase in the gain [37]. Split ring resonators exhibit negative permeability below plasma frequency. In conventional antenna, inclusion of improves the antenna radiation characteristics. Addition of the meta-material layer in a graphene-loaded patch results in improved gain and generated more linear characteristics of graphene. The features of graphene are utilized in many medical imaging applications in the THz spectrum.

4.1.3 Carbon Nanotubes

Carbon nanotubes (CNT) are one-dimensional molecular structure constructed by rolling graphene sheet into cylinder. Depending on the sheets of graphene utilized, it can be classified into single-walled carbon nanotube (SWCNT) and multiwalled carbon nanotube (MWCNT), doped CNT, bundled CNT (BCNT), and hybrid CNT. The electrical properties of CNT are determined by the material's dimension and edge geometry. The skin effect of carbon bonds in CNT becomes highly negligible at the THz frequency band.

The properties of BCNT are scaled version of single CNT and vary on the number of sheets that are over laid. The radiation metrics of the CNT antenna are based on various factors such as the anisotropic surface resistivity, density of the layers, and its wavelength scaling factor. BCNT is a smart alternative for THz and optical frequencies as it has a higher current density. An efficient antenna THz frequency range requires the density of BCNTs to be more than 10^3 (CNTs/μm) [38].

Forming a composite material of CNT and metal, the characteristics of the material changed significantly. The integration of a dipole antenna with the CNT composite material resulted in a performance enhancement [39]. The new CNT antenna had a better efficiency in the THz frequency band with each double-walled carbon nanotube (DWCNT) is surrounded by another material jacket as depicted in Fig. 7 [39]. The graphene-coated jacket of DWCNT decreased the resonant frequency, whereas the copper-coated DWCNT increased the resonant frequency [39]. The radiation performance of the conventional material used in antenna fabrication can be incorporated with the double-layer structures for operation in THz region.

The surface conductivity of the fabricated structure depends on the material composition. The surface conductivity of the single CNT for small radii can be approximated as given Eq. 12

$$\sigma_{\text{cnt}} \cong -j\frac{2{e_0}^2 \nu_F}{\pi^2 h\left(\omega - \frac{j}{\tau}\right)} \tag{12}$$

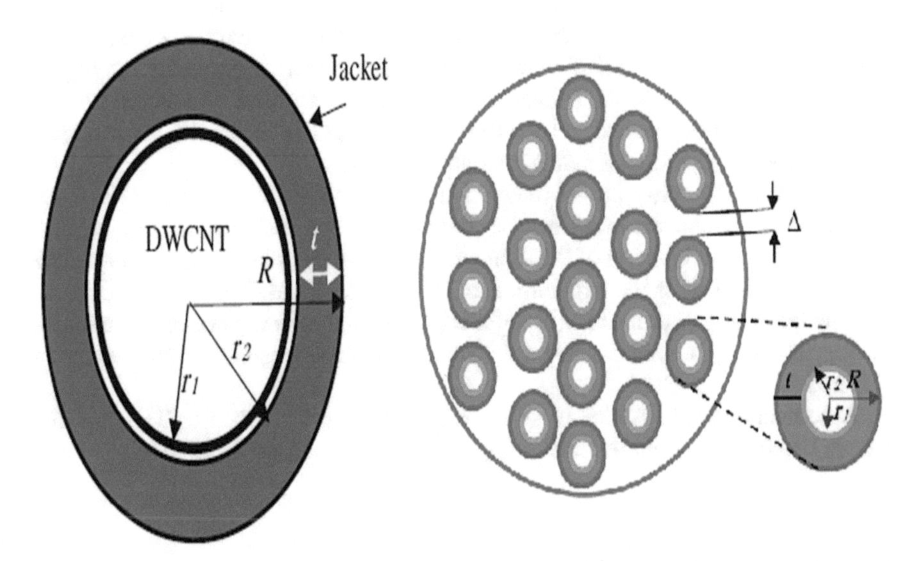

Fig. 7 Double-walled carbon nanotube composite structure ([39], reproduced courtesy of The Electromagnetics Academy)

Table 1 Comparison of material properties

Parameter	Copper	Gold	Graphene	CNT
Tensile strength	587 MPa	124 MPa	1.5 TPa	5–500 GPa
Thermal conductivity (W m^{-1} K^{-1})	400	310	5000	3000
Density (g cm^{-3})	8.92	19.3	2.267	1.40
Electronic mobility (cm^2 V^{-1} S^{-1})	32	42.6	2×10^5	8×10^4
Current density (A cm^{-1})	10^6	10^6	10^9	10^9
Velocity of electron (m S^{-1})	1.57×10^6	1.40×10^6	10^6	10^6

where e_0 is the electron charge, v_F is the Fermi velocity of graphene, and τ is the relaxation time.

The effective conductivity of SWCNT composite material depends on the number of coating material and their conductivity. The effective conductivity with k different materials is given by Eq. 13

$$\sigma_{\text{composite}} = \sum_{j=1}^{k} m_j \sigma_{zj} \tag{13}$$

where m_j represents the volume fraction of the material.

The CNT structures support lower modes of wave propagation resulting in size reduction of the antennas. For the same dimensions, the resistivity of a single CNT is lower than that of a strand of gold resulting in lower fabrication costs. The higher electrical conductivity of CNT gives a greater edge as a choice of material in THz spectrum. Due to its physical and electrical properties, CNT makes a suitable choice in many THz application including wearable antenna.

Apart from the conventional materials used in antenna design, there has been exploration of composite materials for efficient radiation characteristics in the THz spectrum. The combination of the elements creates a variation in the chemical and electrical properties of the structure. From the above discussion, we can conclude that the material properties have a major impact on the performance of the antenna. A comparison of the material properties suitable for THz antenna design is stated in Table 1.

A comparative analysis of the various research works related to THz antenna design based on different material composition is summarized in Table 2.

4.2 *THz Antenna Structures*

The physical dimension and the structural layout of the antenna govern the operational performance of the system. By varying the geometrical attributes, the performance parameters of the antenna such as bandwidth, gain, and directivity can be

Table 2 Comparison of materials-based THz antennas

Ref. No	Year	Structure	Operating frequency	Substrate	Dimension	Gain	Other significances
[32]	2020		0.9–1.2 THz	Polyethylene naphthalate		11.4 dBi	• A hybrid multilayer structure • Performance of composite materials (perovskite, gold and combination of the above) is analyzed
[40]	2021		0.96 and 1.12 THz	Rogers RO4350B p-type doped silicon Silicon dioxide	144 × 72 mm^2		• Independent beam steering dual-band THz antenna incorporating hexagonal active FSS • Frequency reconfigurability is demonstrated with composite multilayer structure
[41]	2021		0.45–0.50 THz	GaAs	0.8 × 0.8 × 0.13 mm^3	6.5 dBi	• A antenna on chip (AoC) based on meta-surface and substrate integrated waveguide (SIW) technologies • Layers arrangement: Cu–SiO_2–Cu–SiO_2–Al–GaAs–Cu • Efficiency: 65% • Fractional Bandwidth: 10.5 GHz (450–500 GHz)

(continued)

Table 2 (continued)

Ref. No	Year	Structure	Operating frequency	Substrate	Dimension	Gain	Other significances
[26]	2020	Multilayer PBG	0.448–10 THz	Teflon	200 × 260 μm^2	21.22 dB	• A UWB graphene-based hexagonal-shaped multilayer structure using of photonic crystal and dielectric grating • Carried out the analysis of effect of varying the shape and airgap of PBG • Impedance bandwidth: 9.552 THz • Efficiency: 93.23% • Minimum Directivity: 21.53 dB
[42]	2019	Graphene-based patch	1.1–1.25 THz	Graphene		5-7-6-5dBi	• Four different antenna design based on reflector model is analyzed • The parameters of antenna are regulated by varying the chemical potentials of the graphene • Reflector-transmission window model and reflector-director model are discussed

enhanced. Addition of slots, layers, and defects in ground plane leads to multiband operation features [27]. The prominent THz antenna structures and their relevance is discussed in this section.

Several high-gain antennae have been reported in the literature in the THz band. In the initial phases, the traditional antenna structures were adapted for operating in the higher frequency. Scaling conventional antenna structures to the terahertz range leads in suboptimal designs.

- *P. Neikirk* et al. [43] demonstrated an imaging application in the THz band incorporating lens antennas arrays.
- *Nagatsuma* et al. [44] implemented a short-range communication link around 120 GHz was using planar dipole, slot antenna, and a silicon lens.
- *Tomohiro Seki* et al. [21] presented a high-gain multilayers antenna on a Teflon substrate for wireless communication.
- *Meng Li* et al. [45] demonstrated a miniaturized meta-material based antenna using SRR with gain of 2.25 dB and VSWR 1.143 was for wireless communication [45].
- *G. Singh* [46] enhanced frequency reuse at THz frequency was using 2D electromagnetic crystal substrate on the rectangular microstrip patch radiator. The patch enhanced with a substrate had a higher gain of 8.248 dB at 852 GHz.
- *Younssi* et al. [47] presented a study on the performance metrics of a RMSA operating at 0.6 to 0.8 THz with and without superstrate. The addition of the superstrate enhanced the radiation efficiency of the antenna. A gain of 10.43 dBi at 0.6929 THz reported makes it an optimal choice for long haul communication.
- *Akyildiz* et al. [20] proposed ultra-massive MIMO (1024 $\times$ 1024) communication in the THz band using graphene-based THz plasmonic antenna. For the lower frequency band, meta-material-based nanoantennas and the upper band deploys graphene material.
- *S. Singhal* [48] depicted an elliptical-shaped antenna on a polyamide substrate with a defective ground plane. The structure radiated an omnidirectional radiation with a 12 dB peak gain and 5 THz impedance bandwidth.
- *Kushwala* et al. [49] proposed a linear scaling of patch antenna from GHz to THz band using CSRR with a bandwidth of 1 THz
- *Goyal* et al. [50] analyzed the impact of variation in dimension of patch and PBG structure. The antenna has an impedance bandwidth: 36.23 GHz. Photonic crystals are incorporated to improve the performance.
- *Zhou* et al. [51] presented a reconfigurable rectangular patch antenna on a glass substrate with an efficiency of 66.71% and bandwidth of 79.01 GHz. The structure has a peak gain of 5.08 dB at 1 THz

A comparative analysis of the various research works related to THz antenna design based on different antenna structures is summarized in Table 3.

Table 3 Comparison of THz antenna structures

Ref. No	Year	Structure	Operating frequency	Substrate	Dimension	Gain	Other significances
[52]	2019		1.02 THz	Quartz	180 × 212 × 10 μm^3	5.75 dB	• A miniaturized meta-material-based rectangular patch antenna with SRR • The impact of return loss is analyzed for variation in SRR width and thickness
[53]	2020		23–40 GHz	Roger 5880	80 × 80 × 1.57 mm^3	12 B	• MIMO antenna with four different arc-shaped stripes • Impedance bandwidth: 23 GHz to 40 GHz • Mean Effective Gain <3 dB • Envelope correlation coefficient <0.5 • Isolation >20 dB • Efficiency >70 %
[54]	2020		12 THz	RT6010LM	24 × 28 × 21.6 mm^3	14.7 dB	• High-gain multilayer antenna used rectangular meta-surface for gain enhancement • Operating bandwidth: 8–13 THz • Gain enhancement: 10.5 dB more than the conventional radiating structure

(continued)

Table 3 (continued)

Ref. No	Year	Structure	Operating frequency	Substrate	Dimension	Gain	Other significances
[55]	2020	Leaf-shaped recursive structure	1.12–10.02 THz	Polyamide	600 × 500 μm^2	18.66 dBi at 7.5 THz	• Utilized trigonometric concept • Six sinusoidal tapered leaves arranged in a plant shape (nature inspired structure) • Bandwidth: 8.9 THz
[56]	2020	Inverted K shape	8.8 THz	Polyamide	600 × 600 μm^2	22.1 dB	• An inverted K-shaped structure • DGS and inclusion of slots for bandwidth enhancement • Wide impedance bandwidth: 0.46–8.84 THz

5 Challenges and Research Scope

The data rates requirements using wireless communication are increasing manifold with each passing day. The unleashed capacity of THz frequency spectrum is the most feasible option to envision the high data rates. Additional band to the existing operational spectrum enhances bandwidth capacity of the wireless system allowing the user to navigate the applications at much higher data rates. THz communication is in the tethering stage, and there are several challenges exist in the implementation of this technology. With the increase in frequency, the attenuation of the signal increases rapidly [3]. For better efficiency, the THz systems must avert the atmospheric path loss encountered in this spectrum and require a robust system design. The antenna design, fabrication, and validation issues are discussed in this section.

5.1 Antenna Design Parameters

For application in the RF band, a high-gain narrow band or wide band moderate gain antenna are utilized. The antenna design is focused on achieving the optimal parameters of gain, band width, size, and cost. Achieving a combination of these parameters generally results in conflict of interest.

The bandwidth achieved by a radiating structure depends on the size of the antenna. Ultra-wide band antenna leads to a larger size of the antenna [57]. Conventional antennae achieve larger bandwidth with an increase in dimensions of the structure that in turn increases the fabrication cost. In general, narrow band antenna have high gain while ultra-wideband antenna has low gain. THz spectrum yields smaller dimension structure, and gain can be improved with the increase in aperture. Achieving a wide band characteristic with high directivity is not possible without an enlargement in the dimensions of the antenna. Therefore, in compact low-profile antenna, there is an always a compromise with the directivity and wideband attained.

In THz antenna, we need more stringent features which are a combination of high gain, ultra-wide band/multiband functionality, small size and conformal. Designing an antenna incorporating with all these performance metrics demands a more innovative design procedure.

5.2 Antenna Fabrication

5.2.1 Fabrication Material

Conventional antennas are fabricated using substrate like FR-4, Roger and similar dielectric polymers. For THz band, the loss associated with these materials is high causing a decrease in gain of antenna. Deficient matching of substrate material can lead to inefficient radiation. The challenge is to enhance to conventional elements to match the THz band or discover alternate material to meet the requirements.

The design of antenna using composite material combining already proven materials like copper with graphene/Teflon leads to new innovative materials. These structures achieve better performance than single element radiators [32]. Variants of CNT materials with additional layers improve the radiation characteristics and enable dimension reduction [39]. The creation of new materials will revolutionize the THz antenna utilization.

For THz antenna, graphene is a promising material for simple antenna shapes. Complex structures and 3D fabrication are yet to be explored [35]. There is very minimal measured statistics for analyzing the performance of graphene. The low refractive index and absorption properties of polymer materials in the THz band help to achieve enhanced bandwidth and efficiency [58]. Meta-materials are another potential element for THz antenna [48]. For application requiring beam steering features, reconfigurable meta-surface is a viable solution. The addition of superstrate layer improves the performance. The material compatibility of the different layers is also an area to be explored.

5.2.2 Fabrication Technology

Due to their shorter wavelength, THz antenna is smaller in dimensions causing difficulties in the fabrication process. At higher frequencies, the texture of the material varies. THz spectrum antenna must have a smooth surface finish that can be fabricates only with precision machinery with low tolerance. These specifications cannot be met with the existing technology used for conventional antenna fabrication. The fabrication cost depends on the material and shape of the antenna. For wide usage of the antenna, the cost must be minimum.

The existing manufacturing approaches for THz band reported in literature are printed circuit broad (PCB) technology, 3D printer technology [34], micromachining microelectromechanical system (MEMS) technology, focused ion beam technology, and nanofabrication technology [59–61].

At present, the THz fabrication is limited to 1 THz. More unique and robust fabricating methods are required for high precision THz antenna. Availability of affordable, reliable high precision fabrication techniques will facilitate for more utilization of THz antenna.

5.3 THz Measurement

There are many reported study on the simulation of high-performance THz antenna; however, the fabrication and measurement are major concerns in the antenna validation [62]. The measurement of antenna performance metrics is done in the far field region by providing uniform illumination of the antenna. It is more convenient to

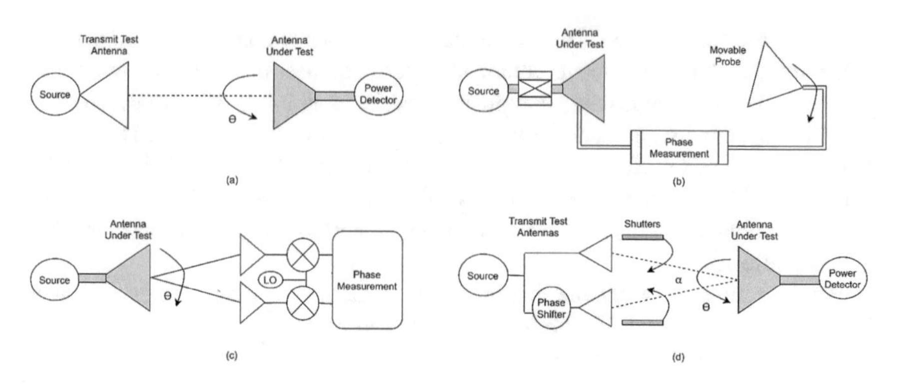

Fig. 8 Antenna measurement **a** Power pattern measurement reception. **b** Amplitude measurement. **c** Phase measurement. **d** Interferometric amplitude and phase measurement

measure the metrics by featuring the test antenna as a receiver configuration. By virtue of reciprocity theorem, the parameters are identical for transmitter mode. In THz spectrum antenna measurement technique, it is more expedient to model the antenna under test (AUT) in reception rather than transmission. The power measurements are more prominently done in this spectrum.

In general, the AUT is rotated in the receiving mode to capture the radiation measurements [63]. The arrangement can be used for amplitude and phase measurement as shown in Fig. 8 [63].

A compact test range (CATR) based on geometric optics has been validated for testing horn antenna in the THz range [64]. For testing micromachined-based components, a two hologram-based antenna test ranges have been demonstrated.

There is a dearth in the availability of measurement equipment for validation of THz antenna due to availability of measurement equipment, such as vector network analyzer operating in the entire spectrum of THz band. The measurement facilities in the THz band are still an evolving research area by itself.

5.4 Future Research Scope

The salient research scope in THz antenna is briefed below:

- Redesign the conventional metals antenna to suit THz band.
- Optimize the antenna geometry UWB operation with high gain and directivity.
- Develop antenna using graphene/meta-material/meta-surfaces.
- Design low-cost miniaturized antenna for specific applications.
- Create innovative fabrication technologies.
- Build testing range for THz band.
- Integrate THz antenna on a chip.

6 Conclusion

THz band is a prospective solution for ultra-high-speed communication in future wireless applications. In this chapter, a comprehensive review of the THz band antenna is presented. The salient features and applications of THz band are discussed. The antenna design in THz band based on material and antenna structure available in literature is presented. The difficulties in antenna fabrication and measurement available in the THz band are highlighted. Finally, we have put forward some open research issues in THZ antenna design.

References

1. M.Z. Chowdhury, M. Shahjalal, S. Ahmed, Y.M. Jang, 6G wireless communication systems: applications, requirements, technologies, challenges, and research directions. IEEE Open J. Commun. Soc. **1**, 957–975 (2020). https://doi.org/10.1109/ojcoms.2020.3010270
2. Technology Trends of Active Services in the Frequency Range 275–3000 GHz (2015). https://www.itu.int/dms_pub/itu-r/opb/rep/R-REP-SM.2352-2015-PDF-E.pdf
3. I.F. Akyildiz, J.M. Jornet, C. Han, Terahertz band: next frontier for wireless communications. Phys. Commun. **12**, 16–32 (2014). https://doi.org/10.1016/j.phycom.2014.01.006
4. Z. Hajiyat, A. Ismail, A. Sali, M.N. Hamidon, Antenna in 6G wireless communication system: specifications, challenges, and research directions. Optik **231**, 166415 (2021). https://doi.org/10.1016/j.ijleo.2021.166415
5. P. Siegel, Terahertz technology. IEEE Trans. Microw. Theory Tech. **50**(3), 910–928 (2002). https://doi.org/10.1109/22.989974
6. N. Horiuchi, Terahertz technology: endless applications. National Photon. **4**(3), 140–140 (2010)
7. A.M.A. Sabaawi, C.C. Tsimenidis, B.S. Sharif, Planar bowtie nanoarray for THz energy detection. IEEE Trans. Terahertz Sci. Technol. **3**(5), 524–531 (2013). https://doi.org/10.1109/tthz.2013.2271833
8. Y. Kadoya, THz wave propagation on strip lines: devices, properties, and applications, in *2007 19th International Conference on Applied Electromagnetics and Communications* (2007). Published. https://doi.org/10.1109/icecom.2007.4544441
9. D. Woolard, R. Brown, M. Pepper, M. Kemp, Terahertz frequency sensing and imaging: a time of reckoning future applications. Proc. IEEE **93**(10), 1722–1743 (2005). https://doi.org/10.1109/jproc.2005.853539
10. A. Srithar, K. Ruby, C. Manickam, C. Mekala, Terahertz imaging patch antenna for cancer diagnosis applications. Int. J. Appl. Eng. Res. **10**, 15232–15236 (2015)
11. D. Crawley, C. Longbottom, V.P. Wallace, B. Cole, D. Arnone, M. Pepper, Three-dimensional terahertz pulse imaging of dental tissue. J. Biomed. Opt. **8**(2), 303 (2003). https://doi.org/10.1117/1.1559059
12. M.V. Hidayat, C. Apriono, Design of 0.312 THz microstrip linear array antenna for breast cancer imaging application, in *2018 International Conference on Signals and Systems (ICSigSys)* (IEEE, 2018), pp. 224–228
13. H. Ben Krid, Z. Houaneb, H. Zairi, Reconfigurable graphene annular ring antenna for medical and imaging applications. Progress Electromagn. Res. M. **89**, 53–62 (2020). https://doi.org/10.2528/PIERM19110803
14. D.F. Plusquellic, K. Siegrist, E.J. Heilweil, O. Esenturk, Applications of terahertz spectroscopy in biosystems. ChemPhysChem **8**(17), 2412–2431 (2007)
15. L. Yu, L. Hao, T. Meiqiong, H. Jiaoqi, L. Wei, D. Jinying, Z. Yang, et al., The medical application of terahertz technology in non-invasive detection of cells and tissues: opportunities and challenges. RSC Adv. **9**(17), 9354–9363 (2019)

16. Y.C. Shen, Terahertz pulsed spectroscopy and imaging for pharmaceutical applications: a review. Int. J. Pharm. **417**(1–2), 48–60 (2011)
17. T. Nagatsuma, Photonic generation of terahertz waves for communications and sensing, in *Proceedings of the 2nd International Conference on Telecommunications and Remote Sensing* (2013). Published. https://doi.org/10.5220/0004785000430048
18. K. Tekbıyık, A.R. Ekti, G.K. Kurt, A. Görçin, Terahertz band communication systems: challenges, novelties and standardization efforts. Phys. Commun. **35**, 100700 (2019)
19. J. Li, A. Salandrino, N. Engheta, Shaping light beams in the nanometer scale: a Yagi-Uda nanoantenna in the optical domain. Phys. Rev. B **76**(24), 245403 (2007)
20. I.F. Akyildiz, J.M. Jornet, Realizing ultra-massive MIMO (1024×1024) communication in the (0.06–10) terahertz band. Nano Commun. Networks **8**, 46–54 (2016)
21. T. Seki, N. Honma, K. Nishikawa, K. Tsunekawa, Millimeter-wave high-efficiency multilayer parasitic microstrip antenna array on teflon substrate. IEEE Trans. Microw. Theory Tech. **53**(6), 2101–2106 (2005)
22. K.M. Luk, S.F. Zhou, Y.J. Li et al., A microfabricated low-profile wideband antenna array for terahertz communications. Sci. Rep. **7**, 1268 (2017). https://doi.org/10.1038/s41598-017-01276-4
23. J.F. Harvey, M.B. Steer, T.S. Rappaport, Exploiting High millimeter wave bands for military communications, applications, and design. IEEE Access **7**, 52350–52359 (2019). https://doi.org/10.1109/access.2019.2911675; K. Sengupta, A. Hajimiri, A 0.28 THz power-generation and beam-steering array in CMOS based on distributed active radiators. IEEE J. Solid-State Circ. **47**(12), 3013–3031 (2012)
24. M.C. Kemp, P.F. Taday, B.E. Cole, J.A. Cluff, A.J. Fitzgerald, W.R. Tribe, Terahertz for military and security applications. Int. Soc. Opt. Photon. 44–53 (2003)
25. S.U. Hwu, K.B. deSilva, C.T. Jih, Terahertz (THz) wireless systems for space applications, in *2013 IEEE Sensors Applications Symposium Proceedings* (2013). https://doi.org/10.1109/sas.2013.6493580
26. M.A.K. Khan, M.I. Ullah, M.A. Alim, High-gain and ultrawide-band graphene patch antenna with photonic crystal covering 96.48% of the terahertz band. Optik **227**, 166056 (2021)
27. K. Anusha, Design techniques for compact microstrip antennas. Int. J. Eng. Adv. Technol. (IJEAT) **8**(2S), 388–390 (2018)
28. N. Laman, D. Grischkowsky, Terahertz conductivity of thin metal films. Appl. Phys. Lett. **93**(5), 051105 (2008)
29. S. Dittakavi, I. Abdel-Motaleb, Indium tin oxide/barium strontium titanate THz sensor antenna, in *NAECON 2018-IEEE National Aerospace and Electronics Conference* (IEEE, 2018), pp. 622–625
30. G. Rosolen, Fabrication of terahertz coupling structures by electron beam lithography. PIERS Online **4**(4), 441–444 (2008)
31. A.A. Abohmra, H.T. Abbas, J.U.R. Kazim, M. Saqib Rabbani, C. Li, A. Alomainy, M. Imran, Q.H. Abbasi, An ultrawideband microfabricated gold-based antenna array for terahertz communication. IEEE Antennas Wirel. Propag. Lett. 1. https://doi.org/10.1109/lawp.2021.3072562
32. A. Abohmra, H. Abbas, M. Al-Hasan, I.B. Mabrouk, A. Alomainy, M.A. Imran, Q.H. Abbasi, Terahertz antenna array based on a hybrid perovskite structure. IEEE Open J. Antennas Propag. **1**, 464–471 (2020)
33. K.S. Novoselov, A.K. Geim, S.V. Morozov, D. Jiang, Y. Zhang, S.V. Dubonos, A.A. Firsov, Electric field effect in atomically thin carbon films. Science **306**(5696), 666–669 (2004)
34. T.A. Elwi, H.M. Al-Rizzo, D.G. Rucker, E. Dervishi, Z. Li, A.S. Biris, Multi-walled carbon nanotube-based RF antennas. Nanotechnology **21**(4), 045301 (2009). https://doi.org/10.1088/0957-4484/21/4/045301
35. M. Dashti, J.D. Carey, Graphene microstrip patch ultrawide band antennas for THz communications. Adv. Func. Mater. **28**(11), 1705925 (2018)
36. M.A.K. Khan, M.I. Ullah, R. Kabir, M.A. Alim, High-performance graphene patch antenna with superstrate cover for terahertz band application. Plasmonics **15**, 1719–1727 (2020)

37. M.M. Seyedsharbaty, R.A. Sadeghzadeh, Antenna gain enhancement by using metamaterial radome at THz band with reconfigurable characteristics based on graphene load. Opt. Quant. Electron. **49**(6), 221 (2017)
38. S. Choi, K. Sarabandi, Performance assessment of bundled carbon nanotube for antenna applications at terahertz frequencies and higher. IEEE Trans. Antennas Propag. **59**(3), 802–809 (2010)
39. Y.N. Jurn, S.A. Mahmood, I.Q. Habeeb, Performance prediction of bundle double-walled carbon nanotube-composite materials for dipole antennas at terahertz frequency range. Progr. Electromagn. Res. M **88**, 179–189 (2020). https://doi.org/10.2528/pierm19101604
40. Y.J. Yang, B. Wu, Y.T. Zhao, Dual-band beam steering THz antenna using active frequency selective surface based on graphene. EPJ Appl. Metamater. **8**, 12 (2021)
41. A.A. Althuwayb, M. Alibakhshikenari, B.S. Virdee, H. Benetatos, F. Falcone, E. Limiti, Antenna on chip (AoC) design using metasurface and SIW technologies for THz wireless applications. Electronics **10**(9), 1120 (2021)
42. Y. Luo et al., Graphene-based multi-beam reconfigurable THz antennas. IEEE Access **7**, 30802–30808 (2019). https://doi.org/10.1109/ACCESS.2019.2903135
43. P.P. Tong, D.P. Neikirk, P.E. Young, W.A. Peebles, N.C. Luhmann, D.B. Rutledge, Imaging polarimeter arrays for near-millimeter waves. IEEE Trans. Microw. Theory Tech. **32**(5), 507–512 (1984)
44. T. Nagatsuma, A. Hirata, Y. Royter, M. Shinagawa, T. Furuta, T. Ishibashi, H. Ito, A 120-GHz integrated photonic transmitter, in *International Topical Meeting on Microwave Photonics MWP 2000 (Cat. No. 00EX430)* (IEEE, 2000), pp. 225–228
45. M. Li, X.Q. Lin, J.Y. Chin, R. Liu, T.J. Cui, A novel miniaturized printed planar antenna using split-ring resonator. IEEE Antennas Wirel. Propag. Lett. **7**, 629–631 (2008)
46. G. Singh, Design considerations for rectangular microstrip patch antenna on electromagnetic crystal substrate at terahertz frequency. Infrared Phys. Technol. **53**(1), 17–22 (2010)
47. M. Younssi, A. Jaoujal, M.D. Yaccoub, A. El Moussaoui, N. Aknin, Study of a microstrip antenna with and without superstrate for terahertz frequency. Int. J. Innov. Appl. Stud. **2**(4), 369–371 (2013)
48. S. Singhal, Ultrawideband elliptical microstrip antenna for terahertz applications. Microw. Opt. Technol. Lett. **61**(10), 2366–2373 (2019). https://doi.org/10.1002/mop.31910
49. K. Bhattacharyya, S. Goswami, K. Sarmah, S. Baruah, A linear-scaling technique for designing a THz antenna from a GHz microstrip antenna or slot antenna. Optik **199**, 163331 (2019). ISSN 0030-4026. https://doi.org/10.1016/j.ijleo.2019.163331
50. R. Goyal, D.K. Vishwakarma, Design of a graphene-based patch antenna on glass substrate for high-speed terahertz communications. Microw. Opt. Technol. Lett. **60**(7), 1594–1600 (2018)
51. T. Zhou, Z. Cheng, H. Zhang, M.L. Berre, L. Militaru, F. Calmon, Miniaturized tunable terahertz antenna based on graphene. Microw. Opt Technol. Lett. **56**(8), 1792–1794 (2014)
52. A. Devapriya, S. Robinson, Investigation on metamaterial antenna for terahertz applications. J. Microw. Optoelectron. Electromagn. Appl. **18**, 377–389 (2019). https://doi.org/10.1590/2179-10742019v18i31577
53. D.A. Sehrai, M. Abdullah, A. Altaf, S.H. Kiani, F. Muhammad, M. Tufail, S. Rahman, A novel high gain wideband MIMO antenna for 5G millimeter wave applications. Electronics **9**(6), 1031 (2020)
54. K.K. Farzad, D.Z. Yasaman, High-gain multi-layer antenna using metasurface for application in terahertz communication systems. Int. J. Electron. Dev. Phys. **4**(1). https://doi.org/10.35840/2631-5041/1707
55. U.S Keshwala, S. Rawat, K. Ray, Plant shaped antenna with trigonometric half sine tapered leaves for THz applications. Optik **223**, 165648 (2020). ISSN 0030-4026. https://doi.org/10.1016/j.ijleo.2020
56. U. Keshwala, S. Rawat, K. Ray, Inverted K-shaped antenna with partial ground for THz applications. Optik **219**, 165092 (2020)
57. R. Cicchetti, E. Miozzi, O. Testa, Wideband and UWB antennas for wireless applications: a comprehensive review. Int. J. Antennas Propag. (2017)

58. H. Mao, L. Xia, X. Rao, H. Cui, S. Wang, Y. Deng, D. Wei, J. Shen, H. Xu, C. Du, A terahertz polarizer based on multilayer metal grating filled in polyimide film. IEEE Photon. J. **8**(1), 1–6 (2016). https://doi.org/10.1109/jphot.2015.2511093
59. B. Zhang, W. Chen, Y. Wu, K. Ding, R. Li, Review of 3D printed millimeter-wave and terahertz passive devices. Int. J. Antennas Propag. (2017)
60. V.M. Lubecke, K. Mizuno, G.M. Rebeiz, Micromachining for terahertz applications. IEEE Trans. Microw. Theory Tech. **46**(11), 1821–1831 (1998)
61. L. Guo, H. Meng, L. Zhang, J. Ge, Design of MEMS on-chip helical antenna for THz application, in *2016 IEEE MTT-S International Microwave Workshop Series on Advanced Materials and Processes for RF and THz Applications (IMWS-AMP)* (IEEE, 2016), pp. 1–4
62. T.J. Reck, C. Jung-Kubiak, J. Gill, G. Chattopadhyay, Measurement of silicon micromachined waveguide components at 500–750 GHz. IEEE Trans. Terahertz Sci. Technol. **4**(1), 33–38 (2014). https://doi.org/10.1109/tthz.2013.2282534
63. Z. Popovic, E.N. Grossman, THz metrology and instrumentation. IEEE Trans. Terahertz Sci. Technol. **1**(1), 133–144 (2011). https://doi.org/10.1109/TTHZ.2011.2159553
64. R. Shan, Y. Yao, J. Yu, X. Chen, Design of tri-reflector compact antenna test range for THz antenna measurement, in *2012 International Conference on Microwave and Millimeter Wave Technology (ICMMT)* (2012). https://doi.org/10.1109/icmmt.2012.6230080

THz Meta-Atoms Versus Lattice to Non-invasively Sense MDAMB 231 Cells in Near Field

Abhirupa Saha, Sanjib Sil, Srikanta Pal, Bhaskar Gupta, and Piyali Basak

Abstract In this chapter, the utility of THz metasurface versus a single unit cell in sensing MDAMB 231 breast cancer cells at 0.75 THz has been demonstrated with a sensitivity of 43 GHz per RIU. A meta-atom is sufficient to exhibit the properties of a left-handed material as in simultaneously negative permeability and a negative permittivity, thus leading to a negative refraction with the cancer cell suspension measured under room temperature condition. It is much easier to drop the minute quantity of the analyte and incubate the cultures in batches by dedicating single unit cells to detect carcinoma condition with time. This problem of choosing a single versus an array of cells is solved using FDTD computational results that guarantee accurate outcomes in the presence of a greater number of inductors and capacitors to enhance the shift and improve the Q factor to track any tender cellular metabolic changes in space and time. Capacity of reflectance and transmittance extracted meta-scattered information using wave ports replaced the challenges of the TDS setup of a terahertz spectroscopy for a single cell. Comparisons with seeded cells were done with that of bare meta-units to further study the overall refractive index enhancement and see their impact on the efficacy of the sensors. The novel single-atom approach gives more information via percentage reflectance rather than conventional metasurface that takes up a lot of area; hence, a single surface can be used to sense different types of cells and their metabolism with time.

A. Saha (✉) · P. Basak
School of Bioscience and Engineering, Jadavpur University, Kolkata 700032, India

S. Sil
Department of Electronics and Communication Engineering, Calcutta Institute of Engineering and Management, Kolkata 700040, India

S. Pal
Department of Electronics and Communication Engineering, Birla Institute of Technology, Mesra, Ranchi 835215, India

B. Gupta
Department of Electronics and Telecommunication Engineering, Jadavpur University, Kolkata 700032, India

S. Das et al. (eds.), *Advances in Terahertz Technology and Its Applications*,
https://doi.org/10.1007/978-981-16-5731-3_20

Keywords Breast cancer · Metasurfaces · Unit cell · Plasmonic metamaterials · Terahertz · Dielectric measurement · Split ring resonator · Apoptosis · Biosensing

1 Introduction

Terahertz bio-sensing via metamaterials is an ongoing trend in the biomedical industry [1]. There have been permutations and combinations of active and passive elements along with tuning the Fermi energy [2] of the metallic element to enhance the sensitivity toward the intended molecule. Micro-fluidic approaches have been gaining quite a lot of exposure, and here, mostly one unit cell is quite enough [3] rather than the two-dimensional pattern of cells. THz window spectral signatures can actually better match to that of the inherent vibrations of the dipoles in the non-ionized molecular aggregates. Artificial dielectrics were first introduced during the Second World War by J. C. Bose while attempting to rotate electromagnetic waves by using twisted fibers, and then, lightweight lenses were designed by Winston E. Kock in Bell Laboratories. The molecules making up these dielectrics are small metallic patterns printed on a substrate; the size is decided by the frequency, conventionally to behave as sub -wavelength lattice units. Wolfgang J. R. Hoefer et al [4] have presented the embedded elements that are reactive and add to the real network nodes to form a negative refractive index verified by measuring the scattering angle [5], and bioengineers cultivate Fano resonance [6] that aids in tracing cell metabolism changes, and thus live cell imaging. Metamaterials have thus become non-invasive chiral detectors of electric as well as magnetic properties of a given analyte. In the present chapter, reflectance-based spectroscopy-aided localized surface plasmons overcome impedance mismatches due to electromagnetic field enhancements of a metasurface absorber cavity [7]. Although there are several reviews of THz-based biological applications of various nanostructures [8], there is a lack of delving into an array preference; Sect. 2 elaborates this along with Floquet theory in near field. Section 3 describes the cell culture aspect, followed by Brillouin zone in Sect. 4, Sect. 5 illustrates the sensor design, setup in Sect. 6, and finally, the unit sensor results are discussed in Sect. 7.

The designs of the units [9] differ in their intended mode generation and further functionality that serves the application. Labyrinth type patterns activate certain areas of the units across the surface and the analyte in their case a fungus has to be seeded at the specific locations where the field confinement takes place. There have been many utilizations of dielectric spectroscopy [10], but using them to intervene with the sensing capabilities is rare. And then, associating this with overcoming with the drawbacks of THz is further interesting. In this chapter, we are going to understand the necessity of a single unit cell versus an entire array of cells. If we delve into the antenna engineering analogy, the increase in the number of meta-atoms enhances the directivity of the entire sensing platform that is reflecting the sensed information to the receiving antenna that acts as the port to measure the scattering parameters. The SRR and their variations have been a major player in the sensing league, and their contribution to react smartly under THz has been useful for detecting various bio-molecules with their spectral signature exactly matching under the irradiation of THz window. The material science and chemistry involved with photonic transaction have made the bioengineers join hands with the microwave community, and this calls for a setup where the biological cells can be sensed using the fabricated platform without any kind of contamination and the treated cells can be immediately incubated for further analysis.

2 Metasurface Versus Unit Atom

Let us start by thinking why we need to mold passive structures to manipulate their wave matter interaction such that they have a negative group velocity and negative refractive index. The main topic of concern is the need for an array of unit meta-atoms to weave a platform or to just leave it at one lattice atom. The author has done simulations comparing the use of Floquet ports to that of the use of wave port to excite only one cell. In both the cases, the THz water absorption problem is same. But the difference is in the effective circuit response controlling the sensitivity in terms of resonant frequency shift due to subtle changes in the refractive index of the biological element to be monitored. The use of unit cell can be beneficial for micro-fluidic setup and an IC with other modules to acquire and process the data before storing it at a cloud database to make an entire system out of the metamaterial sensor flow. The cells form a continuous effective medium whose series impedance Z_s and shunt capacitance Y_p are given by the following equations:

$$Z_s = j2\omega L_l^{'}\Delta l + \tfrac{1}{j\omega C_0} = j\omega\left(2L_l^{'}\Delta l - \tfrac{1}{\omega^2 C_0}\right) = j2\omega\mu_m\Delta l \tag{1}$$

$$Y_p = j2\omega C_l^{'}\Delta l + \frac{1}{j\omega L_0} = j\omega\left(2C_l^{'}\Delta l - \frac{1}{\omega^2 L_0}\right) = j2\omega\varepsilon_m\Delta l \tag{2}$$

2.1 Floquet Theory Bloch Waves

During simulation of an array, the concept of Bloch waves come up. The use of Floquet is to shine the incoming THz light along the cells very uniformly so that each and every cell gets equal amount of power that the unit cell aloe would have got. The fermions should be actively coupled in forming the surface plasmon polariton which is a periodic excess and shortage of electrons along the metal dielectric interface of the meta-surface containing the analyte and the incoming radiation; we have considered near-field versus far-field analogy as well in the next sections of the chapter. In order to have non-invasive prototype, the far-field is a pragmatic approach where the patient or the target can be located at a distance $2D^2/\lambda$.

2.2 Near-Field Theory

Apart from plasmonic effect in 0.1–1.1 THz operating frequency range, reducing the distance between the sensing platforms and the source of test radiation prevents isotropic loss due to the vague presence of distance and other environmental factors. We also compare the effect of varying refractive index under near field for a seeded unit cell, in terms of sensitivity, and will also be varying the distance of the test ports from the live cells under diagnosis in the next sections. The power intensity reaching the material overcomes the free space or isotropic loss as the distance between the feeding antenna and the sensor platform is gradually reduced such that there is an inductive coupling that feeds power from one cell to the other medium. The electric field remains parallel to the surface of the asymmetrical split ring resonator with a backward strip from where the measurements are taken. The near field is such that the meta-atoms are adhered on the surface facing the cells and the strip is serving as the inductor where the resonances due to polarized dipoles and L-C systems interfere to create locally confined hot spots that act as absorbers in some frequencies and reflectors in the neighborhood (Fig. 2).

2.3 Transmission Line Model for Unit Cells and an Array

As implied from Fig. 1b, the only difference due to the addition of a greater number of cells is in the capacitative loading due to the presence of large number of gaps between the corresponding units. Inductive loading and periodicity affect the generation of

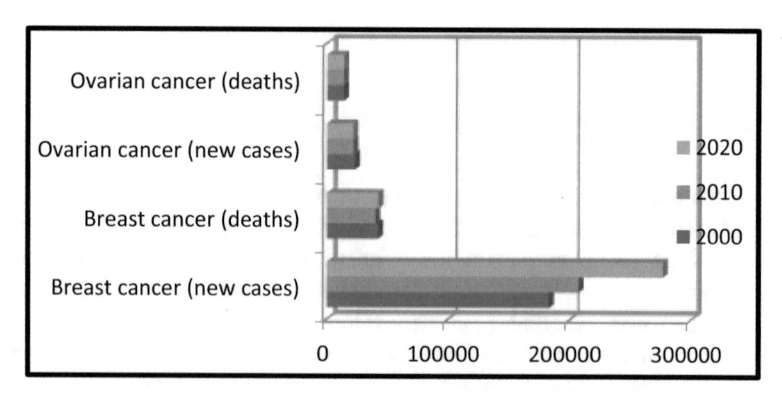

Fig. 1 Over the years, breast cancer is still a prevalent problem in women than ovarian carcinoma

Fig. 2 **a** Radiation experiment to check the health of the cells while testing. **b** Fabricated structures and transceiving patch antennas

dipole responsible for radiation. In the chapter objective, the sensing capabilities are compared, and hence for the metasurface or cell to behave as a detector, non-radiative losses are highly important in the presence of forward waves and the backward waves that get induced.

3 Cell Culture

Preparing the cells to be diagnosed under live conditions is very important for testing the sensor. Protocols for culturing MDAB 231 cells in Gibco Dulbecco's Modified Eagle Medium (DMEM) were conducted as per conventional methodology in a sterilized environment. Experiments have been done to make sure that the temperature and power density and range of radiation do not disrupt the cell health. In fact, the morphology does not change and no cell wall perforation took place [10]. Dielectric properties were measured in the C band and coded to the THz regime (Fig. 3).

4 Brillouin Zone in THz Excitation

Thermal changes were not observed while treating the cells, yet an array periodicity affects folding the Brillouin zone; hence, the resonating points and this determine the choice of the material to be used for the patterns to responsively transmit useful information after encountering the cells to be measured. If the use of multiple unit cells makes the real part of the refractive index shift beyond the edges of the Brilluoin zone, that means the use of negative refractive index [11, 12]-based Q factor Fano delays are redundant and hence unit cells are better. Furthermore, if the wavelength

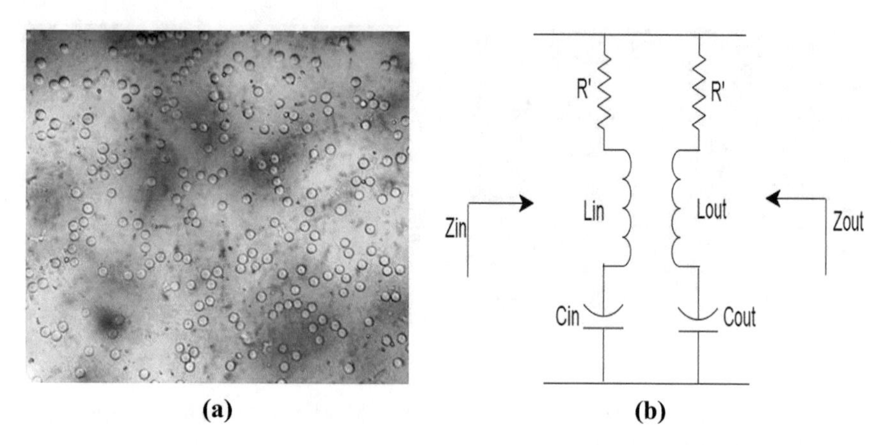

Fig. 3 **a** Cultured MDAMB 231 dispersed in DMEM medium. **b** Equivalent circuit of the cell on MS

of the impinged radiation deceases beyond the lattice dimension, here each unit cell is a meta-atom, then the sub-wavelength characteristics of the medium perishes.

5 Sensor Methodology

The spade structure has been chosen [13] to form the atoms. This geometry has been simulated to give the best results in terms of quality factor, and hence a smart material detector. The workflow starts with the source of THz light, and this can be either antenna that emits the required working frequency and then the sensor and then the receiving antenna to acquire and serve as the second port of the system to understand the transmittance and absorbance property of the unit cell vs an entire surface. This is different than the array of atom-based sensors that have specific microlocation for microorganisms like *E. Coli* [14], yet again the health of the cells under test have not been monitored although such experiments exist [15]. This setup of non-invasively testing a unit cell is much convenient than dipping the unit cell in glucose or analyte solution [16].

5.1 Design of Unit Cell

The asymmetric split-based unit cell is uniquely defined as a continuation the field and port analysis are focused in this chapter although holography can also be carried out as the oblique angle of incidence of the irradiating C band radiation is varied [17]. If we study the pattern of the metal, it has a broken asymmetry at two the sides of the arms that are diagonally opposite to each other, and hence, due to the non-radiative fields at these gaps, the incoming radiation gets confined at these locations. And there is no ground plane in this case that is replaced by a strip line to resemble the permeability negating metallic coupler. Thus, the magnetic field permeates through the broken asymmetry split spades while the back strips create the polarization of electric field along them instead of a ground plane. Gaps imply capacitance resulting in magnetic polarization while electric polarization is caused by inductive strips:

$$\nabla \times \vec{H} = j\omega\varepsilon_0 \vec{E} + j\omega \vec{P_e} = j\omega\varepsilon_0\omega \vec{E} - j\frac{\vec{E}}{\omega L_0 d} \tag{3}$$

$$\nabla \times \vec{E} = -j\omega\mu_0 \vec{H} - j\omega\mu_0 \vec{P_e} = -j\omega\mu_0 \vec{H} + j\frac{\vec{E}}{\omega C_0 d} \tag{4}$$

where the effective material permittivity and permeability are expressed in terms of ε_0 and μ_0, respectively.

6 Experiments for Comparing Seeded Atoms

The setup is done in CST Studio Suite® initially that involves a layer of dielectric 1.6 mm thick which is generally the conventional FR4 and a pattern of metal generally copper that is 0.02 mm thick and the cells seeded uniformly on top of the layer on which the material under test is seeded or in housed adhered to the culture flasks (Fig. 4).

For the simulations, at first, the Floquet port tests were carried out on the seeded membrane, and then, only one meta-unit was taken with the perfect electric and perfect magnetic wall boundaries across the x and y boundary walls depending on the impinging wave polarization. The z-direction is always open both in case of periodic and single unit. The position of the wave port is also varied to study the impact of one cell and find out if it can compensate the efficacy of the membrane altogether,

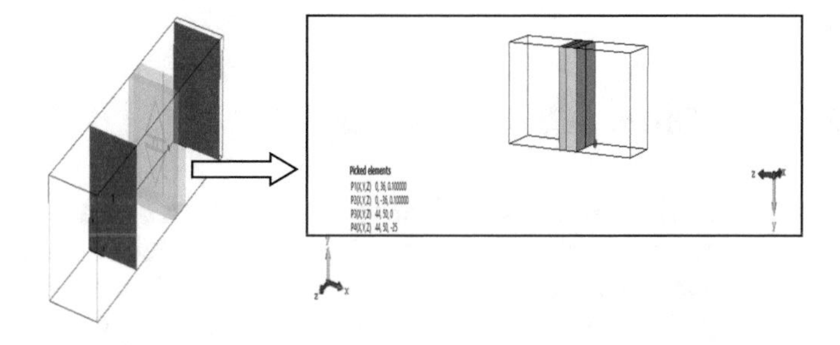

Fig. 4 CST Microwave Studio simulation setup, with two wave ports shown in red, and blue region showing cells

and the effect on scattered waves is also interpreted. The important aspect of this experiment is to find if there is any change in the sensitivity due to the presence of the periodic array. The periodicity is kept as a constant value. Thermal and mechanical parameters were also kept unperturbed. Normal incidence with TM mode excitation is done by choosing a single mode in the wave ports and the Floquet mode. This technique can be used to tinker the dielectric properties instead of using a THz TDS [18] or facing the problems of open-ended coaxial probe technique [19, 20]. All of the biophysics have their roots in Maxwell's equations [20, 21, 22].

7 Resulting Nature of Biosensor

7.1 Unit Cell Results

The unit meta-atom acts as a reflector after 0.15 THz from the reflectance parameters, and from the transmittance, it is also evident that our atom works as a broadband sensing reflector. This is a similar behavior in case of a bunch of cells housed together, where the Q factor is much higher; this contributes to the transmission resonance at a precise point that shifts with the change of the refractive index of the cells. SAR was observed to be same for both cases.

a. **Comparing single meta-atom with and without MDAMB layers** (Fig. 5).

b. **Varying the near-field proximity of the meta-units from wave ports** (Figs. 6, 7 and 8).

In the figures, the z1 implies distance between two wave ports. As the distance between the ports is increased, the reflectance becomes more and more perfect, with a 100% reflectance at 0.3 THz and a transmission at 0.1 THz that reduces with the increase. A typical gap of 120 mm from our meta-atom creates an unusual response, whereas the scattering reflectance from the second port behind the surface is well responsive to the distance of the target from the ports, and thus can be used as a proximity sensor too. The simulations were done in MATLAB.

7.2 Meta Surface-Based Array Results

Periodic boundary conditions were imposed along the xz- and yz-directions, the simulation results to which the progress has meshing errors, but the Floquet port approach is more flexible with the specification of TE and TM modes in the frequency solver settings. The point is to investigate whether the unit cell and the array give the same results in terms of resonant frequency and in terms of more frequency shift and higher Q factor that is higher return loss at the point of resonance and also the change in the amplitude of the reflected signal.

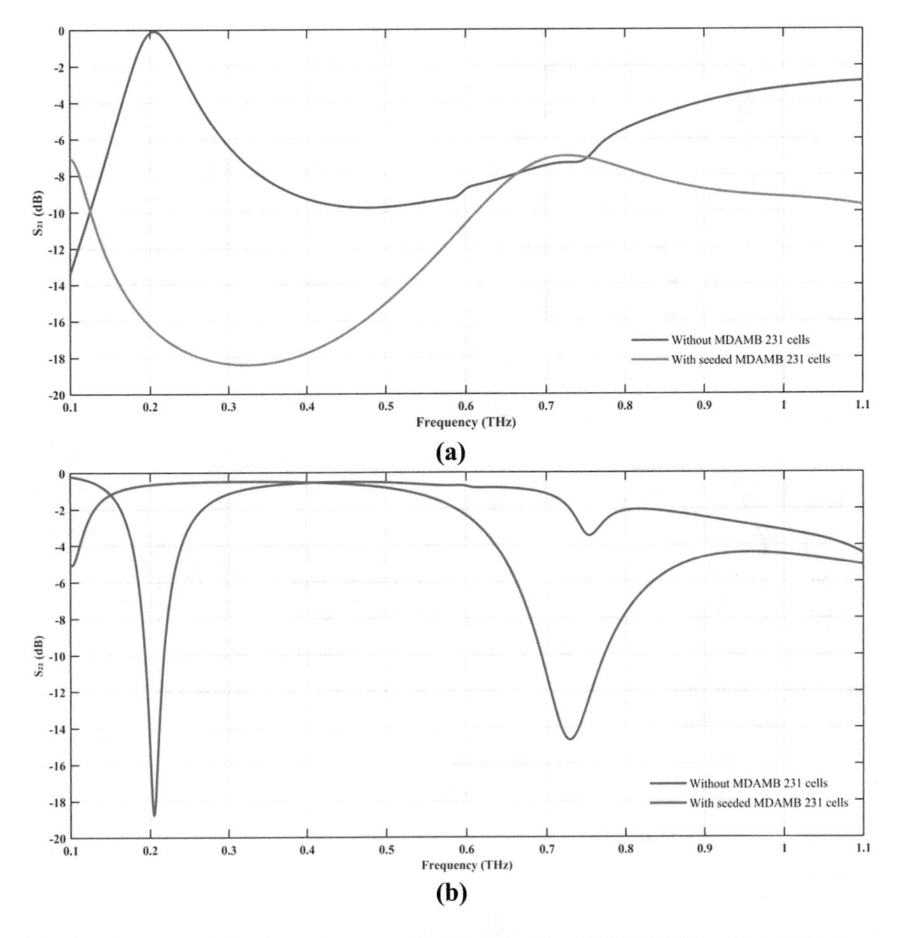

Fig. 5 **a** Transmitted signals on the meta-unit with and without MDAMB 231 cells. **b** Reflectance from the meta-atom with and without cells

a. **Transmittance and reflectance for cells versus no cells on an array** (Fig. 9).

b. **S11 array versus unit cell** (Fig. 10).

From the figure, it is evident that the amplitude of reflected waves is increased from −16 to −7.49 dB when the cells are introduced to the array, along with a frequency shift of −188.54 GHz. Finally, it is noted that the unit cell behaves as an efficient filter as compared to that of the array that had a few frequency points where the array behaves as a complete absorber and thus might affect the sensor performance. Comparison of refractive index-based identification capability for array and unit cell in Fig. 11 proves that a unit cell is sufficient to even sub-categorize cancer.

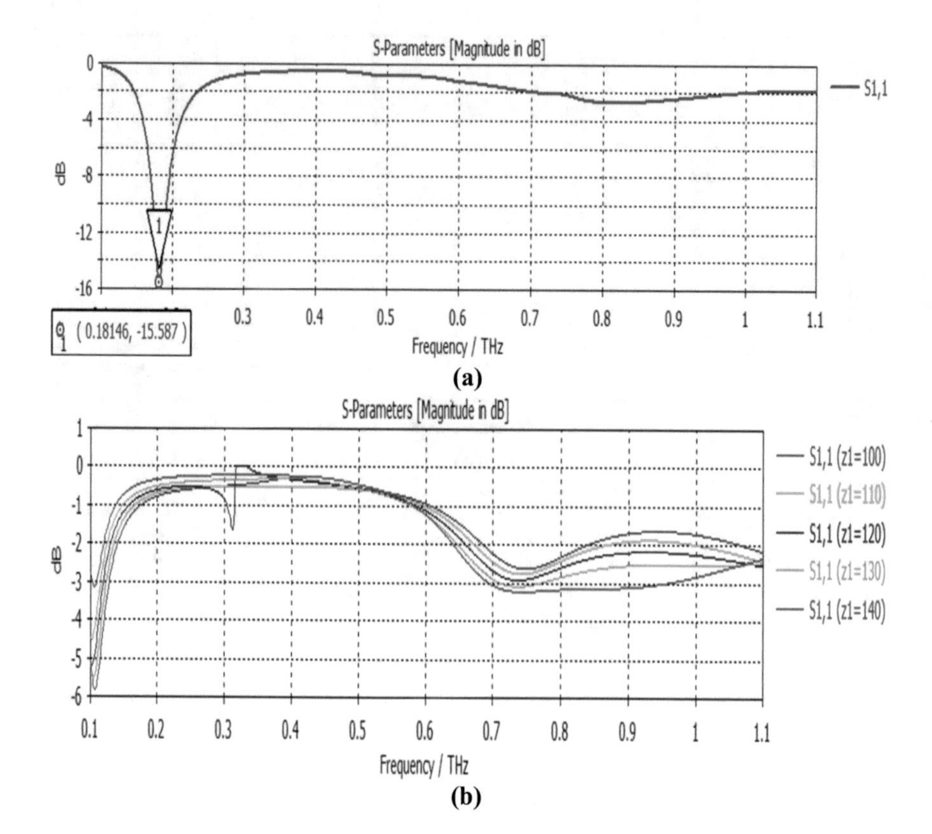

Fig. 6 **a** S11 parameters when no cells are present. **b** S11 parameters as a result of port to sensor gap variation

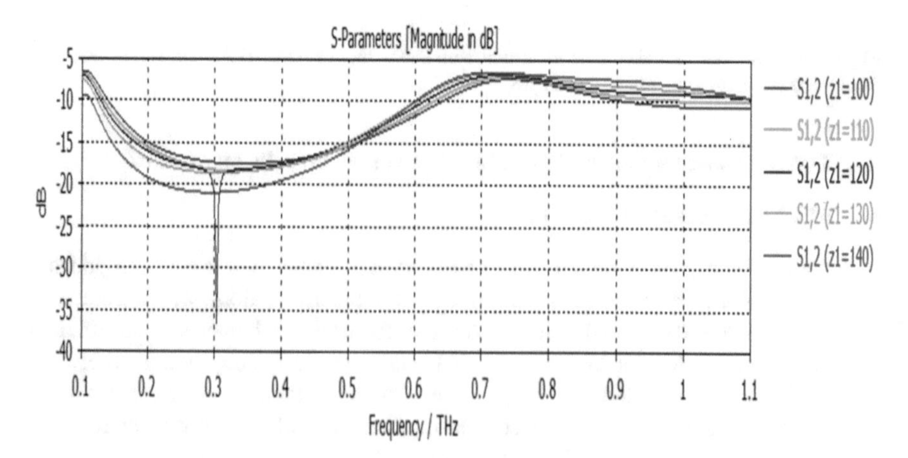

Fig. 7 Effect of port distance on S12 and complete reflectance observed at 0.3 THz

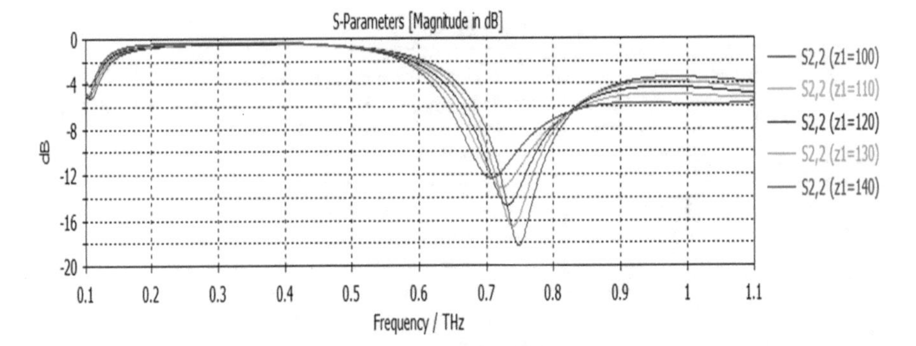

Fig. 8 S22 parameters as a result of port to sensor distance variation

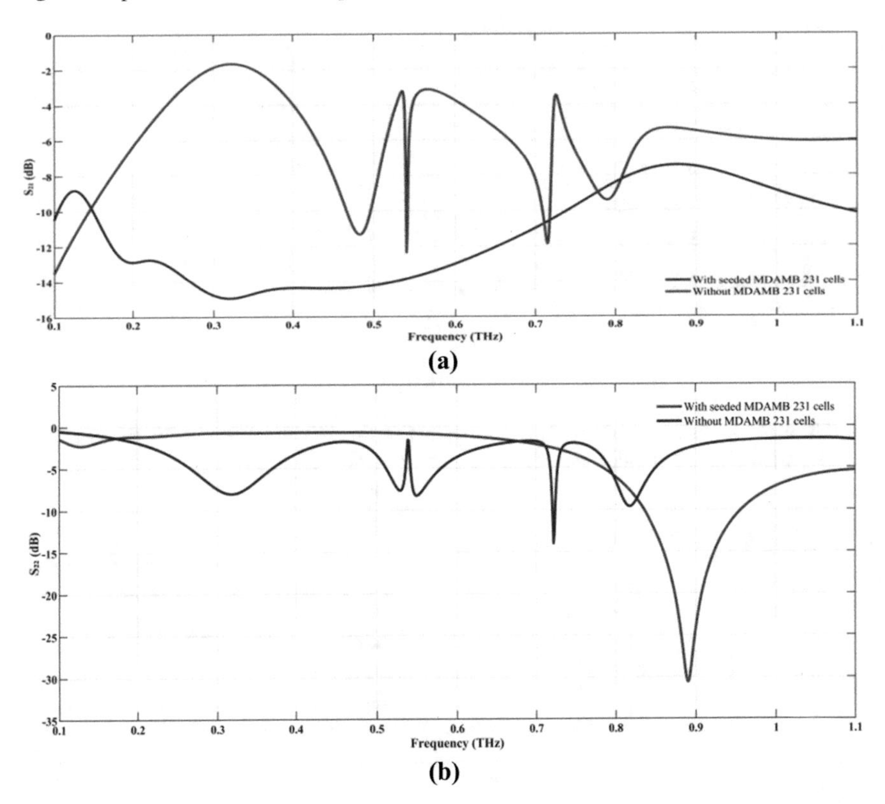

Fig. 9 **a** S21 parameter comparing the bare and the cell seeded metasurface. **b** S22 parameter comparing the bare and the cell seeded metasurface

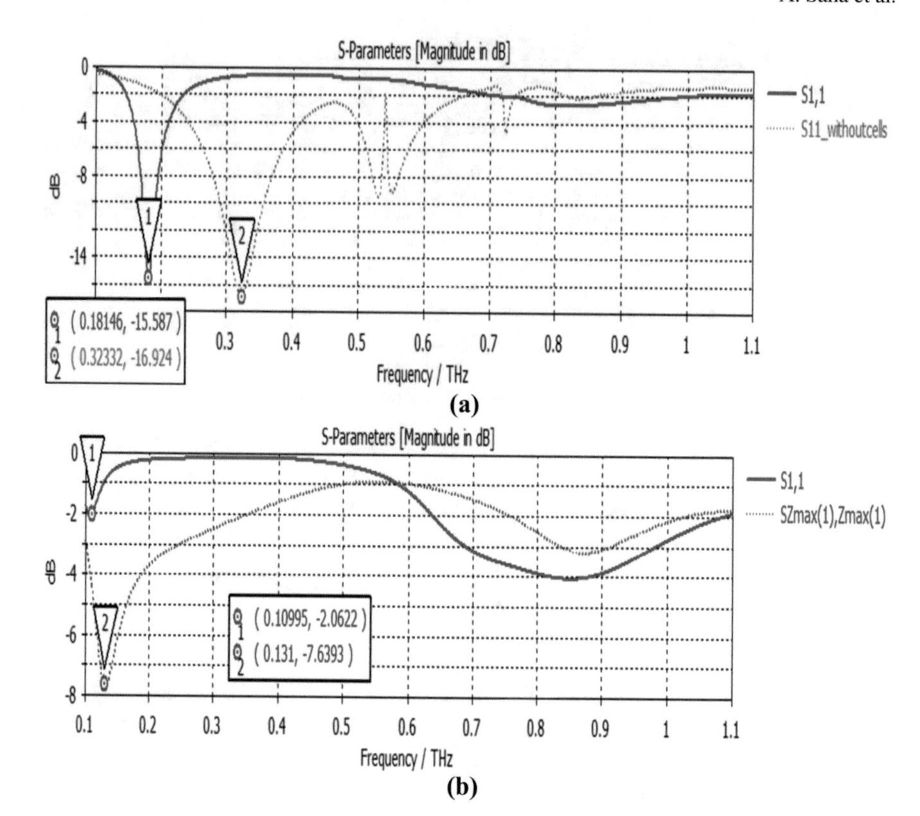

Fig. 10 **a** Comparing S11 curves, blue—array, and red—unit cell without cells. **b** Seeded unit versus seeded array S11 comparison, deep red—unit cell, dotted—array

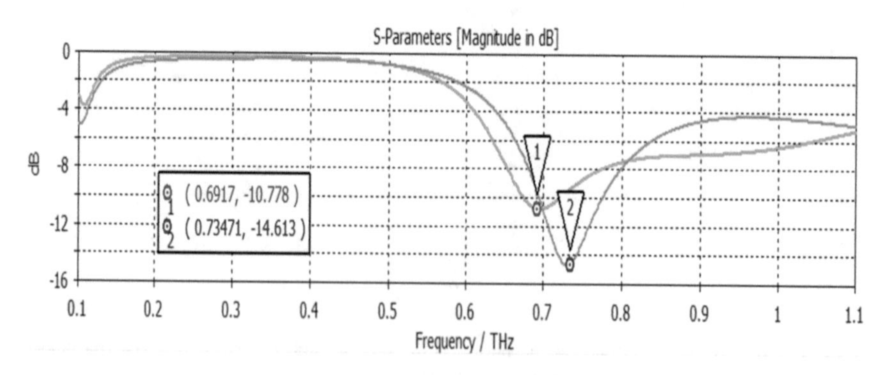

Fig. 11 43 GHz delay after varying the RIU by 1 unit, green: epsilon 64.57, orange: 92.98

8 Conclusion

The need for a single unit cell that suffices bio-sensing requirements is claimed after comparing the results from their corresponding infinite array, and this happened to be an effective deciding factor to design smart THz communication platform for future biotelemetry-based sensor IC design. Unit cells give better capability to differentiate sub-types, while an array has the benefit of high Q deciding factor. The unit cell has a sensitivity of 43 GHz per unit increase in refractive index. Results infer that the single atoms provide better sensitivity when it comes to categorizing sub-types with a red shift of 188.54 GHz; even if the impedance spectrum is calculated from the results, the difference in the amplitude of reflected wave can be another significant parameter as proven from the results, considering a single unit will of course lead to a compact packaging capability to accommodate other components like signal processing and power amplifier circuitry. It is found that the reflection is much more enhanced in case of a single cell but that does not aid to sensitivity for which the membrane gives us an appreciable resonant frequency blue shift. The resonant frequency can be explicitly tuned by changing the chemical potential if a material like graphene is used instead of titanium, which excites surface plasmons in the THz spectra.

References

1. X. Yan, The terahertz electromagnetically induced transparency-like metamaterials for sensitive biosensors in the detection of cancer cells. Biosens. Bioelectron. **126**, 485–492 (2019)
2. W. Xu, L. Xie, J. Zhu, L. Tang, R. Singh, C. Wang, Y. Ma, H.T. Chen, Y. Ying, Terahertz biosensing with a graphene-metamaterial heterostructure platform. Carbon **141**, 247–252 (2019)
3. Z. Geng, X. Zhang, Z. Fan, X. Lv, H. Chen, A route to terahertz metamaterial biosensor integrated with microfluidics for liver cancer biomarker testing in early stage. Sci. Rep. **7**(1), 1–11 (2017)
4. W.J. Hoefer, P.P. So, D. Thompson, M.M. Tentzeris, Topology and design of wideband 3D metamaterials made of periodically loaded transmission line arrays, in *IEEE MTT-S International Microwave Symposium Digest* (Long Beach, CA, 2005), pp. 12–17
5. R.A. Shelby, D.R. Smith, S. Schultz, Experimental verification of a negative index of refraction. Science **292**(5514), 77–79 (2001)
6. A. Farmani, A. Mir, M. Bazgir, F.B. Zarrabi, Highly sensitive nano-scale plasmonic biosensor utilizing Fano resonance metasurface in THz range: numerical study. Phys. E. **104**, 233–240 (2018)
7. L. Cong, S. Tan, R. Yahiaoui, F. Yan, W. Zhang, R. Singh, Experimental demonstration of ultrasensitive sensing with terahertz metamaterial absorbers: a comparison with the metasurfaces. Appl. Phys. Lett. **106**(3), 031107 (2015)
8. R. Zhou, C. Wang, W. Xu, L. Xie, Biological applications of terahertz technology based on nanomaterials and nanostructures. Nanoscale **11**(8), 3445–3457 (2019)
9. J.B. Pendry, Negative refraction makes a perfect lens. Phys. Rev. Lett. **85**(18), 3966 (2000)
10. S. Gabriel, R.W. Lau, C. Gabriel, The dielectric properties of biological tissues: III. Parametric models for the dielectric spectrum of tissues. Phys. Med. Biol. **41**(11), 2271 (1996)
11. H. Li, A. Denzi, X. Ma, X. Du, Y. Ning, X. Cheng, F. Apollonio, M. Liberti, J.C. Hwang, Distributed effect in high-frequency electroporation of biological cells. IEEE Trans. Microw. Theory Tech. **65**(9), 3503–3511 (2017)

12. V.G. Veselago, The electrodynamics of substances with simultaneously negative values of ε and μ. Sov. Phys. Usp. **10**, 509–514 (1968)
13. M. Choi et al., A terahertz metamaterial with unnaturally high refractive index. Nature **470**, 369–373 (2011)
14. A. Saha, S. Sil, .S. Pal, B. Gupta, P. Basak, Meta sensing ovarian cancer cells at THz from C band radiation biophysics, in *Advances in Smart Communication Technology and Information Processing: OPTRONIX* (Springer Singapore, 2020), pp. 111–122
15. W. Wang, C. Li, R. Qiu et al., Modelling of cellular survival following radiation-induced DNA double-strand breaks. Sci Rep **8**, 16202 (2018)
16. M. Beruete, I. Jáuregui-López, Terahertz sensing based on metasurfaces. Adv. Opt. Mater. **8**(3), 1900721 (2020)
17. S. Afroz, S.W. Thomas, G. Mumcu, S.E. Saddow, Implantable SiC based RF antenna biosensor for continuous glucose monitoring, in *SENSORS* (IEEE, 2013), pp. 1–4
18. V. Kumari, G. Sheoran, T. Kanumuri, SAR analysis of directive antenna on anatomically real breast phantoms for microwave holography. Microw. Opt. Technol. Lett. **62**(1), 466–473 (2020)
19. M.M. Nazarov, O.P. Cherkasova, A.P. Shkurinov, Study of the dielectric function of aqueous solutions of glucose and albumin by THz time-domain spectroscopy. Quantum Electron. **46**(6), 488 (2016)
20. A. La Gioia, E. Porter, I. Merunka, A. Shahzad, S. Salahuddin, M. Jones, M. O'Halloran, Open-ended coaxial probe technique for dielectric measurement of biological tissues: challenges and common practices. Diagnostics **8**(2), 40 (2018)
21. M. David, Microwave Engineering (Wiley, Inc., 1998), p. 11
22. R.E. Collin, *Field Theory of Guided Waves* (IEEE Press, Piscataway, N.J., 1990), Chap. 12

Printed in the United States
by Baker & Taylor Publisher Services